DEVELOPMENT AND INTEGRATION OF MICROGRIDS

Edited by **Wen-Ping Cao** and **Jin Yang**

Development and Integration of Microgrids
http://dx.doi.org/10.5772/65582
Edited by Wen-Ping Cao and Jin Yang

Contributors

Shahbaz Siddiqui, Manoj Fozdar, Narayanan K, John Fletcher, Daming Zhang, Youmin Zhang, Oliver Probst, Luis Ismael Minchala-Avila, Andrew Harrison Hubble, Taha Selim Ustun, Reza Sabzehgar, Bouna Mohamed Cisse, Jing Wang, Fabio Mottola, Daniela Proto, Angela Russo, Guido Carpinelli, Toufique Ahmed Soomro, Du Guanhao, Syed Abid Ali Shah Bukhari, Jon Andoni Barrena, Eneko Unamuno, Luisa Alfieri, Caramia Pierluigi, Antonio Bracale, Chen Xu, Fei Xiong, Runhai Jiao, Zhenyu Zhou

Published by InTech
Janeza Trdine 9, 51000 Rijeka, Croatia

Publishing Process Manager Maja Bozicevic
Technical Editor SPi Global
Cover InTech Design team

First published August, 2017
Printed in Croatia
Legal deposit, Croatia: National and University Library in Zagreb

Additional hard copies can be obtained from orders@intechopen.com

Development and Integration of Microgrids, Edited by Wen-Ping Cao and Jin Yang
p. cm.
Print ISBN 978-953-51-3399-5
Online ISBN 978-953-51-3400-8

For EU product safety concerns:
IN TECH d.o.o., Prolaz Marije Krucifikse Kozulić 3, 51000 Rijeka, Croatia,
info@intechopen.com or visit our website at intechopen.com.

Meet the editor

Editor, Dr. Wen-Ping Cao received his BEng degree in electrical engineering from Beijing Jiaotong University, China, in 1991, and the PhD degree in electrical machines and drives from the University of Nottingham, UK, in 2004. He is currently a Chair Professor of Electrical Power Engineering and the Head of Power Electronics, Machines and Power System (PEMPS) Group at Aston University, Birmingham, UK. His research interests include design, fault analysis and condition monitoring of power components and electrical systems.

Editor, Dr. Jin Yang received his BEng and MSc degrees from North China Electric Power University, China, in 2003 and 2006, respectively, and his PhD degree from the University of Glasgow, UK, in 2011. He is now a lecturer at the School of Engineering and Applied Science, Aston University, UK. His research interests include operations and optimization of power transmission and distribution networks, protection of renewable power generation systems and smart grid and low-carbon technologies.

Contents

Preface

Microgrids are an indispensable element of "smart grids" when incorporating distributed generation from renewable energy sources (such as wind, solar and bioenergy) into the power grids. Microgrids can be AC, DC or in a hybrid form and can operate in connection with the main grid or in islanding mode. As the penetration of new and renewable energy sources continues to increase especially in developing countries and in rural areas, the control, operation and fault profile of the systems can be technically challenging. This book responds to the latest development of microgrids and attempts to provide some solutions to the control, system integration, fault and disturbance issues when involving renewable power generation (on the generation side) and electric vehicles (on the consumption side). In addition, some chapters also provide some insight into management, operation, data and economic analysis of the microgrid systems to enrich the value of the book.

This book contains 11 chapters and is organized into 3 sections. Section I outlines AC/DC microgrid concepts, design and control considerations. Section II deals with operational and technical aspects of microgrids, followed by Section III to address the issues associated with the interconnection and integration of microgrid systems. The editors would like to acknowledge all the contributing authors and the Publishing Process Manager Ms. Maja Bozicevic.

Dr. Wen-Ping Cao and Dr. Jin Yang
Aston University
Birmingham, United Kingdom

Section 1

AC/DC Microgrid Concepts, Design and Control Considerations

Chapter 1

Overview of Technical Challenges, Available Technologies and Ongoing Developments of AC/DC Microgrids

Reza Sabzehgar

Additional information is available at the end of the chapter

http://dx.doi.org/10.5772/intechopen.69400

Abstract

Gradual depletion of fossil fuel resources, poor energy efficiency of conventional power plants, and environmental pollution have led to a new grid architecture known as smart microgrid. The smart microgrid concept provides a promising solution that enables high penetration of distributed generation from renewable energy sources without requiring to redesign the distribution system, which results in stable operation during faults and disturbances. However, distributed generators/loads and interaction between all nodes within a microgrid will substantially increase the complexity of the power system operation, control, and communications. Many innovative techniques and technologies have been proposed to address the complexity and challenges of microgrids including power quality, power flow balancing, real-time power management, voltage and frequency control, load sharing during islanding, protection, stability, reliability, efficiency, and economical operation. All key issues of the microgrids, different solutions, and available methods and technologies to address such issues are reviewed in this chapter. Pros and cons of each method are discussed. Furthermore, an extensive comprehensive review for researchers and scholars working on microgrid applications is provided in this chapter to help them identify the areas that need improvements and innovative solutions for increasing the efficiency of modern power distribution grid.

Keywords: AC microgrid, DC microgrid, microgrid power quality, power management, microgrid modeling, power electronics, renewable energies

1. Introduction

The smart microgrid concept was first developed in Refs. [1, 2]. The capability of integration of a large number of distributed power generation sources and high penetration of renewable sources such as photovoltaic panels, wind turbines, and fuel cells within a smart microgrid made it very attractive. In fact, management of multiple distributed power generation sources including renewable energy sources in a smart microgrid offers many advantages such as significant reduction in the need for a transmission and high voltage distribution system, satisfying the growing demand of electricity, and improving energy utilization efficiency and reliability. However, interaction between all nodes within a microgrid including both conventional and renewable energy generation sources will not only increase the complexity of the power system considerably, but it will also raise challenging issues of reliability and power quality due to intermittent nature of renewable sources. Therefore, many technical challenges must yet be overcome to ensure safe, secure, reliable, optimized, efficient, and cost effective operation of the microgrid.

Academic scholars and industry experts offered many innovative techniques and technologies to address the challenges of microgrids such as power quality and power flow balancing [3–12], voltage and frequency control [13–35], power management [36–48], optimization [49–62], stability [63–66], reliability and protection [67–77], and dynamic modeling [78–95]. In this chapter, the concept of microgrid is first presented. All key issues of microgrid and the proposed innovative techniques addressing theses issues are then briefly reviewed. Finally, dynamic modeling and control of microgrid in different circumstances is discussed.

The rest of this chapter is organized as follows: Section 2 presents the microgrid concept. Power quality and power flow balancing of microgrid is discussed in Section 3. Section 4 reviews different control methods, power management strategies, and optimization techniques for microgrid. The key issues and technical challenges of microgrid stability, reliability, and protection are then summarized in Section 5. Section 6 discusses dynamic modeling of microgrid. Finally, conclusion and discussion are presented in Section 7.

2. Microgrid concept

A microgrid is a local group of electricity sources and loads that are intelligently managed through power electronics interfaces. A microgrid normally operates in parallel with utility electrical macrogrid, which is denoted as grid-tied mode of operation. It is capable of operating independently in islanded mode. In grid-tied mode, the microgrid becomes part of the main utility grid, and serves the grid with the excess power from distributed renewable generation sources. In this mode of operation, system control and scheduling a transition to islanded mode are done based on the system operation information such as generation outputs, demands, voltages, and the status of protection relays. In islanded mode, however, the microgrid must provide load balance by load shedding and load management. In this mode of operation, some loads might be shutdown to guarantee that there is enough power for

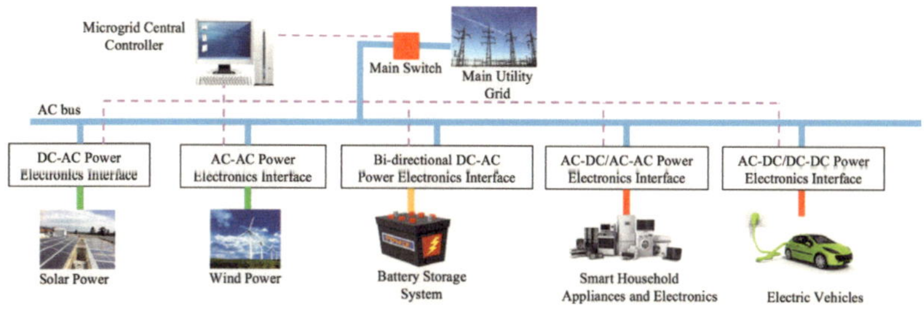

Figure 1. Schematic of an AC microgrid.

critical loads. To restore the transition back to grid-tied mode, the frequency, voltages, and phase angles should be within acceptable limits and must be synchronized.

A microgrid can be generally classified into AC and DC types. The AC microgrid is the main type of microgrid where all distributed power generation sources and loads are connected to a common AC bus through power electronic interfaces as sown in **Figure 1**. However, in DC microgrids a DC bus is used as a common bus especially for small-scale commercial and residential applications, due to higher efficiency and controllability as the extra power conversion stages are eliminated and synchronization and reactive power compensation are not needed in this configuration [96–101].

Integration of various distributed generators/loads and interaction among them within a smart microgrid remain in place regardless of the type of microgrid due to the complexity of power systems technologies, control and communication techniques. In the following sections, many challenges raised by smart microgrid applications such as power quality, protection, stability, reliability, and efficiency are briefly discussed.

3. Microgrid power quality

Power quality is an important issue in a microgrid due to intermittent nature of the integrated distributed renewable energy sources within the microgrid, the transition between grid-tied and islanded modes of microgrid, nonlinear loads, injected current harmonics by power electronic devices, and loads with considerable reactive power demand.

Energy storage, filtering, and proper control schemes are three main categories utilized by scholars to improve the power quality of a microgrid. In Ref. [3], a two-level and a three-level controlled voltage structure were developed as the active power conditioners (APC) to improve the power quality. Their satisfactory results validate the ability of the controlled APCs in improving the power quality even in a weak microgrid context with strongly nonlinear loads and voltage unbalance. A micro-source grid-connected inverter control method based on an advanced *f-P*, *V-Q* droop control was proposed in Ref. [4] for microgrid power quality management. Authors in Ref. [5] employed a digital power processor in a closed loop scheme to remove and compensate for unwanted harmonic content of power system variables such as voltages and currents, and thus improve the power quality. Zhang et al. [6] took advantage of both energy storage and filtering by combining a flywheel energy storage system and an active power filter for power quality improvement of microgrid. Their simulation results show that the combined system can maintain short-term uninterrupted power supply and meet harmonic content standard. Belov et al. [7] created a virtual prototype of a microgrid including AC/DC/AC converters and energy storage devices. They developed a mathematical model of the converter with energy storage device on the basis of the bridge-element (B-element) concept. Their experimental results demonstrate low-voltage distortion caused by the AC/DC/AC converter. In [8], a cooperative control approach is applied to distribute switching power interfaces in a low-voltage residential smart microgrid to optimize exploitation of local energy sources and improve power quality. It is assumed that the low-voltage residential smart microgrid has limited or no communication capability to neighbor units. Cheng et al. [9] employed distributed multiple active filter system to remove power system's harmonics. They used multiple active-filter units instead of a centralized large-rated active filter to reduce the harmonics in the power system and thus improve the power quality. In Ref. [10], Pulse Width Modulation (PWM) converters are utilized to improve the generators' power factor and produce less harmonics. By changing voltage amplitude and phase of PWM converters, generators' active and reactive powers are controlled to meet the required power quality. A combined system of active power filter and static var compensator is employed in Ref. [11] to reduce the harmonic current produced by power electronic devices and load with considerable reactive power demand, and thus improve power quality of microgrid. In Ref. [12], an optimal power control strategy is presented for a microgrid operating in islanded mode. The proposed control strategy is based on particle swarm real-time self-tuning method.

4. Microgrid control strategies, power management, and optimization

Generally, the control strategies of microgrid can be classified in three different levels including primary (local), secondary (power management), and tertiary (optimization) control level [13].

Primary control, which is also known as local or internal control, is simply based on local measurements. In this local control level, no communication is needed [14]. The frequency and voltage deviation in microgrid depend on active and reactive power mismatch, the load characteristic, and the droop model of distributed generation sources. Therefore, frequency/voltage droop control and inverter output control strategies are the general methods being utilized in the primary level of control [15–26].

Unlike the primary control, the secondary control level approaches strongly need fast and reliable communication systems set by IEC 61850 standard [34, 35]. Both centralized and decentralized control approaches can be used in secondary control level [27–30]. Non-model-based fuzzy and neural network controllers [30, 31] and model-based predictive controllers are the examples of centralized control approaches. Multi-agent based control approaches [32, 33] are the examples of decentralized control methods in secondary control level of microgrid. Also, peak shaving, load following, frugal discharge, state of charge set-point, full power/

http://dx.doi.org/10.5772/intechopen.69400

minimum run time, and ideal predictive dispatch strategies [36–38] are other examples of secondary level control algorithms that have been proposed based on the main power management strategies.

Coordination of multiple microgrids based on the requirements of the main utility system is done in the tertiary control level. Tertiary control is not done by the microgrid itself, and is normally considered part of the main utility grid.

In addition to the developed control strategies for secure, reliable, and stable operation of microgrid, many energy management approaches have also been proposed to handle the characteristics of different power generators and storage systems within the microgrid. In Ref. [39], a microcontroller-based power management system is utilized for the online operation of an experimental low-voltage microgrid to control the battery state-of-charge. In Ref. [40], a dynamic energy management strategy is proposed for a photovoltaic (PV)-based micro grid with combined energy storage. The combined storage system includes batteries and super capacitors. Batteries have low power density and their charge and discharge rate are low causing severe stress under quick load fluctuations. Super capacitors, on the other hand, have high power density and they can easily tolerate quick load fluctuations, but cannot be used as an energy storage system alone as they cannot supply load for a longer time. Combining batteries and super capacitors provides a high-power density storage system. In Ref. [41], authors developed an online power energy management strategy for a hybrid fuel cell/battery as another type of energy storage system utilized in a microgrid. Their proposed method includes three layers, where in first layer, all possible operation modes of microgrid are captured. The power split between batteries and fuel cells is done in the second layer by a fuzzy controller. The set points of each subsystem are regulated in the third layer. In Ref. [42], an energy management system is proposed for a microgrid, including advanced PV generators with embedded storage units and a gas microturbine. The power management is done both centrally at the microgrid side and locally at the customer side by exchanging data and order through a communication network. The proposed power management strategy relies on PV power predictions, load forecasting, and distributed battery storage system. In Ref. [43], an overall power management strategy is proposed to manage power flows among the different energy sources and the storage system within the microgrid in which the primary power sources of the system are wind and PV, and a fuel cell-electrolyzer combination is used as a backup and a long-term storage system. Katiraei and Iravani [44] investigated three power management strategies based on voltage-droop characteristic, voltage regulation, and load reactive power compensation to address real and reactive power management of a multiple distributed generation microgrid system.

In addition to control and power management, optimization is the other pillar of efficient and reliable operation of the microgrid. An optimal allocation methodology and economic analysis of energy storage system within a microgrid is proposed in Ref. [45] using genetic algorithm on the basis of net present value (NPV). In Ref. [46], the optimization is done by a coordinated control approach in two layers including the schedule layer and the dispatch layer. The schedule layer ensures the economical operation of microgrid based on forecasting data, and dispatch layer provides power and voltage regulation based on real-time data. The control and coordination between two layers satisfies both the economical benefit and technical constraints of microgrid on long-time operation. Authors in Ref. [47] optimized a microgrid economically by considering a multi-objective cost function including the operational cost of distributed generators, start-up and shut-down costs, and the cost of interrupted loads. A multi-objective optimization is also implemented in Ref. [48] in order to minimize the emissions of the three main pollutants coming from the gas turbines, i.e., CO_2, CO and NO_x. Another multi-objective multi-scenario optimization method is presented in Ref. [49] to evaluate and optimize the performance of the microgrid under various scenarios from different aspects consisting of construction and operation cost, customer outage cost and environment. In Refs. [50, 51], a centralized hierarchical optimization strategy is presented for economic evaluation of a typical microgrid. A decentralized agent-based strategy is also developed in Refs. [52–54] for microgrid power management and optimization. In Ref. [55], an optimization scheme is developed based on the heuristics using a fuzzy neural network. Another neural network-based optimization method is presented in Ref. [56]. Genetic algorithm [57], particle swarm optimization (PSO) [58–61], and ant colony optimization (ACO) [58, 62] are also utilized as intelligent computational methods for microgrid power management and optimization.

Despite all scholarly works and proposed strategies, it is still necessary to develop a comprehensive control and power management strategy to consider all aspects of energy managements such as different modes of microgrid operation and transition modes, voltage and power flow, coordination of controllable units, economical operation, and stability due to the complexity of microgrid.

5. Microgrid stability, reliability, and protection

Stability, reliability, and protection are the key issues of microgrids due to reverse power flows of distributed generation units, local oscillations, transient modes of microgrid, severe frequency deviations in islanded mode operation, and economical and supply-demand uncertainties of microgrid. To address these issues, various strategies have been proposed and developed [63–66]. In Ref. [63], the stability constraints imposed by droop characteristics in an islanded microgrid are identified using a small-signal approach. It was shown that droop gains have significant impact on the stability and microgrids dynamic performance. In Ref. [64], an adaptive feedforward strategy is proposed to change the dynamic coupling between a distributed resource unit and the host microgrid in a way to make the system stability less sensitive and more robust to the droop coefficients and network dynamics. The conflicting goals of proper load sharing and stability of an islanded microgrid is investigated in Ref. [65]. Suitable load sharing requires high values of angle droop specially under weak system conditions. However, overall stability of the system is negatively impacted by high droop gains. To stabilize the system while ensuring proper load sharing, a supplementary control loop around the primary droop control loop is proposed in this chapter. Stability issues of high gains of droop controller are also investigated in Ref. [66]. In this chapter, a reduced-order mathematical model of the microgrid is proposed for prediction of stability of microgrids with large number of inverters. The results show that the stability of the microgrid can be analyzed with each inverter transformed into an equivalent network separately assuming that the interconnection cables are predominantly inductive and the droop laws can be decoupled.

Protection is another major challenge for microgrids. An overview of microgrid's protection techniques and strategies is presented in Refs. [67, 68]. All the proposed protection systems ensure to respond to both main grid (utility) and microgrid faults as fast as possible for proper isolation of the microgrid from the main grid. Fast operation of protection system plays a crucial role in stability of microgrid after transition to islanded operation. Protection against overcurrent is one of the fundamental elements of electricity grid. However, this is a challenging task in microgrid as the total short-circuit current capacity of a microgrid in islanded and grid-tied modes is different. In fact, in grid-tied mode of operation the protection is simplified by the potentially large fault currents, whereas these fault currents may have relatively low values in islanded mode due to integrated power electronics interfaces in microgrid. This low current capacity in islanded mode is not sufficient to trip conventional overcurrent protection. Therefore an adaptive protection system is needed to change relay settings in real time to guarantee that the microgrid is always protected [69, 70]. Another solution is utilizing digital relays equipped with a communication network [71] to protect the microgrid. An easier solution to the protection issue is designing the microgrid to enter islanded mode in a faulty situation before any protection action could take place [72]. To prevent the flow of large line currents during a voltage sag, two current-limiting algorithms are employed in Ref. [73]. These algorithms are used with a voltage-source inverter (VSI) connected in series between the microgrid and main utility grid to imitate a large virtual RL or L impedance for limiting the large line current during utility voltage sags.

To address the issues and limitations of traditional relaying overcurrent protection techniques in microgrid with bidirectional power flow, a voltage-based fault detection and protection strategy is proposed in Ref. [74]. The proposed protection scheme provides reliable and fast detection for different types of faults within the microgrid in which any output disturbance is detected and the strategy for the isolation of the faulty section is initiated. A state observer is utilized in Ref. [75] to detect and identify faults that occur within the protection zone. Fault detection in a DC microgrid and the method for discriminating the stable abnormal operating condition from the faults is presented in Ref. [76]. In Ref. [77], different fault-detection and issues associated with grounding aspect of protection system is discussed.

6. Microgrid modeling and dynamic characteristics

Dynamic characteristics and suitable mathematical model of the microgrid is needed to design an efficient control and power management strategy. However due to high order model of microgrid and its complexity, many studies are performed based on digital computer simulation software packages such as Power Systems Computer Aided Design (PSCAD)/Electromagnetic Transients including DC (EMTDC) and MATLAB/Simulink/Simpowersystem to select appropriate control parameters on a trial and error basis [78–83]. These scholarly works emphasize that simulation-based modeling approaches facilitate a powerful tool to investigate the behavior of microgrid. However, simulation-based models cannot provide a comprehensive prediction of all microgrid scenarios resulting in poor power quality or instability.

To this end, small-signal dynamic model of a microgrid is presented to provide an accurate and valid representation of microgrid for designing and optimizing the proper control strategies [84–87]. To develop an accurate and valid model of microgrid, individual model of each distributed generation source is obtained in its local *dq0* reference frame and then all individual models are transformed to the microgrid global *dq0* frame to form an integrated model of microgrid. For instance, a practical model of energy storage system as one of the most imperative components of the microgrid is developed in Ref. [88].

Although, a comprehensive model of a microgrid provides accurate dynamic characteristics of the grid, it is very complex and increases the computational burden of the designed controller and power management strategy. Therefore, reduced-order model of microgrids is used [89–90] to address the associated issues of complex comprehensive model. It is shown in Ref. [84] that microgrid low frequency modes are highly related to the network configuration and inverter external power loop, whereas the high frequency modes are largely relative to the inverter inner loop, the dynamic characteristics of loads and network. Thus far, a reduced small-signal model of a microgrid operating in islanded mode is derived in Ref. [89] by neglecting the effect of microgrid high-frequency modes. It should be noted that model order reduction and linearization is done around an operating point, assuming small signal variation around the desired operation point, to apply linear system theory to the power grid. However, network variables would not always remain in small neighborhood of the desired operating point. In these cases, nonlinear model [91, 92] or large-scale system model [93] provide a suitable model of the microgrid. In Ref. [93], linearized state variable models of DC-DC converters and system interconnections at different system operating points and changing interconnections are used to develop a large-scale system model of a DC microgrid. Also, a modular and scalable model for a DC microgrid is presented in Ref. [94] which is independent of the type of renewable energy sources. The authors extend the proposed model of 40-house village equipped with PV panels and energy storage system to an islanded DC microgrid. In Ref. [95], the model of a hybrid power generation system is developed using transfer function model of integrated power electronics converters within the microgrid.

7. Conclusion and discussion

The power delivery system has gradually changed from large-scale unidirectional conventional power generation sources to small-scale bidirectional distributed power generation sources/loads including new plugged-in electric vehicles as bidirectional loads and renewable energy sources such as solar photovoltaic and wind energy. The microgrid concept provides promising solution for the transition from a conventional delivery system to the future grid with high penetration of distributed generation from renewable energy sources without requiring to redesign the distribution system. However, distributed generators/loads and interaction between all nodes within a microgrid substantially increases the complexity of power systems technologies, control techniques, and communications among grid's components.

http://dx.doi.org/10.5772/intechopen.69400

Microgrid challenges	Methods/technologies	Comments/examples
Power quality	Energy storage	Electrical batteries, flywheel mechanical storage, thermal storage
	Filtering	Active power conditioners (APC)
	Proper control methods	*f–P*, *V–Q* droop control, optimal power control
	Energy storage + Filtering	Flywheel storage and active power filter
Control strategies	Primary (local) control	Frequency/voltage droop control
	Secondary centralized control	Non-model-based fuzzy and neural network controllers and model-based predictive controllers
	Secondary decentralized control	Multiagent-based control approaches
	Tertiary (optimization) control	Part of the main utility grid and not microgrid itself
Energy management	Combined energy storage	Combined batteries and super capacitors, hybrid fuel cell/battery
	Power generation prediction and load forecasting	Managing power flow among the different energy sources and the storage system within the microgrid
	Voltage-droop characteristic	Voltage regulation, and load reactive power compensation
Energy optimization	Multi-objective optimization using intelligent methods	Genetic algorithm, fuzzy neural networks, particle swarm optimization (PSO), and ant colony optimization (ACO)
Stability	Proper control strategies	Supplementary control loop around the primary droop control loop
Protection	Adaptive protection system	
	Digital relays	
	Current-limiting algorithms	
	Voltage-based fault detection	
	State observer	
Modeling	Software-based model	PSCAD/EMTDC and MATLAB/ Simulink
	Comprehensive small-signal model	Accurate, but very complex
	Reduced small-signal model	Model order reduction and linearization around an operating point
	Nonlinear model	

Table 1. Summary of different microgrid technologies.

This chapter provided an overall vision of microgrids and their main requirements. An overview of several proposed methods and developed competing technologies for seamless deployment of microgrid and their pros and cons were also presented. **Table 1** summarizes these methods and technologies and their key features. Despite the progresses made over the last few years, technologies remain immature and not yet ready for commercial stage as shown by a large number of different methods, strategies, and policies. To transform the current microgrid into fully commercial, reliable, and cost-effective power grid, a combination of targeted research, development, engineering work, and government incentives is still necessary.

Author details

Reza Sabzehgar

Address all correspondence to: rsabzehgar@mail.sdsu.edu

Electrical and Computer Engineering Department, San Diego State University, San Diego, California, USA

References

[1] Lasseter B. Microgrids [distributed power generation]. In: IEEE Power Engineering Society Winter Meeting. Vol. 1. 2001; pp. 146-149

[2] Marnay C, Robio FJ, Siddiqui AS. Shape of the microgrid. In: IEEE Power Engineering Society Winter Meeting. Vol. 1. 2001; pp. 150-153

[3] Balanuta C, Vechiu I, Gurguiatu G. Improving micro-grid power quality using three-phase four-wire active power conditioners. In: IEEE 16th International Conference on System Theory, Control and Computing (ICSTCC). 2012. pp. 1-5

[4] Li Y, An L, Chun Ming T, Fei R, Shuang Jian P. A micro power quality management technology based on grid-connected inverter. In: IEEE International Conference on Electricity Distribution (CICED). 2010. pp. 1-8

[5] Rafiei S, Moallem A, Bakhshai A, Yazdani D. Application of a digital ANF-based power processor for micro-grids power quality enhancement. In: IEEE 29th Annual Applied Power Electronics Conference and Exposition (APEC). 2014. pp. 3055-3059

[6] Zhang L, Li L, Cui W, Li SH. Study on improvement of micro-grid's power quality based on APF and FESS. In: IEEE Innovative Smart Grid Technologies – Asia (ISGT Asia). 2012. pp. 1-6

[7] Belov V, Butkina A, Bolschikov F, Leisner P, Belov I. Power quality and EMC solutions in micro grids with energy-trading capability. In: IEEE International Symposium on Electromagnetic Compatibility (EMC Europe). 2014. pp. 1203-1208

[8] Tenti P, Costabeber A, Mattavelli P. Improving power quality and distribution efficiency in micro-grids by cooperative control of Switching Power Interfaces. In: IEEE International Power Electronics Conference (IPEC). 2010. pp. 472-479

[9] Cheng P-T, Lee T-L. Distributed active filter systems (DAFSs): A new approach to power system harmonics. IEEE Transactions on Industry Applications. 2006;**42**(5):1301-1309

[10] Wei H, Jianhua Z, Qinghua X, Ziping W. The impact on power quality by PWM converter in micro-grid. In: IEEE International Conference on Sustainable Energy Technologies (ICSET). 2008. pp. 239-243

[11] Dong J, Li L, Ma Z. A combined system of APF and SVC for power quality improvement in microgrid. In: IEEE Power Engineering and Automation Conference. 2012. pp. 1-4

[12] Al-Saedi W, Lachowicz SW, Habibi D, Bass O. Power quality improvement in autonomous microgrid operation using particle swarm optimization. In: IEEE PES Innovative Smart Grid Technologies. 2011. pp. 1-6

[13] Olivares DE, Mehrizi-Sani A, Etemadi AH, Caizares CA, Iravani R, Kazerani M, Hajimiragha AH, Gomis-Bellmunt O, Saeedifard M, Palma-Behnke R, Jimnez-Estvez GA, Hatziargyriou ND. Trends in Microgrid Control. IEEE Transactions on Smart Grid. 2014;**5**(4):1905-1919

[14] Karimi H, Nikkhajoei H, Iravani MR. Control of an electronically-coupled distributed resource unit subsequent to an islanding event. IEEE Transactions on Power Delivery. 2008;**23**(1):493-501

[15] De Brabandere K, Bolsens B, Van den Keybus J, Woyte A, Driesen J, Belmans R. A voltage and frequency droop control method for parallel inverters. IEEE Transactions on Power Electronics. 2007;**22**(4):1107-1115

[16] Zeineldin HH, Kirtley JL. Micro-grid operation of inverter based distributed generation with voltage and frequency dependent loads. In: IEEE Power and Energy Society General Meeting. 2009. pp. 1-6

[17] Xiu Y, Xiang Z, Fei Y, Haiyang Z. A research on droop control strategy and simulation for the micro-grid. In: IEEE International Conference on Electrical and Control Engineering. 2011. pp. 5695-5700

[18] Guerrero JM, Vasquez JC, Matas J, de Vicua LG, Castilla M. Hierarchical control of droop-controlled AC and DC microgrids – A general approach towards standardization. IEEE Transactions on Industrial Electronics. 2011;**58**(1):158-172

[19] Mohamed Y, El-Saadany EF. Adaptive decentralized droop controller to preserve power sharing stability of paralleled inverters in distributed generation microgrids. IEEE Transactions on Power Electronics. 2008;**23**(6):2806-2816

[20] Bahrani B, Saeedifard M, Karimi A, Rufer A. A multivariable design methodology for voltage control of a single-DG-unit microgrid. IIEEE Transactions on Industrial Electronics. 2013;**9**(2):589-599

[21] Brabandere KD, Bolsens B, den Keybus JV, Woyte A, Driesen J, Belmans R. A voltage and frequency droop control method for parallel inverters. IEEE Transactions on Power Electronics. 2007;**22**(4):1107-1115

[22] Yu X, Khambadkone AM, Wang H, Terence STS. Control of parallel-connected power converters for low-voltage microgrid – Part I: A hybrid control architecture. IEEE Transactions on Power Electronics. 2010;**25**(12):2962-2970

[23] Li YW, Kao C-N. An accurate power control strategy for power-electronics-interfaced distributed generation units operating in a low-voltage multibus microgrid. IEEE Transactions on Power Electronics. 2009;**24**(12):2977-2988

[24] Vandoorn TL, Meersman B, Degroote L, Renders B, Vande-velde L. A control strategy for islanded microgrids with DC-link voltage control. IEEE Transactions on Power Delivery. 2011;**26**(2):703-713

[25] Marwali MN, Jung J-W, Keyhani A. Control of distributed generation systems - Part II: Load sharing control. IEEE Transactions on Power Electronics. 2004;**19**(6):1551-1561

[26] Delghavi MB, Yazdani A. Islanded-mode control of electronically coupled distributed-resource units under unbalanced and nonlinear load condition. IEEE Transactions on Power Delivery. 2011;**26**(2):661-673

[27] Mehrizi-Sani A, Iravani R. Potential-function based control of a microgrid in islanded and grid-connected modes. IEEE Transactions on Power Systems. 2010;**25**(4):1883-1891

[28] Prodanovic M, Green TC. High-quality power generation through distributed control of a power park microgrids. IIEEE Transactions on Industrial Electronics. 2006;**53**(5):1471-1482

[29] Dimeas AL, Hatziargyriou ND. Operation of a multiagent system for microgrid control. IEEE Transactions on Power Systems. 2005;**20**(3):1447-1455

[30] Pilo F, Pisano G, Soma G. Neural implementation of microgrid central controllers. In: IEEE 5th International Conference on Industrial Informatics. Vol. 2. 2007. pp. 1177-1182

[31] Jia-jun Y, Li-xiao Y, Dong-meng T, Bo L, Dong L. Study on the fuzzy control strategy based on back- to-back micro grid connection. In: EEE Asia Pacific Power and Energy Engineering Conference. 2012. pp. 1-5

[32] Wang Z, Yang R, Wang L. Intelligent multi-agent control for integrated building and micro-grid systems. In: IEEE PES Innovative Smart Grid Technologies. 2011. pp. 1-7

[33] Zheng W-D, Cai J-D. A multi-agent system for distributed energy resources control in microgrid. In: IEEE 5th International Conference on Critical Infrastructure (CRIS). 2010. pp. 1-5

[34] Ustun T, Ozansoy C, Zayegh A. Modeling of a centralized microgrid protection system and distributed energy resources according to IEC 61850-7-420. IEEE Transactions on Power Systems. 2012;**27**(3):1560-1567

[35] Colet-Subirachs A, Ruiz-Alvarez A, Gomis-Bellmunt O, Alvarez- Cuevas-Figuerola F, A. Sudria-Andreu. Centralized and distributed active and reactive power control of a utility connected microgrid using IEC 61850. IEEE Systems Journal. 2012;**6**(1):58-67

[36] Prema V, Rao KU. Predictive models for power management of a hybrid microgrid – A review. In: International Conference on Advances in Energy Conversion Technologies (ICAECT). 2014. pp. 7-12

[37] Dursun E, Kilic O. Comparative evaluation of different power management strategies of a stand-alone PV/Wind/PEMFC hybrid power system. International Journal of Electrical Power and Energy Systems. 2012;**34**:81-89

[38] Madhu sudhakar P, Murali Mohan B. Predictive and optimizing energy management of photo voltaic fuel cell hybrid systems with short time energy storage. International Journal of Engineering Research and Applications. 2012;**2**(6):550-556

[39] Belvedere B, Bianchi M, Borghetti A, Nucci CA, Paolone M, Peretto A. A microcontroller-based power management system for standalone microgrids with hybrid power supply. IEEE Transactions on Sustainable Energy. 2012;**3**(3):422-431

[40] Satishkumar R, Kollimalla SK, Mishra MK. Dynamic energy management of micro grids using battery super capacitor combined storage. In: Annual IEEE India Conference. 2012. pp. 1078-1083

[41] Hajizadeh A, Golkar MA. Intelligent power management strategy of hybrid distributed generation system. Elsevier Electrical Power and Energy Systems. 2007;**29**:783-795

[42] Kanchev H, Lu D, Colas F, Lazarov V, Francois B. Energy management and operational planning of a microgrid with a PV-based active generator for smart grid applications. IEEE Transactions on Industrial Electronics. 2011;**58**(10):4583-4592

[43] Wang C, Hashem M. Power management of a stand-alone wind/photovoltaic/fuel cell energy system. IEEE Transactions on Energy Conversion. 2007;**23**(3):957-967

[44] Katiraei F, Iravani MR. Power management strategies for a microgrid with multiple distributed generation units. IEEE Transactions on Power Systems. 2006;**21**(4):1821-1831

[45] Chen C, Duan S, Cai T, Liu B, Hu G. Optimal allocation and economic analysis of energy storage system in microgrids. IEEE Transactions on Power Electronics. 2011; **26**(10):2762-2773

[46] Jiang Q, Xue M, Geng G. Energy management of microgrid in grid-connected and stand-alone modes. IEEE Transactions on Power Systems. 2013;**28**(3):3380-3389

[47] Nasrolahpour E, Doostizadeh M, Ghasemi H. Optimal management of micro grid in restructured environment. In: IEEE Second Iranian Conference on Renewable Energy and Distributed Generation. 2012. pp. 116-120

[48] Kanchev H, Lu D, Francois B, Lazarov V. Smart monitoring of a microgrid including gas turbines and a dispatched PV-based active generator for energy management and emissions

reduction. In: IEEE Innovative Smart Grid Technology Conference Europe (ISGT Europe). 2010. pp. 1-8

[49] Wang C-S, Yu B, Xiao J, Guo L. Multi- scenario, multi-objective optimization of grid-parallel microgrid. In: IEEE 4th International Conference on Electric Utility Deregulation and Restructuring and Power Technologies. 2011. pp. 1638-1646

[50] Tsikalakis AG, Hatziargyriou ND. Centralized control for optimizing microgrids operation. IEEE Transactions on Energy Conversion. 2008;**23**(1):241-248

[51] Milo A, Gaztaaga H, Etxeberria-Otadui I, Bilbao E, Rodriguez P. Optimization of an experimental hybrid microgrid operation: Reliability and economic issues. In: IEEE International Conference on Power Technology. 2009. pp. 1-6

[52] Oyarzabal J, Jimeno J, Ruela J, Engler A, Hardt C. Agent based micro grid management system. In: IEEE International Conference on Future Power Systems. 2005. pp. 6-11

[53] Hatziargyriou ND, Dimeas H. Operation of a multi-agent system for microgrid control. IEEE Transactions on Power Systems. 2005;**20**(3):1447-1455

[54] Eddy FYS, Gooi HB. Multi-agent system for optimization of microgrids. In: IEEE 8th International Conference on Power Electronics. 2011. pp. 2374-2381

[55] Chakraborty S, Weiss MD, Simes MG. Distributed intelligent energy management system for a single-phase high-frequency AC microgrid. IEEE Transactions on Industrial Electronics. 2007;**54**(1):97-109

[56] Celli G, Pilo F, Pisano G, Soma GG. Optimal participation of a microgrid to the energy market with an intelligent EMS. In: IEEE 7th International Power Engineering Conference, Vol. 2. 2005. pp. 663-668

[57] Deng Q, Gao X, Zhou H, Hu W. System modeling and optimization of microgrid using genetic algorithm. In: IEEE 2nd International Conference on Intelligent Control and Information Processing. 2011. pp. 1540-544

[58] Colson CM, Nehrir MH, Pourmousavi SA. Towards real-time microgrid power management using computational intelligence methods. In: IEEE Power and Energy Society General Meeting. 2010. pp. 1-8

[59] Chung Il-Y, Liu W, Cartes DA, Schoder K. Control parameter optimization for a microgrid system using particle swarm optimization. In: IEEE International Conference on Sustainable Energy Technologies (ICSET). 2008. pp. 837-842

[60] Hassan MA, Abido MA. Optimal design of microgrids in autonomous and grid-connected modes using particle swarm optimization. IEEE Transactions on Power Electronics. 2011;**26**(3):775-769

[61] Prommee W, Pongprapunt N, Ongsaku W. Improved reliability model and optimal protective device placement in micro grid by improved binary particle swarm optimization. In: IEEE 8th International Conference on Power Electronics. 2011. pp. 1514-1519

[62] Colson C, Nehrir M, Wang C. Ant colony optimization for microgrid multi-objective power management. In: IEEE Power Systems Conference and Exposition (PSCE). 2009. pp. 1-7

[63] Barklund E, Pogaku N, Prodanovic M, Hernandez-Aramburo C, Green TC. Energy management in autonomous microgrid using stability-constrained droop control of inverters. IEEE Transactions on Power Electronronics. 2008;**23**(5):2346-2352

[64] Delghavi MB, Yazdani A. An adaptive feedforward compensation for stability enhancement in droop-controlled inverter-based microgrids. IEEE Transactions on Power Delivery. 2011;**26**(3):1764-1773

[65] Majumder R, Chaudhuri B, Ghosh A, Majumder R, Ledwich G, Zare F, Improvement of stability and load sharing in an autonomous microgrid using supplementary droop control loop. IEEE Transactions on Power Systems. 2010;**25**(2):796-808

[66] Iyer SV, Belur MN, Chandorkar MC. A generalized computational method to determine stability of a multi-inverter microgrid. IEEE Transactions on Power Electronronics. 2010;**25**(9):2420-2432

[67] Wei J, Zheng-you H, Zhi-qian B. The overview of research on microgrid protection development. In: IEEE International Conference on Intelligent System Design and Engineering Application (ISDEA). 2010. pp. 692-697

[68] Laaksonen H. Protection principles for future microgrids. IEEE Transactions on Power Electronics. 2010;**25**(12):2910-2918

[69] Oudalov A, Fidigatti A, Degner T, Valov B, Hardt C, Yarza JM, Li R, Jenkins N, Awad B, van Overbeeke F, Hatziargyriou N, Lorentzou M. Advanced architectures and control concepts for more microgrids: Novel protection systems for microgrids. IDeliverable DC2. 2009

[70] Islam M, Gabbar HA. Study of micro grid safety & protection strategies with control system infrastructures. Scientific Research, Smart Grid and Renewable Energy. 2012;**3**(1):1-9

[71] Sortomme E, Venkata SS, Mitra J. Microgrid protection using communication-assisted digital relays. IEEE Transactions on Power Delivery. 2009;**25**(4):2789-2796

[72] Nikkhajoei H, Lasseter RH. Microgrid fault protection based on symmetrical and differential current components. In: Prepared for Public Interest Energy Research, California Energy Commission. 2006

[73] Vilathgamuwa DM, Chiang LP, Wei LY. Protection of microgrids during utility voltage sags. IEEE Transactions on Industrial Electronics. 2006;**53**(5):1427-36

[74] Al-Nasseri H Redfern MA, Li F. A voltage based protection for micro-grids containing power electronic converters. In: IEEE Power Engineering Society General Meeting. 2006. pp. 1-7

[75] Esreraig M, Mitra J. An observer-based protection system for microgrids. In: IEEE Power and Energy Society General Meeting. 2011. pp. 1-7

[76] Lee W-S, Kang S-H. Protection for distributed generations in the DC micro-grid. In: IEEE International Conference and Exhibition on Innovative Smart Grid Technologies. 2011. pp. 1-5

[77] Salomonsson D, Sannino A. Protection of low voltage DC microgrids. IEEE Transactions on Power Delivery. 2009;**24**(3):1045-1053

[78] Katiraei F, Iravani MR, Lehn PW. Micro-grid autonomous operation during and subsequent to islanding process. IEEE Transactions on Power Delivery. 2005;**20**(1):248-257

[79] Zhou J, Chen Y, Huang M, Tong Y. Study on energy control strategies in microgrid-modeling and simulation. In: IEEE Power and Energy Engineering Conference (APPEEC). 2012. pp. 1-4

[80] Brissette A, Hoke A, Maksimovic D, Pratt A. A microgrid modeling and simulation platform for system evaluation on a range of time scales. In: IEEE Energy Conversion Congress and Exposition (ECCE). 2011. pp. 968-976

[81] Mok KY, Norman CFT, Lau WH, Leung MC. Experiment-based simulation for distortion behavior in LV networks for microgrid modeling. In: IEEE Power and Energy Society General Meeting. 2011. pp. 1-6

[82] Chang GW, Zeng GF, Su HJ, Hsu LY, Chang YR, Lee YD, Lin CH. Modelling and simulation for INER AC microgrid control. In: IEEE PES General Meeting Conference and Exposition. 2014. pp. 1-5

[83] Bae S, Kwasinski A. Dynamic modeling and operation strategy for a microgrid with wind and photovoltaic resources. IEEE Transactions on Smart Grid. 2012;**3**(4):1867-1876

[84] Katiraei F, Iravani MR, Lehn PW. Small-signal dynamic model of a micro-grid including conventional and electronically interfaced distributed resources. IET Generation, Transmission and Distribution. 2007;**1**(3):369-378

[85] Raju PESN, Jain T. Small signal modelling and stability analysis of an islanded AC microgrid with inverter interfaced DGs. In: IEEE International Conference on Smart Electric Grid (ISEG). 2014. pp. 1-8

[86] Valdivia V, Diaz D, Gonzalez-Espin F, Foley R, Chang NC. Systematic small signal modeling and stability analysis of a microgrid. In: IEEE 5th International Symposium on Power Electronics for Distributed Generation Systems (PEDG). 2014. pp. 1-5

[87] Zhu M, Li H, Li X. Improved state-space model and analysis of islanding inverter-based microgrid. In: IEEE International Symposium on Industrial Electronics (ISIE). 2013. pp. 1-5

[88] Bahramirad S, Camm E. Practical modeling of Smart Grid SMS™ storage management system in a microgrid. In: IEEE PES Transmission and Distribution Conference and Exposition. 2012. pp. 1-7

[89] Wang Y, Lu Z, Min Y, Wang Z. Small signal analysis of microgrid with multiple micro sources based on reduced order model in islanding operation. In: IEEE Power and Energy Society General Meeting. 2011. pp. 1-9

[90] Rasheduzzaman MD, Mueller JA, Kimball JW. Reduced-order small-signal model of microgrid systems. IEEE Transactions on Sustainable Energy, Early Access, 2015

[91] Nazaripouya H, Mehraeen S. Modeling and nonlinear optimal control of weak/islanded grids using FACTS device in a game theoretic approach. IEEE Transactions on Control Systems, Early Access. 2015

[92] Meza C, Biel D, Jeltsema D, Scherpen JMA. Lyapunov-based control scheme for single-phase grid-connected PV central inverters. IEEE Transactions on Control System Technology. 2012;**20**(2):520-529

[93] Tulpule P, Yurkovich S, Wang J, Rizzoni G. Hybrid large scale system model for a DC microgrid. In: American Control Conference (ACC). 2011. pp. 389-3904

[94] Farooq R, Mateen L, Ahmad M, Akbar SQ, Khan HA, Zaffar NA. Smart DC microgrids: Modeling and power flow analysis of a DC Microgrid for off-grid and weak-grid connected communities. In: IEEE PES Asia-Pacific Power and Energy Engineering Conference (APPEEC). 2014. pp. 1-6

[95] Kim SK, Jeon JH, Cho CH, Ahn JB, Kwon SH. Dynamic modeling and control of a grid-connected hybrid generation system with versatile power transfer. IEEE Transactions on Industrial Electronics. 2008;**55**(4):1677-1688

[96] Gu Y, Xiang X, Li W, He X. Mode-adaptive decentralized control for renewable DC microgrid with enhanced reliability and flexibility. IEEE Transactions on Power Electronics. 2014;**29**(9):5072-5080

[97] Ahmadi R, Ferdowsi M. Improving the performance of a line regulating converter in a converter-dominated DC microgrid system. IEEE Transactions on Smart Grid. 2014;**5**(5):2553-2563

[98] Kumar M, Srivastava SC, Singh SN. Control strategies of a DC microgrid for grid Connected and islanded operations. IEEE Transactions on Smart Grid. 2015;**6**(4):1588-1601

[99] Gu Y, Li W, He X. Frequency-coordinating virtual impedance for autonomous power management of DC microgrid. IEEE Transactions on Power Electronics. 2015;**30**(4):2328-2337

[100] Shadmand MB, Balog RS. Multi-objective optimization and design of photovoltaic-wind hybrid system for community smart DC microgrid. IEEE Transactions on Smart Grid. 2014;**5**(5):2635-2643

[101] Yu X, She X, Zhou X, Huang AQ. Power management for DC microgrid enabled by solid-state transformer. IEEE Transactions on Smart Grid. 2014;**5**(2):954-965

Chapter 2

Distributed Energy Resources to Improve the Power Quality and to Reduce Energy Costs of a Hybrid AC/DC Microgrid

Luisa Alfieri, Antonio Bracale, Pierluigi Caramia and Guido Carpinelli

Additional information is available at the end of the chapter

http://dx.doi.org/10.5772/intechopen.68766

Abstract

This chapter deals with microgrids (μGs), i.e., a group of interconnected loads and distributed energy resources that act as a single controllable entity with respect to the grid. The μGs can be classified into AC and DC μGs depending on the characteristics of the supply voltage, with both solutions characterized by advantages and challenges. Recently, hybrid AC/DC μGs have been developed with the aim to exploit the advantages of both AC and DC solutions. Hybrid μGs require being properly controlled to guarantee their optimal behavior, in both grid-connected and islanding mode. In this chapter, we propose an optimal control strategy for a hybrid μG to be realized in an actual Italian industrial facility. The control strategy operates with the aim to simultaneously minimize the energy costs and to compensate waveform distortions. The key result of the chapter consists in evidencing the technical and economic advantages of the proposed solution by means of real-time simulations of the hybrid μG performed through Matlab/Simulink development tool in the different conditions (grid-connected and islanding mode).

Keywords: hybrid microgrids, distributed energy resources, power quality, optimal control strategies

1. Introduction

Nowadays distributed generators, storage systems, and controllable loads much more actively and simultaneously contribute to an optimal operation of the electrical distribution systems in the frame of the new concepts of smart grids (SGs) and microgrids (μGs).

In this exciting context, new technologies and services are being introduced to make the electrical networks more reliable, efficient, secure, and environmentally friendly.

This chapter deals with μGs that CIGRÉ C6.22 Working Group defines as electricity distribution systems containing loads and distributed energy resources that can be operated in a controlled, coordinated way either while they are connected to the main power network (grid-connected mode), or when operate in islanding mode [1].

As well known, μGs can be AC and DC μGs on the basis of the nature of the supplied voltage [2–4]. AC μGs have the advantage of utilizing existing AC technologies as well as DC μGs seem particularly suitable for supporting the current needs because most distributed generation sources, such as photovoltaic plants, fuel cells, and storage systems, generate DC power directly. Moreover, DC grids can also guarantee high power quality levels to AC sensible loads, even though this provision requires the presence of additional DC/AC converters.

Recently, hybrid AC/DC μGs have been developed with the aim to exploit the advantages of both AC and DC solutions and, in particular, to accelerate the integration process of DC power technologies into the existing consolidated AC systems [5–10].

As well known, several static converters are installed in the hybrid μGs, either to connect renewable generation and storage systems to the AC or DC grids, or as an interface between AC and DC grids. In this context, optimization strategies are required to perform the converters optimal control with the aim of efficiently operating the whole hybrid μG. These control strategies should take into account all goals and operating constraints of both DC and AC sections.

Motivated by the above requirements, in the following sections of this chapter, we propose an optimal control strategy for a hybrid μG to be realized in an actual Italian industrial facility. In particular, the static converters of the hybrid μG are controlled with the aim to simultaneously minimize the energy costs and to compensate waveform distortions, thus ensuring an optimal technical and economical behavior of the whole μG.

The remainder of the chapter is organized as follows. The electrical system of the actual Italian industrial facility is briefly recalled in Section 2, where also the structure of the proposed hybrid microgrid is shown. Section 3 deals with the proposed control strategy. Numerical applications are shown in Section 4, Section 5 provides our conclusions, and some data of the test system under study are explained in Appendix.

2. The electrical industrial system under study

The electrical industrial system under study was a hybrid AC/DC microgrid obtained by the modification of the existing LV distribution system of an actual Italian industrial facility where transformers are assembled. The simplified electrical scheme before the modification is reported in **Figure 1**. The existing facility's electrical distribution system is connected to the MV grid through a 20/0.4–630 kVA transformer and includes four low-voltage feeders; each feeder is dedicated to a different manufacturing process, i.e.: (i) tanks and boxes manufactory, (ii) assembly, (iii) winding and coils and (iv) test. An automatically switched 120 kVAr capacitor bank with 10 kVAr step is installed at bus 3 to guarantee a power factor at PCC (bus 2) equal to 0.95.

http://dx.doi.org/10.5772/intechopen.68766

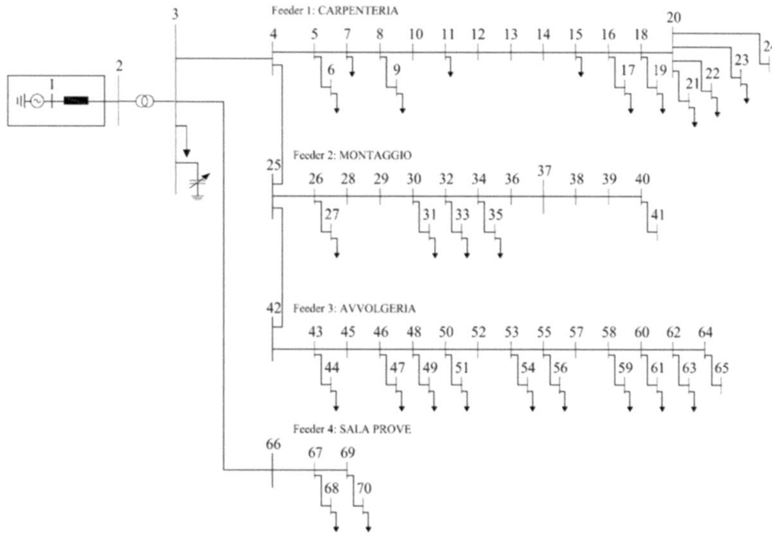

Figure 1. Electrical simplified scheme of the existing electrical distribution system of the industrial plant under study.

The data of the lines, transformers and loads are reported in Appendix.

The modifications on the existing electrical system of **Figure 1** and the applied control strategy implemented in the centralized control system (CCS) are finalized (i) to improve the continuity and quality of the energy delivered to the loads and (ii) to optimize the exchange of energy between industrial plant and MV distribution grid.

The following modifications are effected (**Figure 2**): (i) a DC micro grid (DCμG) is connected at AC bus 16 through an AC/DC grids interfacing converter; (ii) a Battery Energy Storage System (BESS) is installed at AC bus 3 through a DC/AC static converter; and (iii) a dispatchable micro-turbine is connected at AC bus 20 through an AC/AC static converter.[1]

The DC micro grid (DCμG) consists of a photovoltaic (PV) plant equipped with a maximum power point tracker control system and connected to the DC grid through a DC/DC static converter and three sensitive AC loads (Folding walls island robot, Sandblasting machine and PLCs used to control the automation of the whole electromechanical process[2]) moved from the original electrical scheme of **Figure 1**.

The sensitive loads are controllable in terms of start-up times; in particular, their working time intervals are obtained by a day-ahead scheduling aimed at achieving the minimization of the costs for the electricity purchase of the hybrid AC/DC μG [12].[3]

---

[1]In the following we will use the term micro-turbine, BESS and PV generators to indicate the system constituted by the micro-turbine, BESS and PV panels together with their interfacing static converters.

[2]These loads were chosen in accordance with the technical operators of the industrial facility on the basis of a deep study on the industrial process [11]

[3]An Uninterruptible Power Supply (UPS) system is dedicated for PLC power backup to provide a soft shut-down of the load control devices in case AC μG out of service.

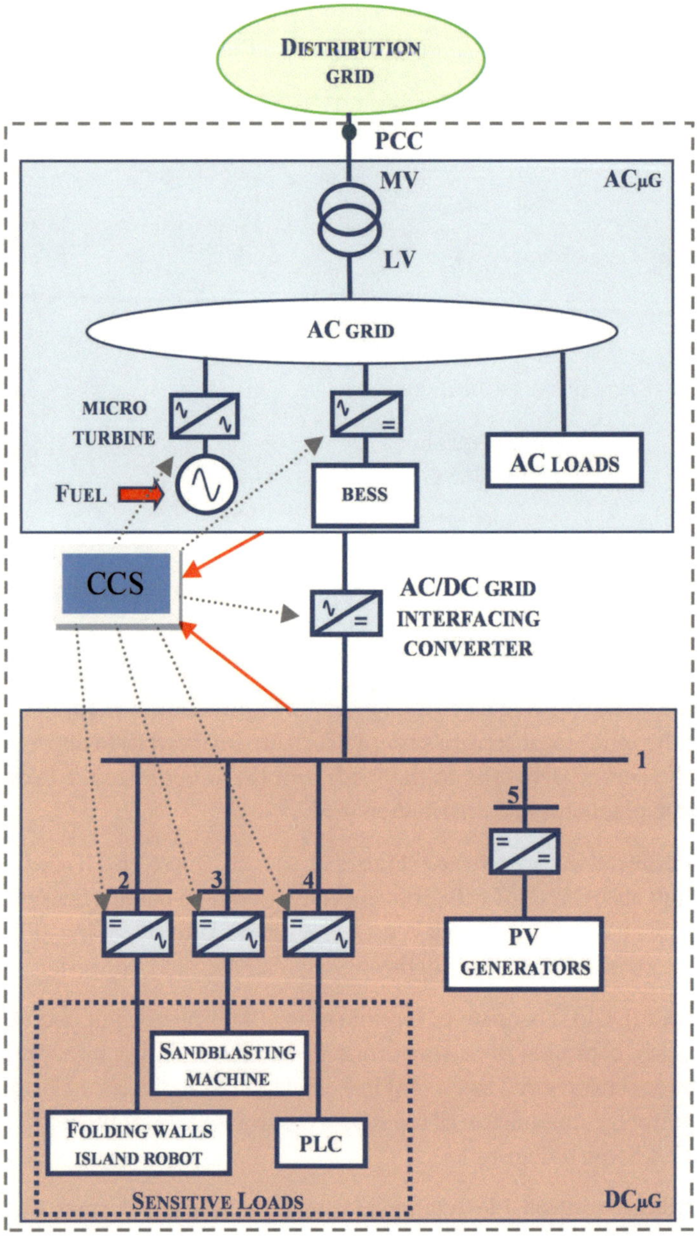

Figure 2. Scheme of the hybrid AC/DC μG.

The above modifications lead to a hybrid μG, which includes AC buses (ACμG) and DC buses (DCμG).

The data of the newly added components are reported in Appendix.

http://dx.doi.org/10.5772/intechopen.68766

3. The proposed control strategy

The proposed control strategy is based on the solution of optimization problems performed at fundamental and at harmonic frequencies, where the efficient operation of the DC and AC grids is guaranteed by a proper control of the static converters installed in the hybrid μG, either to connect micro-turbine and storage system to the AC grid, or as interface between AC and DC grids.

The control strategy allows the hybrid AC/DC μG to operate both in grid-connected and in islanding mode as described in details in the following subsections.

3.1. Grid-connected mode

When the hybrid μG is in *grid-connected mode*, the converters are controlled with the aim to simultaneously minimize: (i) the energy costs sustained for the electrical energy imported from the main distribution network and the production cost of the micro-turbine and (ii) the waveform distortions of the AC bus voltages.

The converter sizes have to be appropriately chosen to correctly perform both the required services.

Two different optimization models are formulated and solved by CCS to provide the reference signals for the energy cost minimization (reference signals at fundamental frequency) and for the waveform distortion compensation (reference signals at harmonic frequencies), as shown in **Figure 3**.

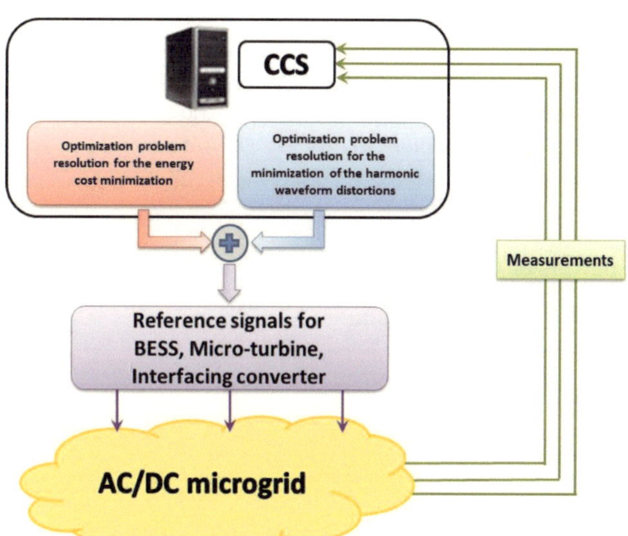

Figure 3. Optimal control strategy for grid-connected mode.

3.1.1. Energy cost minimization

At the fundamental frequency, the real-time operation of the hybrid AC/DC μG is optimized by controlling the active and reactive power of the BESS, the active and reactive power of the micro-turbine and the reactive power provided by the AC/DC grids interfacing converter. The reference signals for the converters, which minimize the total energy cost, are obtained solving a non-linear constrained optimization problem, such as:

$$\min f_{obj}(\mathbf{x}) \tag{1}$$

s.t.

$$\beta_h(\mathbf{x}) = 0 \; h = 1, \ldots, N_{eq} \tag{2}$$

$$\gamma_k(\mathbf{x}) \leq 0 \; k = 1, \ldots, N_{ineq} \tag{3}$$

where f_{obj} is the objective function and β_h and γ_k are the hth equality and the kth inequality constraints to be met, respectively. The vector **x** includes the state and control variables.

Dividing the day into N_t time slots of the same duration Δt, the optimization problem in Eqs. (1)–(3) is repeatedly solved at each ith time slot and its solution furnishes the reference signals needed at the successive time slot $(i + 1)$th for the controllers of micro-turbine, BESS and AC/DC grids interfacing converters.

Input and output data, the objective function and the constraints structure are listed in the following.[4]

The input and output data of the optimization problem are:

Input data:

- the state of charge of the BESS at the end of the ith time slot and at the end of the day (i.e., at the end of time slot N_t). In particular, the value at the end of the ith time slot is obtained as result of the optimization problem in Eqs. (1)–(3) solved at the previous $(i$-1)th time slot; the value at the end of the day is fixed on the basis of the desired value at the beginning of the successive day;
- the forecasted powers of the PV system and of the non-controllable loads, all provided by proper forecasting tools for all the time slots from $(i + 1)$th to N_t [13–15];
- the power required by the controllable loads[5];
- the energy charge;

[4]It should be noted that a fixed value is assigned to the energy stored in the battery at the last time slot of the day (Nt), thus influencing the battery power profile over the whole day; then, the optimization problem becomes multi-period with the consequence that, at the ith time slot, objective function and constraints must be formulated along all the rest of the day (i.e., time slots included in $\{i+1, \ldots, N_t\}$).

[5]The real time control procedure follows a day ahead scheduling procedure (performed off-line), as the one proposed in [12], finalized to schedule the powers of DCμG controllable loads.

http://dx.doi.org/10.5772/intechopen.68766

- the charge and discharge periods of the BESS in ACμG which are fixed a-priori in function of the energy hourly tariff (typically, the discharge period is during the hours of peak energy tariff).

Output data:

- the active and reactive power of the BESS and the corresponding State of Charge (SOC);
- the active and reactive powers of the microturbines;
- the reactive power of the interfacing power converter.

All the output data are calculated from the $(i + 1)$th time slot to the end of the day, but, obviously, only the values of the $(i + 1)$th time slot are sent to the controllers of the ACμG BESS, micro-turbine and AC/DC grids interfacing converter.

The *objective function* in Eq. (1) to be minimized is the daily total costs sustained by the industrial facility for the electrical energy. This cost is the sum of the cost for the energy imported from the MV electrical distribution grid and of the production cost of the micro-turbine. The objective function $f_{obj}(\mathbf{x})$ results in:

$$f_{obj}(\mathbf{x}) = \sum\nolimits_{k=i+1}^{N_t} (\mathrm{Pr}_{E,k}\, P_k^{grid} \Delta t + C_k^{mTG} \Delta t) \tag{4}$$

In Eq. (4), P_k^{grid} is the active power furnished by the distribution grid at the kth time interval, $\mathrm{Pr}_{E,k}$ and C_k^{mTG} are the energy charge and the production cost of the micro-turbine generator at the kth time interval, expressed as [16]:

$$C_k^{mTG} = a_{mTG} \cdot (P_{20,k}^{mTG})^2 + b_{mTG,n} P_{20,k}^{mTG} + c_{mTG}\, u(k), k = \ i+1, \ldots, \ Nt \tag{5}$$

In Eq. (5), a_{mTG}, b_{mTG} and c_{mTG} are the cost coefficients of the micro-turbine, $P_{20,k}^{mTG}$ is the power injected by the micro-turbine connected at busbar 20, $u(k)$ is equal to 1 (if the micro-turbine is switched ON) or 0 (if the micro-turbine is switched OFF). The start-up cost is not considered in Eq. (5) [17].

The *equality and inequality constraints* related to the DCμG, the ACμG, and the AC/DC grids interfacing converter are illustrated in **Tables 1–3**, respectively.

With reference to the DCμG constituted by M_{DC} buses (**Table 1**), the equality constraints are the DC load flow Eq. (6), while the inequality constraints refer to limits on bus voltage magnitudes Eq. (7) and line currents Eq. (8).

In Eq. (6), $P_{s,k}^{DC}$ is the active power injected in the sth DC bus at the kth time slot, $V_{s,k}^{DC}$ is the sth bus voltage magnitude in the kth time slot, G_{sd} is the s-d element of the conductance matrix.[6]

[6]The terms of the conductance matrix don't depend on the time interval, since no time variation of the DCμG topology is assumed.

Load flow equations

$$P_{s,k}^{DC} = \sum_{d=1}^{M_{DC}} V_{s,k}^{DC} V_{d,k}^{DC} G_{sd}\,, \quad k = i+1, \ldots, N_t, s = 1, \ldots, M_{DC} \tag{6}$$

Voltage magnitudes

$$V_{s_{LB}}^{DC} \leq V_{s,k}^{DC} \leq V_{s_{UB}}^{DC}, \quad k = i+1, \ldots, N_t, s = 2, \ldots, M_{DC} \tag{7}$$

Line currents

$$I_{l,k}^{DC} \leq I_{l,\,max}^{DC}, \quad k = i+1, \ldots, N_t,\; l = 1, \ldots, L_{DC} \tag{8}$$

Table 1. DCμG's constraints.

In particular, $P_{s,k}^{DC}$ with $s = 2,\ldots, M_{DC}$ is the scheduled active power of the controllable loads at load busbars and the active power injected by PV generator. In particular, at busbars 2, 3 and 4 only loads are connected and their active powers are the ones reported in Table A. VI of Appendix (with minus sign) corresponding to the busbars 17, 9 and 22 of the actual industrial system shown in **Figure 1**; at bus 5, only PV generator is connected and the active power is equal to the value obtained by forecasted PV solar irradiance; at bus 1, the voltage $V_{1,k}^{DC} = V_{sp}^{DC}$, for each kth time slot, since AC/DC interfacing converter controls the DC voltage at the constant value V_{sp}^{DC}.

In Eq. (7), $V_{s_{LB}}^{DC}$ and $V_{s_{UB}}^{DC}$ are the lower and upper bounds of voltage amplitude of the sth DC bus. In Eq. (8), $I_{l,k}^{DC}$ is the value of current flowing in the lth line during the kth time slot, L_{DC} is the number of the DC grid lines and $I_{l,\,max}^{DC}$ is the ampacity of the lth line.

With reference to the ACμG constituted by M_{AC} buses (**Table 2**), the equality constraints are the AC load flow in Eq. (9) and the balance equation of the energy stored in the BESS at the end of the day (Eq. (16)), while the inequality constraints refer to limits on bus voltage magnitudes (Eq. (10)), line currents (Eq. (11)), active and reactive power at PCC (Eq. (12)), MV/LV transformers apparent power (Eq. (13)), micro-turbine active and apparent power (Eqs. (14) and (15)), energy stored, charging/discharging powers and apparent power of BESS in the generic time slot (Eqs. (17)–(19)).

In Eq. (9), $P_{s,k}^{AC}$ ($Q_{s,k}^{AC}$) is the active (reactive) power injected in busbar s at the kth time slot; $V_{s,k}$ $\vartheta_{s,k}$ are the magnitude and argument of the voltage of sth AC bus at the kth time slot, respectively, $G_{s,j}$ ($B_{s,j}$) is the s-j term of the conductance (susceptance) matrix.[7] The first busbar of the grid in

[7]The terms of the conductance and susceptance matrix don't depend on the time interval, since no time variation of the AC microgrid topology is assumed.

http://dx.doi.org/10.5772/intechopen.68766

AC grid

Load flow equations

$$P_{s,k}^{AC} = V_{s,k}^{AC}\sum_{j=1}^{M_{AC}} V_{j,k}^{AC}[G_{s,j}\cos(\vartheta_{s,k} - \vartheta_{j,k}) + B_{s,j}\sin(\vartheta_{s,k} - \vartheta_{j,k})]$$

$$Q_{s,k}^{AC} = V_{s,k}^{AC}\sum_{j=1}^{M_{AC}} V_{j,k}^{AC}[G_{s,j}\sin(\vartheta_{s,k} - \vartheta_{j,k}) - B_{s,j}\cos(\vartheta_{s,k} - \vartheta_{j,k})] \qquad (9)$$

$$k = i+1, \ldots, N_t \; s = 1, \ldots, M_{AC}$$

Voltage magnitudes

$$V_{s_{LB}}^{AC} \le V_{s,k}^{AC} \le V_{s_{UB}}^{AC}, \quad k = i+1, \ldots, N_t \; s = 2, \ldots, M_{AC} \qquad (10)$$

Line currents

$$I_{l,k}^{AC} \le I_{l,max}^{AC}, \quad k = i+1, \ldots, N_t \; l = 1, \ldots, L_{AC} \qquad (11)$$

P_{PCC} powers

$$P_{2-3,k}^{AC} \le P_{PCC}^{Max} \qquad (12)$$

$$Q_{2-3,k}^{AC} \ge 0, \quad k = i+1, \ldots, N_t$$

MV/LV transformer apparent power

$$\sqrt{P_{tr,k}^2 + Q_{tr,k}^2} \le S_{TR}, \quad k = i+1, \ldots, N_t \qquad (13)$$

Micro-turbine

Active and apparent power

$$P_{min}^{mTG}\, u(k) \le P_{20,\,k}^{mTG} \le P_{max}^{mTG}\, u(k), \quad k = i+1, \ldots, N_t \qquad (14)$$

$$\sqrt{P_{20,\,k}^{mTG2} + Q_{20,\,k}^{mTG2}} \le S_{nom}^{mTG}, \quad k = i+1, \ldots, N_t \qquad (15)$$

BESS

Energy stored at the end of the day

$$E_i - \Delta t \sum_{k=i+1}^{N_t} \gamma_k\, P_{3,k}^{BESS} = E_{sp}^{in}, \quad \gamma_k = \begin{cases} \dfrac{1}{\eta_{bess}} & if\ P_{3,k}^{BESS} \ge 0 \\ \eta_{bess} & if\ P_{3,k}^{BESS} < 0 \end{cases} \qquad (16)$$

AC grid	
Energy stored in the generic time slot	
$$E_{sp}^{lb} \leq E_{k-1} - \Delta t \, P_{3,k}^{BESS} \leq E_{sp}^{ub}, \quad k = i+1, \ldots, (N_t - 1).$$	(17)
Charging and discharging power in the generic time slot	
$$\begin{array}{ll} -P_{max}^{ch} \leq P_{3,k}^{BESS} \leq 0 & k \in \Omega_{ch,k} \\ 0 \leq P_{3,k}^{BESS} \leq P_{max}^{dch} & k \in \Omega_{dch,k} \end{array}$$	(18)
Apparent power	
$$\sqrt{P_{3,k}^{BESS2} + Q_{3,k}^{BESS2}} \leq S_{nom}^{BESS}, \quad k = i+1, \ldots, N_t$$	(19)

Table 2. ACμG's constraints.

Figure 1 is the slack busbar. Then, $V_{1,k}^{AC} = V_{sp}^{AC}$ and $\vartheta_{1,k} = 0$, for each kth time slot, where V_{sp}^{AC} is the fixed slack voltage magnitude. In particular, with reference to each kth time slot:

- at the load bus bars, $P_{s,k}^{AC}$ and $Q_{s,k}^{AC}$ are the forecasted active and reactive powers of the non-controllable loads, whose rated powers are reported in Table A. VI;
- at bus 3, $P_{3,k}^{AC}$ is the sum of the active power of the BESS $(P_{3,k}^{BESS})$ and the forecasted active power of the non-controllable load connected at bus 3 while $Q_{3,k}^{AC}$ is the sum of the reactive power of the BESS $(Q_{3,k}^{BESS})$, the forecasted reactive power of the non-controllable load connected at bus 3 and the reactive power injected by capacitor bank;
- at bus 20, $P_{20,k}^{AC}$ and $Q_{20,k}^{AC}$ are, respectively, the active $(P_{20,k}^{mTG})$ and reactive $(Q_{20,k}^{mTG})$ powers of the micro-turbine;
- at bus 16, $P_{16,k}^{AC}$ and $Q_{16,k}^{AC}$ are the active and reactive power injected by the AC/DC grids interfacing converter;
- at buses, where no generation or loads are connected, the $P_{s,k}^{AC}$ and $Q_{s,k}^{AC}$ values are zero.

In Eq. (10), $V_{S_{LB}}^{AC}$ and $V_{S_{UB}}^{AC}$ are the values of lower and upper bounds of voltage at the sth AC bus, respectively.

In Eq. (11), $I_{l,k}^{AC}$ is the value of current flowing in the lth line at the kth time interval, L_{AC} is the number of lines of the AC grid and $I_{l,max}^{AC}$ is the ampacity of the lth line.

In Eq. (12), the active power $P_{2-3,k}^{AC}$ at the PCC busbar during the kth time interval (coincident with P_k^{grid} in Eq. (4)) is limited by contractual active power P_{PCC}^{Max} and the reactive power $Q_{2-3,k}^{AC}$

at the PCC busbar during the kth time interval has to be positive since reactive power cannot be injected into distribution network. The power flows $P^{AC}_{2-3,k}$ and $Q^{AC}_{2-3,k}$ can be expressed in a simple way using state variables.

In Eq. (13), $P_{tr,k}$ and $Q_{tr,k}$ are the values of active and reactive power flowing through the transformer at the kth time interval which can be expressed in function of state variables; S_{TR} is the size of transformer (630 kVA).

In Eq. (14), the active power produced by micro-turbine ($P^{mTG}_{20,k}$) is within a normal operating interval $[P^{mTG}_{min}, P^{mTG}_{max}]$.

In Eq. (15), $Q^{mTG}_{20,k}$ is the reactive power injected by converter of the micro-turbine and S^{mTG}_{nom} is its nominal apparent power.

In Eq. (16), E_i is the energy stored in the BESS at the ith time interval, $P^{BESS}_{3,k}$ is the BESS power at the kth time slot, E^{in}_{sp} is the desired energy stored in the BESS at the end of the day that is equal to the desired energy at the beginning of the successive day, η_{bess} is the battery efficiency in charging/discharging mode.

In Eq. (17), the upper value E^{ub}_{sp} is the battery size while the lower value E^{lb}_{sp} must be specified on the basis of the maximum allowable depth of discharge, thus preserving an adequate value of the battery lifetime [18]. E_{k-1} is BESS energy stored at the end of (k-1)th time slot.

In Eq. (18), P^{ch}_{max} and P^{dch}_{max} are the maximum power rates in charging and discharging modes; $\Omega_{dch,k}$ and $\Omega_{ch,k}$ are the set of time slots from $(i + 1)$th to N_tth in which the BESS is allowed to discharge and charge, respectively.

In Eq. (19), $Q^{BESS}_{3,k}$ is the reactive power injected by converter of the BESS and S^{BESS}_{nom} is its nominal apparent power.

With reference to AC/DC grids interfacing converter (**Table 3**), the equality constraints are the DC and AC voltage balance equation (Eq. (20)) and the DC and AC active power balance equation (Eq. (21)), while an inequality constraint (Eq. (22)) refers to a limit on apparent power flowing through the AC/DC grids interfacing converter.

In Eq. (20), $m_{a,k}$ is the amplitude modulation ratio of the PWM interfacing converter, $V^{AC}_{16,k}$ is the RMS voltage of the bus at AC side of converter and $V^{DC}_{1,k}$ is the voltage amplitude of the bus at DC side of converter (the first node of the DCμG) at the kth time slot. As before mentioned, $V^{DC}_{1,k}$, for each kth time slot, is equal to V^{DC}_{sp}.

In Eq. (21), η_{rect} (η_{inv}) is the converter efficiency in rectifier (inverter) operation of the power converter.

In Eq. (22), the active ($P^{AC}_{16,k}$) and reactive ($Q^{AC}_{16,k}$) powers injected by the converter from DCμG to AC μG are limited by its size (S^{conv}_{nom}) at the kth time slot.

Interfacing converter	
DC and AC voltage	
$$V_{1,k}^{DC} = \frac{2\sqrt{2}}{m_{a,k}\sqrt{3}} V_{16,k}^{AC}, \quad k = i+1, \ldots, N_t$$	(20)
DC and AC active power	
$$P_{16,k}^{AC} = -\xi_k P_{1,k}^{DC} \qquad \xi_k = \begin{cases} \frac{1}{\eta_{rect}} & if\ P_{16,k}^{AC} \leq 0 \\ \eta_{inv} & if\ P_{16,k}^{AC} > 0 \end{cases}, k = i+1, \ldots, N_t$$	(21)
Apparent power	
$$\sqrt{P_{16,k}^{AC\ 2} + Q_{16,k}^{AC\ 2}} \leq S_{nom}^{conv}, \quad k = i+1, \ldots, N_t$$	(22)

Table 3. Constraints on AC/DC grids interfacing converter.

3.1.2. Waveform distortion compensation

At harmonic frequencies, the centralized control system furnishes reference current signals that ACμ*G* power converters have to be injected in order to compensate waveform distortions at AC busbars. These reference currents are obtained as the solution of an optimization problem in the frequency domain, based on the optimal control theory [19–21]. The optimization problem consists of minimizing the voltage harmonics at AC busbars without significantly oversizing of the converters employed for the compensation.

The Fourier transform $\overline{\boldsymbol{U}}$ of the converters reference currents to be injected is obtained by minimizing the following system index:

$$J = \int_{-\infty}^{+\infty} (\overline{\boldsymbol{Y}}^* \cdot \boldsymbol{Q} \cdot \overline{\boldsymbol{Y}} + \overline{\boldsymbol{U}}^* \cdot \boldsymbol{R} \cdot \overline{\boldsymbol{U}}) d\omega \tag{23}$$

In Eq. (23), $\overline{\boldsymbol{Y}}$ is the Fourier transform of the complex output vector $\boldsymbol{y}$, which includes the harmonic voltages at all network buses and at all considered harmonic frequencies; $\boldsymbol{Q}$ and $\boldsymbol{R}$ are weight matrices and the symbol * indicates the complex conjugate.

The analytical Fourier transform of the vector of the injected currents $\overline{\boldsymbol{U}}$ corresponding to the minimization of index of Eq. (23) is given by [20]:

$$\overline{\boldsymbol{U}} = -\dot{\boldsymbol{G}}^{-1} \cdot \dot{\boldsymbol{P}} \cdot \overline{\boldsymbol{D}} \tag{24}$$

where $\dot{G} = (\boldsymbol{L} \cdot \boldsymbol{C} \cdot \dot{\boldsymbol{\Phi}} \cdot \boldsymbol{B})^* \cdot \mathrm{L} \cdot \boldsymbol{C} \cdot \dot{\boldsymbol{\Phi}} \cdot \boldsymbol{B} + \boldsymbol{R}$ and $\dot{P} = (\boldsymbol{L} \cdot \mathrm{C} \cdot \dot{\boldsymbol{\Phi}} \cdot \boldsymbol{B})^* \cdot \boldsymbol{L} \cdot \boldsymbol{C} \cdot \dot{\boldsymbol{\Phi}} \cdot \boldsymbol{F}$.

$\overline{\boldsymbol{D}}$ is the vector of the disturbances (harmonic currents injected at buses where non-linear loads are present at all considered harmonic frequencies) and $\overline{\boldsymbol{\Phi}} = (\mathrm{j}\omega\boldsymbol{I} - \boldsymbol{A})^{-1}$ being $\boldsymbol{I}$ the identity matrix and $\boldsymbol{Q}$ factorized in $\boldsymbol{Q} = \boldsymbol{L}^*\boldsymbol{L}$.

The matrices ***A, B, C,*** and ***F*** are the system state equation matrices obtained from the following linear state space model, derived from the application of the Kirchoff's laws in the time domain to the ACμG [21]:

$$\begin{cases} \dfrac{dx}{dt} = \boldsymbol{A}\boldsymbol{x} + \boldsymbol{B}\boldsymbol{u} + \boldsymbol{F}\boldsymbol{d} \\ y = \boldsymbol{C}\boldsymbol{x} \end{cases} \tag{25}$$

where $\boldsymbol{x}$ is the state vector (inductor currents and capacitor voltages), $\boldsymbol{d}$ is the vector of the disturbances, $\boldsymbol{u}$ is the vector of the compensating currents injected by the converters for voltage harmonic compensation, and $\boldsymbol{y}$ is the output vector. ***A*, B, C,** and ***F*** depend on the parameters of the distribution network (matrices ***A*** and ***C***) and on the busbars where the disturbing loads and the converters participating to system harmonic compensation are connected (matrices ***F*** and ***B***).

It should be noted that the values of the element of vector $\overline{\boldsymbol{D}}$ elements are estimated thanks to a disturbance-estimation algorithm based on a Kalman Filter [22, 23].

It should be also noted that the values of the weight matrix $\boldsymbol{Q}$ elements depend on targets assigned to the waveform distortion compensation levels at the nodes, with the rule that the values of the weights are assigned greater in the busbars where particularly sensitive loads are connected (area compensation). The values of the weights of matrix $\boldsymbol{R}$ elements are assigned on the basis of targets on the rate of converter sizes used for the harmonic compensation: greater values of the $\boldsymbol{R}$ elements correspond to the smaller rate of the converter and vice versa [19]. With reference to the considered industrial study system, the converters installed at buses 3, 16 and 20 participate, using a part of their size, to harmonic compensation.

3.2. Islanding mode

In addition, the control strategy applied by CCS is able to disconnect the hybrid μG from the distribution network when abnormal conditions occur. We will consider the behavior of the AC/DC μG in the case of supply interruptions. In such conditions, local energy resources and loads are optimally controlled in order to guarantee high power quality levels to priority loads. As priority loads, we refer to electrical loads requiring high levels of continuity of supply.

In particular, in islanding mode, the so-called "Single Mater Operation" control strategy is adopted [24]. In fact, while in grid-connected mode all inverters installed into the hybrid μG operate in PQ mode (inverter is used to supply a given active P and reactive Q power set point), in islanding mode the BESS interface converter operates feeding the μG with predefined values of voltage and frequency (V-f control mode), as well as the remaining inverters, on the contrary, continue to operate in PQ mode.

As soon as the hybrid μG is in the islanding mode, all loads are disconnected except the priority loads whose power demand can be provided by the BESS and micro-turbine (the PV system is not considered in the power balance because its availability depends on the uncertain presence of the primary source). As an example, in the considered case study, the priority loads are assumed to be the sensitive loads connected to the DCμG. In this case, the maximum power required by the priority loads is 72 kW (corresponding to the power absorbed when all sensitive loads are working). Also in the worst conditions (long interruption arising when BESS charge is E_{sp}^{lb} = 102 kWh), the energy that should be generated by BESS and micro-turbine in 1 hour[8] (equal to 72 kWh) is significantly lower than the energy they can provide (equal to 132 kWh) .[9]

Eventually, in islanding mode, the proposed control strategy starts to supply the priority loads. Anyway, according to a predefined list of load priorities, some of the disconnected loads can be reconnected taking into account of: (i) the available power from PV generator; (ii) the scheduled power required by sensitive loads; (iii) the maximum time fixed for the autonomous operation of the hybrid μG; and (iv) the value of the SOC of battery. The available energy for the islanding operation can be easily calculated from the knowledge of the SOC of battery, the forecasted PV power generation in the time intervals successive to the disconnection and the rated power of micro-turbine.

4. Numerical applications

A simulation of the hybrid micro grid of Section 2 using *Matlab® Simulink®* toolbox was implemented in order to show the feasibility of the proposed control strategy both in grid-connected and in islanding mode. Several case studies were performed, and, for sake of conciseness, the results relative to a typical working day are shown. In particular, in the following subsections, results are presented for the two considered operation modes, separately.

Remind that, in the grid-connected mode, the converters participating in the energy cost minimization problem are: (i) the BESS converter at bus 3; (ii) the AC/DC grids interfacing converter at bus 16; and (iii) the micro-turbine converter at bus 20. In particular, the AC/DC grids interfacing converter (micro-turbine) participates optimally exchanging only reactive (active) power with the ACμG.

[8]The maximum time interval in islanding mode operation of hybrid μG was fixed to 60 minutes considering the mean value of the time duration of long interruptions in the area where the industrial facility is located.

[9]We assume that in the worst conditions the PV production is not available and the SOC of battery is initially equal to its minimum value (20%). In these conditions, in order to supply the sensitive loads (72kWh), the micro-turbine provides 30kWh and the BESS has to furnish only 42 of the 102kWh available. This, however, determines a further discharge of the BESS and a final SOC equal to 11.76%, causing a reduction of the expected battery life. However, since 60 minutes long interruptions occur in distribution networks only few times during the battery life and taking into account that the worst conditions are infrequent, the actual life reduction of the BESS can be considered largely tolerable.

The converters participating in the harmonic compensation are: (i) the AC/DC grids interfacing converter at bus 16 and (ii) the micro-turbine converter at bus 20. Note also that sizes of the converters, shown in Table A. VIII, are chosen according to the aforesaid operating conditions. Specifically, while the 120 kVA of the BESS converter is fully utilized for the energy cost minimization, only 80 kVA of the AC/DC grids interfacing converter and 30 kVA of the micro-turbine converter are addressed for the same operation. The remaining 20 kVA of the AC/DC grids interfacing converter and 10 kVA of the micro-turbine converter are used for the harmonic compensation.

4.1. Grid-connected mode

The considered day was divided into 72-time slot of equal duration (20 min), and the inputs of the proposed control strategy for the energy cost minimization are:

- The forecasted powers of the PV system and of the non-controllable loads, both obtained before each time slot for all the successive time slots of the considered day. As an example, Figures 4 show the forecasted power of the PV system (**Figure 4a**) and the forecasted active power (solid line), reactive power (dotted line) of the non-controllable load at bus 33 (**Figure 4b**) obtained at the first time slot;
- The scheduled power profile of the controllable loads, as illustrated in **Figure 5**;
- The energy charge reported in **Figure 6** [25]. This tariff is applicable to medium and small industrial customers and for service in the common areas of multi-family complexes. It was selected since it is characterized by a large spread between on peak and off peak prices; this is a necessary condition to make it possible a fruitful use of storage systems nowadays;
- The SOC of the BESS at bus 3 equal to 43.36% at the beginning of the day. The same value is required also at the end of the considered day according to scheduling procedure.

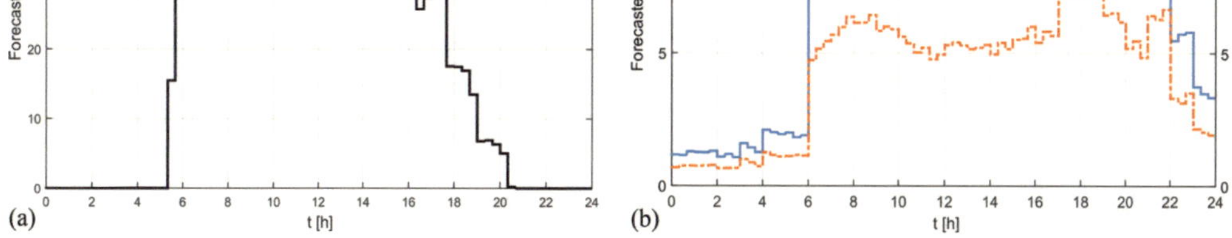

Figure 4. Forecasted power profile for the first time slot: (a) power production of the PV system; (b) active (solid line) and reactive (dotted line) power of the non-controllable load at bus 33.

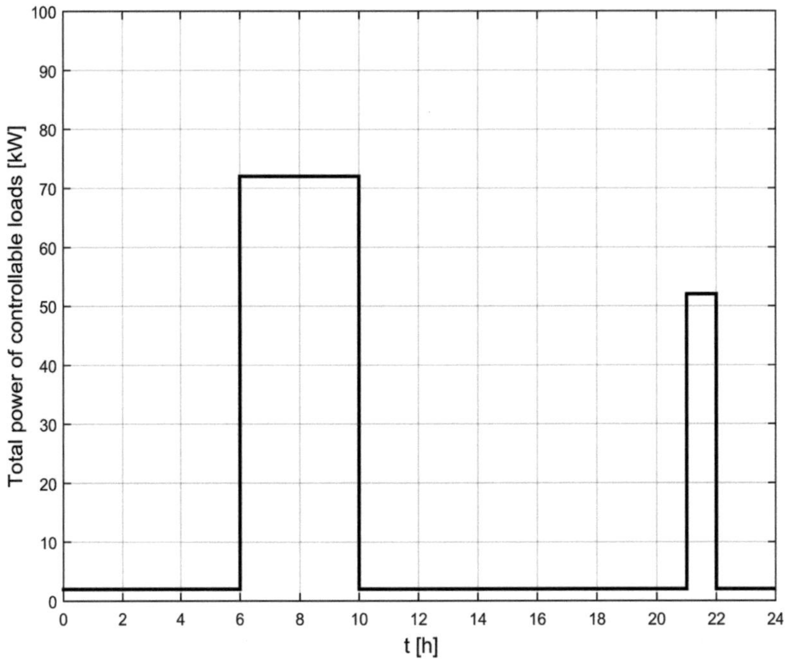

Figure 5. Scheduled profile of the power of the controllable loads.

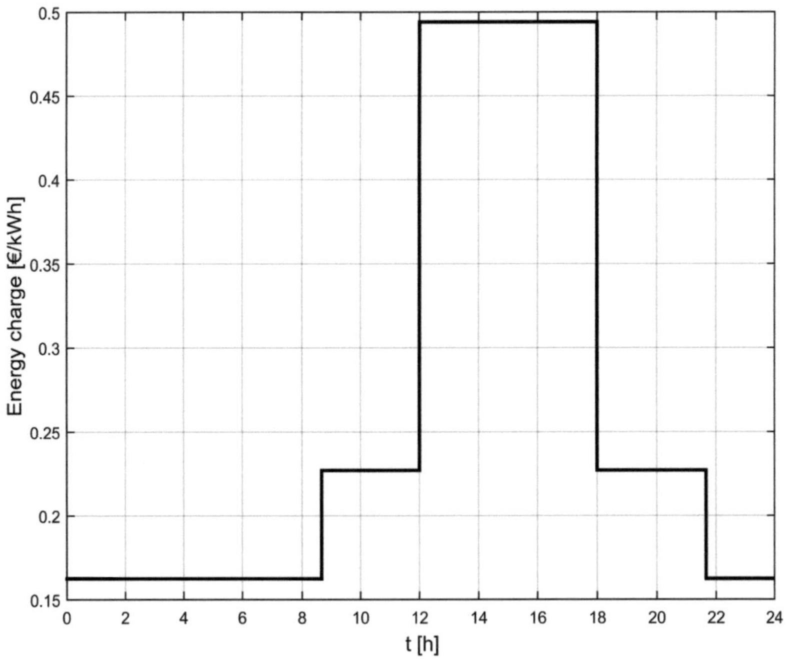

Figure 6. Energy charge.

The time behavior of the main quantities of interest is illustrated in the following relatively to the considered working day.

Figure 7 shows the active (solid line) and reactive (dotted line) power profiles of the BESS converter. In **Figure 8**, the active (solid line) and reactive (dotted line) power profiles of the AC/DC grids interfacing converter are plotted. **Figure 9** shows the reactive power supplied by the controllable capacitor bank. Eventually, in **Figure 10**, the active (solid line) and reactive (dotted line) power profiles at the PCC are reported. The active power of the micro-turbine is not reported since it always operates at a constant value of 30 kW.

From the analysis of **Figures 7–10**, the following considerations can arise:

- the μG operates practically at unitary power factor all over the day (**Figure 10**); the measured power factor of the industrial system (without the hybrid μG), instead, was 0.85 on average. Indeed, the forecasting error on PV power generation and on non-controllable load powers causes an operation of network slightly different to unity power factor and, as a consequence, the reactive power absorbed by the hybrid μG at the PCC is not exactly zero;
- the reactive power furnished by BESS and capacitor bank is maximized in the hours corresponding to load peak power (**Figures 7** and **9**);
- the active power provided by the BESS during the discharge from 12:00 to 14:00 is limited by the converter size (in this interval also reactive power is furnished by the converter);
- the BESS discharge reduction at 16:20 and the residual BESS discharge at 17:20 are justified by the numerical deviation between the BESS set point and the corresponding simulation response (**Figure 7**).

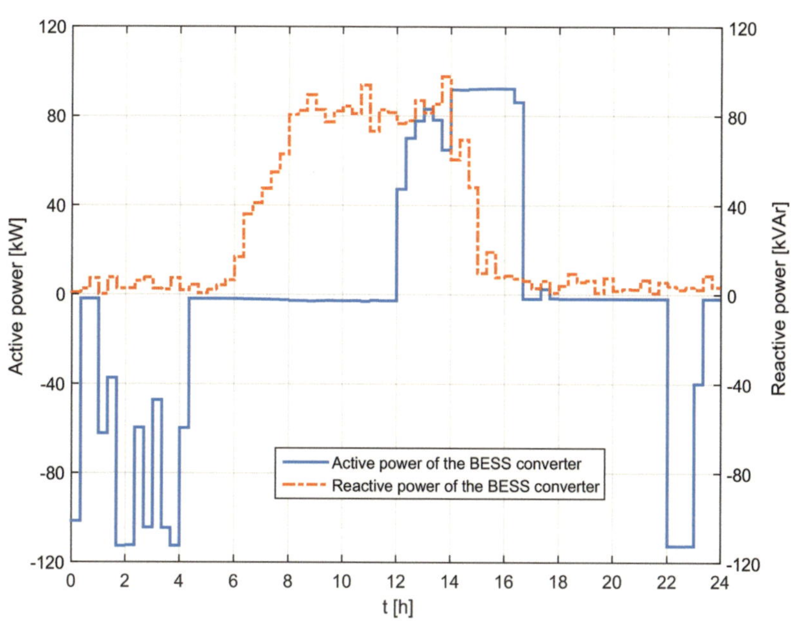

Figure 7. Active (solid line) and reactive (dotted line) power profiles related to the BESS converter at bus 3.

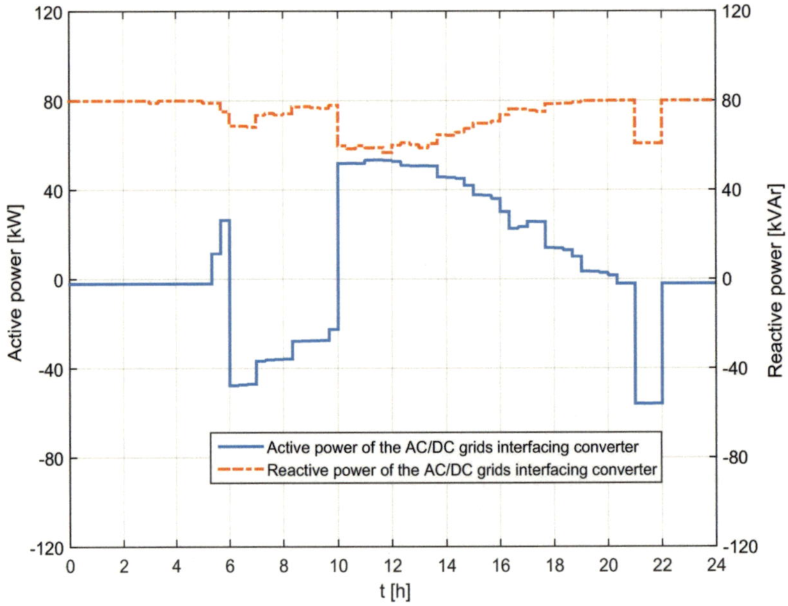

Figure 8. Active (solid line) and reactive (dotted line) power profiles related to the AC/DC grids interfacing converter at bus 16.

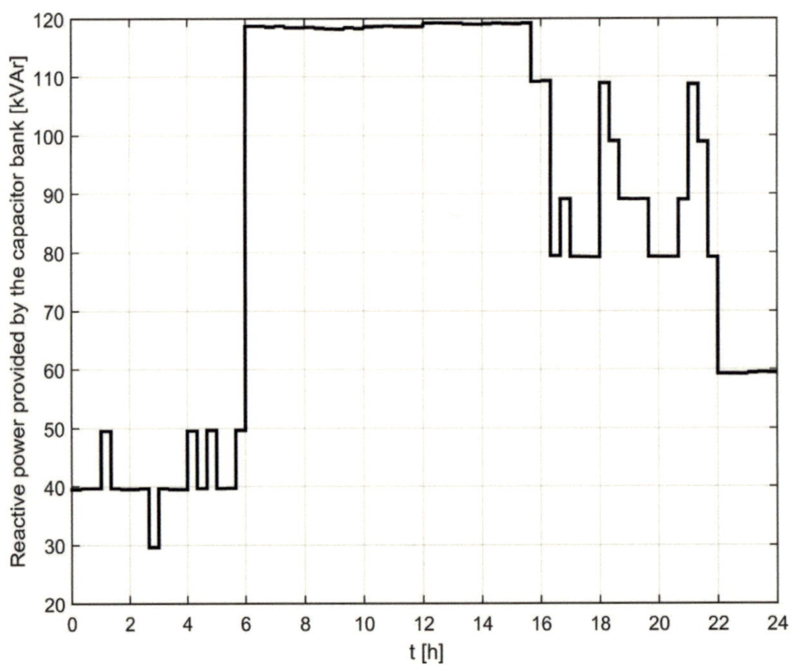

Figure 9. Reactive power profile of the capacitor bank at bus 3.

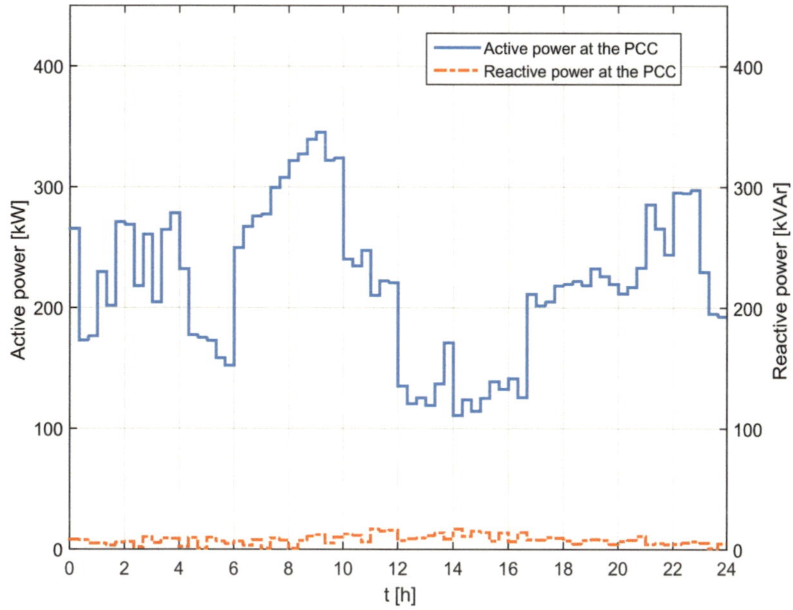

Figure 10. Active (solid line) and reactive (dotted line) power profiles at the PCC.

Anyway, the BESS totally discharges (i.e. the SOC reaches 20%) in the interval where the tariff is higher, maximizing in this way the economic benefit.

Note also that the reactive power production inside the μG allows not only the reduction of the power losses but also the maximization of the internal active power generation, determining the minimization of the objective function. In fact, an increasing of the mere internal active power production determines a corresponding decreasing of power factor at the PCC; therefore, a production of reactive power inside the hybrid grid is needed to guarantee an adequate power factor at the PCC.

In **Figure 11**, a comparison among the energy charge profile (dotted line) and the active power (solid line) of BESS converter (**Figure 11a**) and the active power of AC/DC grids interfacing converter (**Figure 11b**) is reported. **Figure 12** shows a comparison among the energy charge profile (dotted line), the active power profile at the PCC of the hybrid μG (solid line) and of the actual industrial facility (dashed line).

The analysis of these last figures clearly evidences that:

- to maximize the economic benefits, the maximum allowable discharge for the BESS arises in the hours of the maximum value of tariff;
- the active power flowing from DC to AC μG is maximized in the hours of the maximum value of tariff, according to the day-ahead scheduling procedure that shifted the controllable loads far from the maximum tariff price and the PV generation available during those hours;

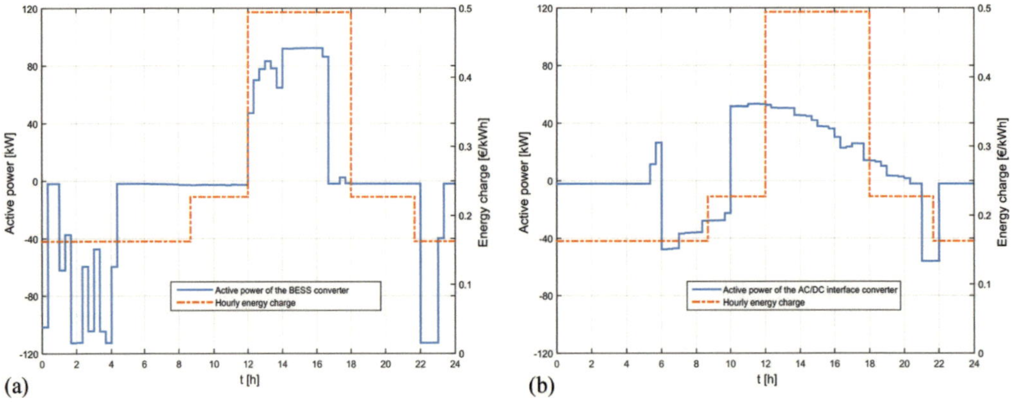

Figure 11. Comparison among the energy charge profile (dotted line), the active power profiles (solid line) related to: (a) the BESS converter at bus 3 and (b) to the AC/DC grids interfacing converter at bus 16.

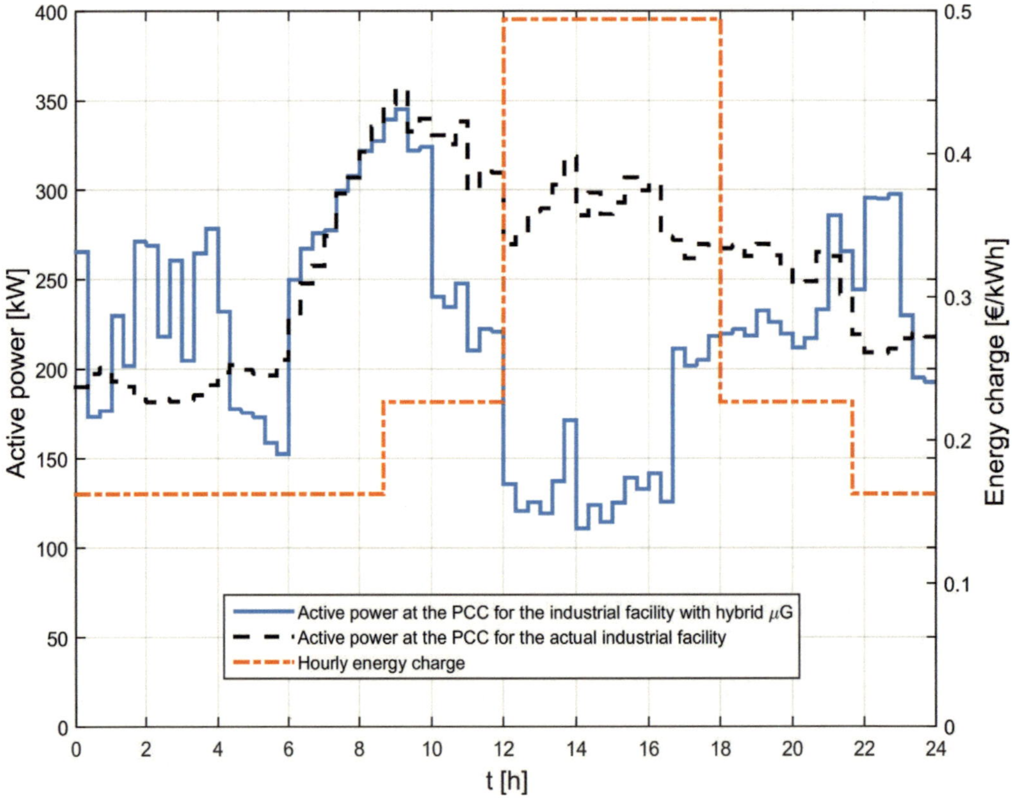

Figure 12. Comparison among the energy charge profile (dotted line), the active power profile at the PCC of the industrial system with μG (solid line) and of the actual industrial system (dashed line).

- the active power at the PCC of hybrid μG increases respect to the value of the existing industrial facility when the tariff price is minimum, due to: (i) the BESS charge profile and

(ii) the shifting of the controllable loads; on the other hand, the active power at the PCC of hybrid μG decreases when the tariff price assumes higher values, due to: (i) the shifting of the controllable loads; (ii) the PV system production; and (iii) the BESS discharge profile.

The aforesaid observations justify the reduction, shown in **Figure 13**, of the hourly costs to be sustained by the hybrid μG (solid line) respect to the ones sustained by the existing industrial facility (dotted line). In particular:

- from 00:00 to 04:00 and from 22:00 to 23:00, the hybrid μG has costs slightly greater than the existing industrial system, cause of the BESS charge;
- from 06:00 to 07:00 and from 21:00 to 22:00, the hybrid μG has costs slightly greater than the existing industrial system, cause of the shifting of controllable loads;
- from 10:00 to 18:00, the hybrid μG has cost significantly lower than the existing industrial system, cause of both the PV system production and BESS discharge;
- from 18:00 to 21:00, the hybrid μG has costs lower than the existing industrial system, mainly cause of the micro-turbine operation, whose production costs are lower than the considered energy tariff, and also due to the PV system production.

Globally, the proposed procedure allows a daily reduction of the total costs for the energy purchase of about 26%.

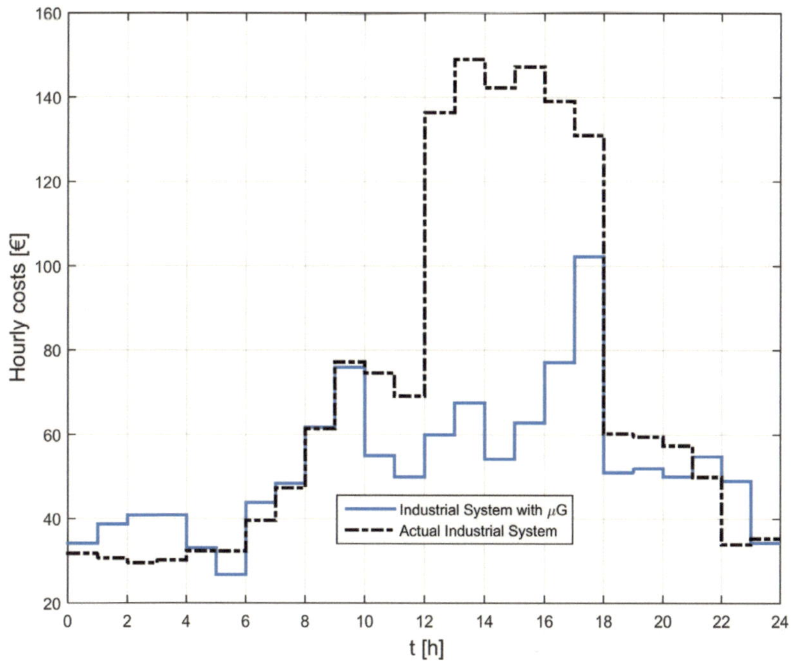

Figure 13. Hourly costs to be sustained by the industrial system with hybrid μG (solid line) and by the actual industrial system (dotted line).

With reference to the waveform distortion compensation, an analysis of the load currents in the existing industrial system revealed that main contributions to voltage distortions were due to the harmonic of order h = 5, 7, 11 and 13. So, the procedure described in Section 3.1.2 was applied only for the compensation of the aforesaid harmonic orders. The compensation action is applied in a working day using the AC/DC grids interfacing converter and the micro-turbine converter.

As illustrated in Section 3, the inputs needed to solve the minimization problem and to obtain the reference currents of converters (see Eq. (23)) are the estimated harmonic disturbances $\overline{\boldsymbol{D}}$ and the **Q** and **R** matrices.

The loads producing waveform distortion are marked with an asterisk in Table A. VI. As an example, **Figures 14** and **15** report the harmonic currents amplitudes injected by disturbing loads at the buses 23 and 33 during the time slot from 16:20 to 16:40.

The terms of the weight matrix **R** are obtained from the knowledge of the size of the converters dedicated to harmonic compensation, and giving priority to the compensation of the harmonic with the lowest orders. As an example, the value of **R** term related to the compensation of the 5th harmonic order is 0.59, while the value of **R** term related to the compensation of the 13th harmonic order is 0.23.

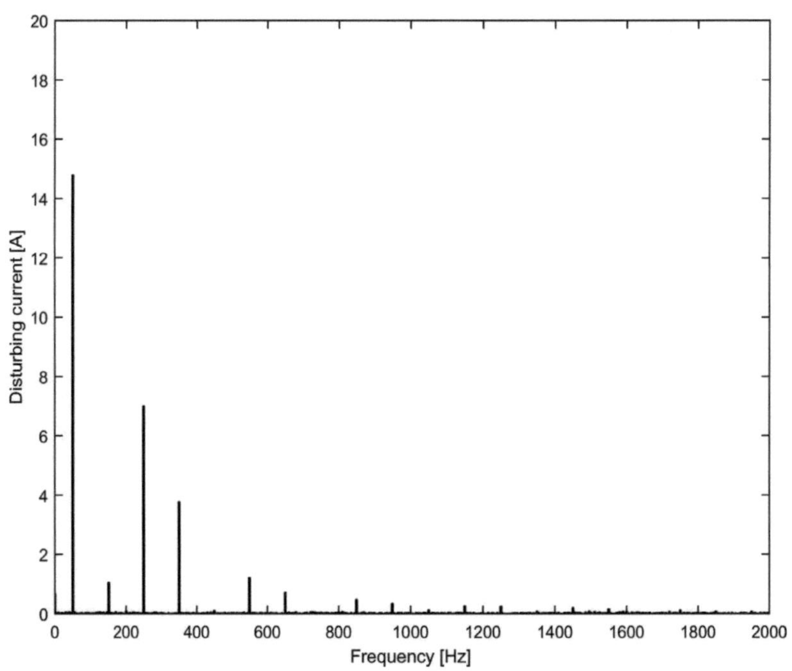

Figure 14. DFT estimation of harmonic currents amplitudes injected by disturbing loads at the bus 23 during the time slot from 16:20 to 16:40.

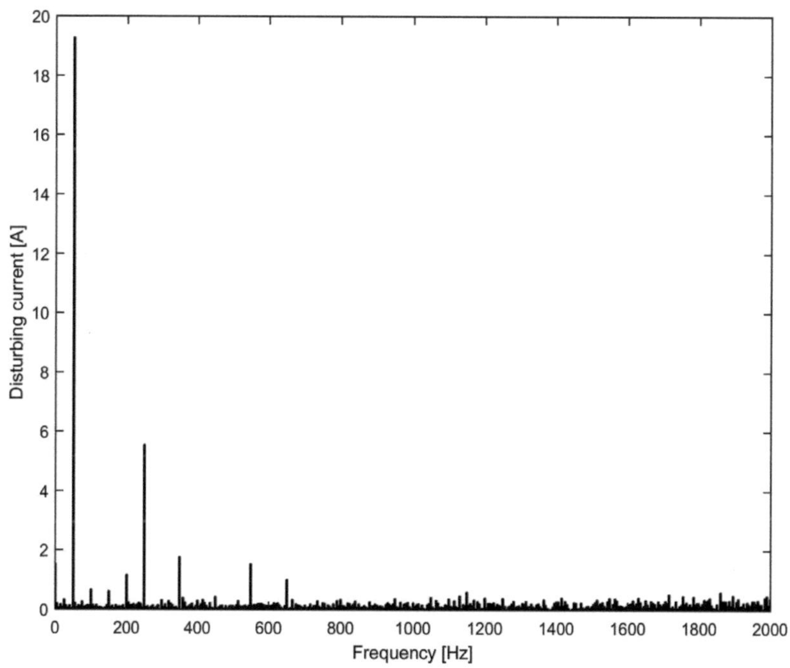

Figure 15. DFT estimation of harmonic currents amplitudes injected by disturbing loads at the bus 33 during the time slot from 16:20 to 16:40.

The terms of weight matrix **Q** are set giving priority to the harmonic compensation for the feeder "tanks and boxes manufactory"; thus the elements of **Q** corresponding to nodes of this feeder are 50 times greater than those of the other feeders.

Figures 16–18 show the comparison of total harmonic distortion (THD) with (solid bars) and without (empty bars) the proposed compensation procedure. Specifically, **Figure 16** shows the THD comparison over the buses in the time slot corresponding to the time interval from 16:20 to 16:40, while **Figures 17** and **18** show the THD comparison, in the whole day, for the buses 23 and 35, respectively. The benefits in term of THD reduction with the application of the proposed procedure are evident and significant, since, as observed in **Figures 16–18**, the THD values in all of the buses of the hybrid μG in time stayed always under the 0.6% with the waveform distortion compensation. Without compensation, the THD overcame the 4.5% in some buses for some time slots.

4.2. Islanding mode

To simulate islanding mode of operation of the hybrid μG, an interruption in the MV distribution network supplying the industrial facility was simulated. The interruption determines the switching from grid-connected mode to islanding mode. The control strategy described in Section 3.2 is applied; the BESS converter switches from P-Q control mode to V-f control mode

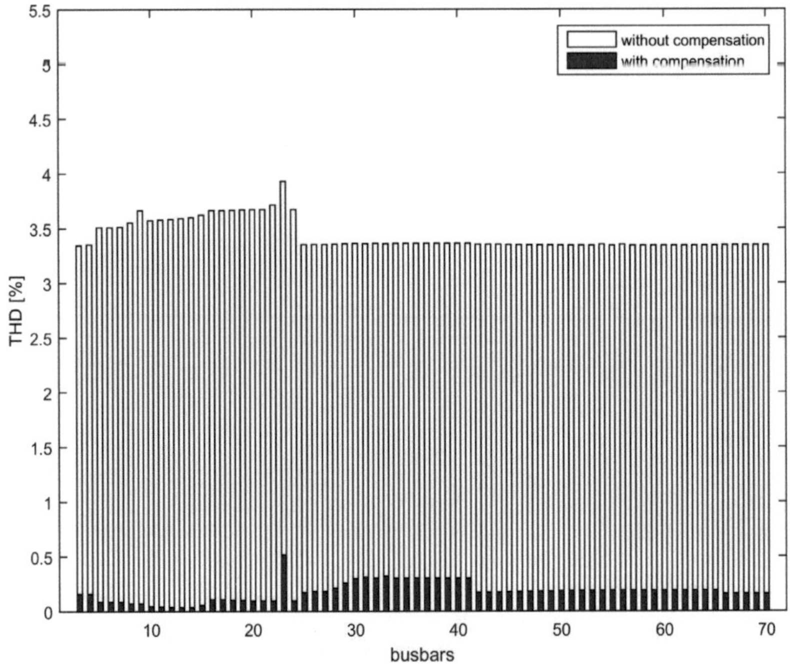

Figure 16. THD comparison with (solid bars) and without (empty bars) the proposed compensation procedure corresponding to the time interval from 16:20 to 16:40.

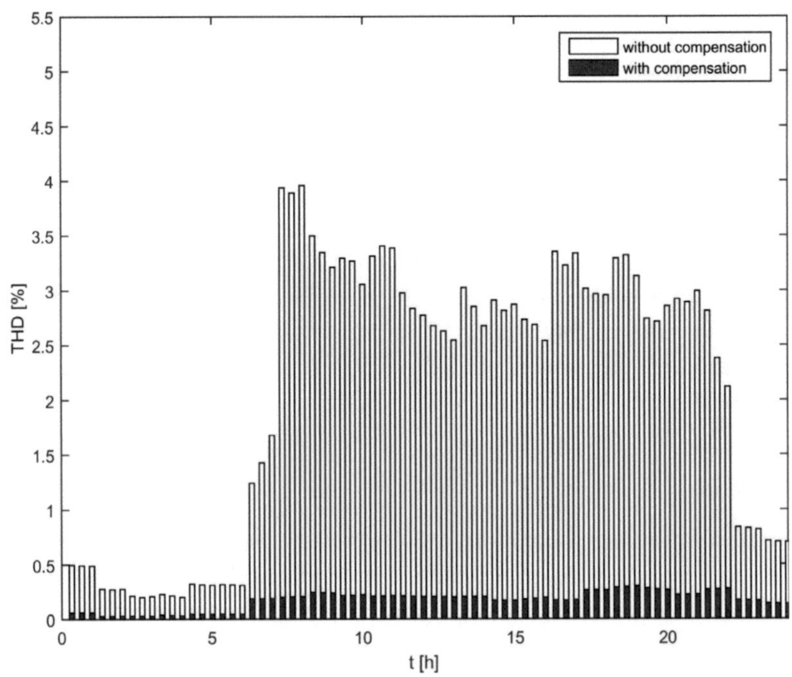

Figure 17. THD comparison with (solid bars) and without (empty bars) the proposed compensation procedure corresponding to the bus 23.

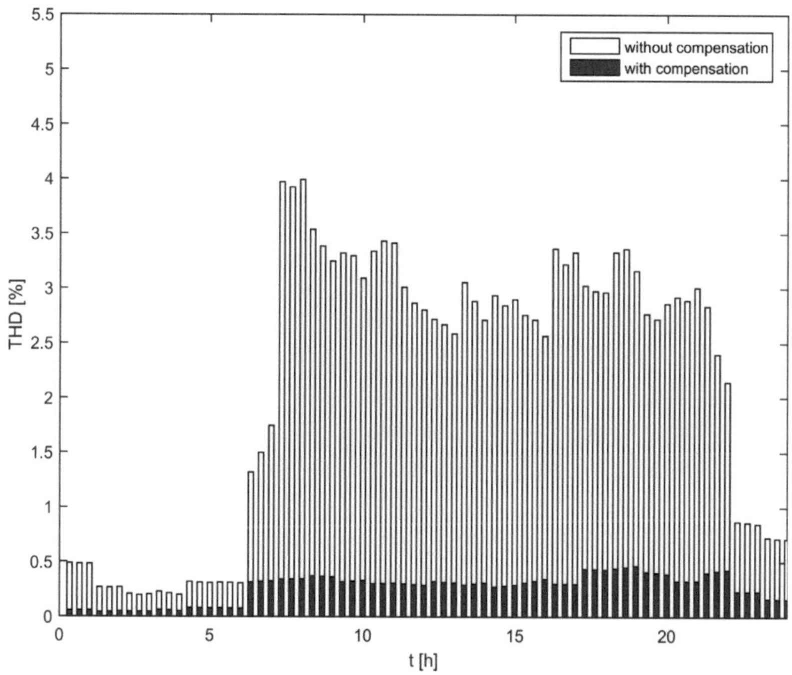

Figure 18. THD comparison with (solid bars) and without (empty bars) the proposed compensation procedure corresponding to the bus 35.

since the RMS voltage at PCC becomes lower than 5% of nominal voltage [26], that is the threshold for the switching of the BESS converter control strategy. The remaining inverters, on the contrary, continue to operate in P-Q mode. The interruption arises at 6 h 25′ 1″ of the working day analyzed in Section 4.1. As soon as the islanding condition arises, all loads of the hybrid μG are disconnected except the sensitive loads present in the DC μG. As illustrated in **Figure 5**, at 6 h 25′ all sensitive loads are working and absorbing 72 kW, micro-turbine and photovoltaic generators produce 30 and 31 kW, respectively. BESS is characterized by a SOC of about 100%.

In **Figure 19**, the voltage profile at bus 3 versus time is reported. In **Figure 20**, the active power injected by BESS and AC/DC grids interfacing converters are shown. **Figures 19** show how the proposed approach allows to supply the hybrid μG during the interruption with obvious advantages in terms of continuity of the supplied energy.

Figure 20a shows the increasing of BESS active power from the value at 6 h 25′ in grid-connected mode (see also **Figures 7** and **11**) to the value needed to supply the hybrid μG in islanded mode. In particular, this value, in the considered application, is related to the grid active power losses and active power required by the sensitive loads connected to DCμG reduced by the active power furnished by micro-turbine and PV system. As expected, the active power flowing in the AC/DC grids interfacing converter remains constant to the value of grid-connected mode (see also **Figures 8** and **11**).

In **Figure 20b**, reactive power variations from the values at 6 h 25′ in grid-connected mode (see also **Figures 7** and **8**) to the values needed to provide the reactive power required by the hybrid

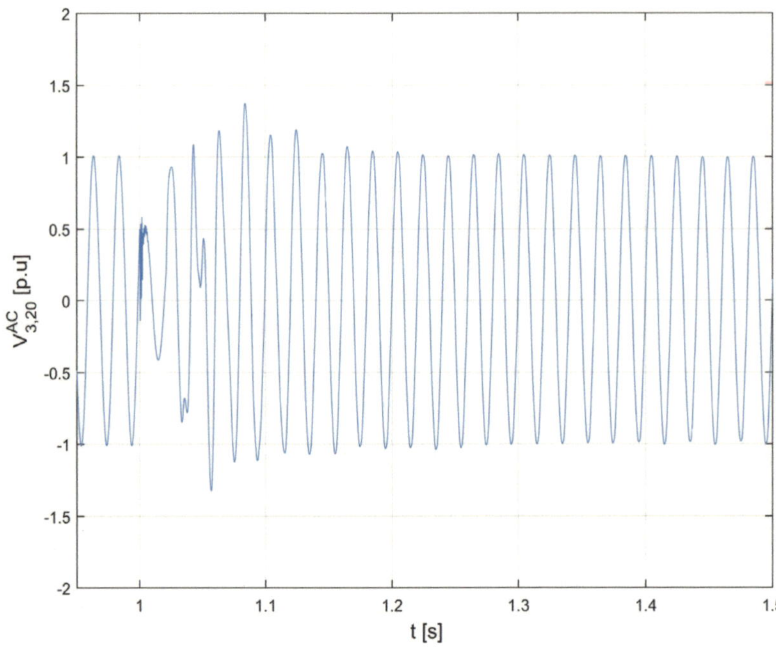

Figure 19. Phase-A voltage at bus 3 versus time starting from 6 h 25′.

μG in islanding mode is shown. In particular, since only the sensitive loads connected to DC μG are working, only reactive power losses are required and provided by BESS converter.

Note that the active and reactive power transients shown in **Figure 20** are due to the switching of the BESS converter regulator from P-Q control mode to voltage control mode. Moreover, although the other converters in the hybrid μG continue to be controlled in P-Q mode, also their responses are influenced by the transient at bus 3.

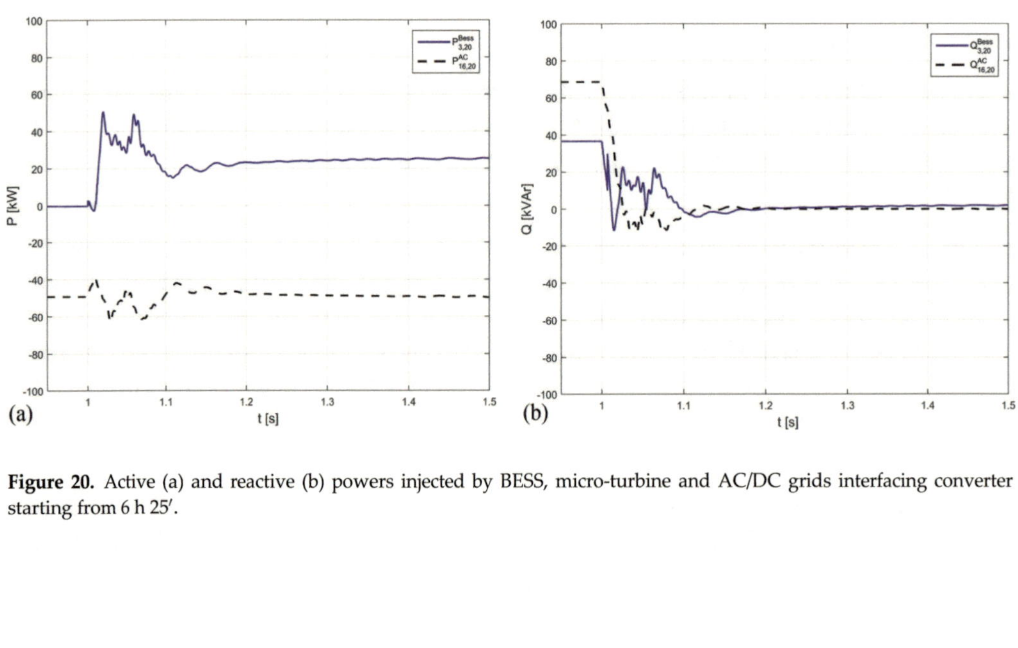

Figure 20. Active (a) and reactive (b) powers injected by BESS, micro-turbine and AC/DC grids interfacing converter starting from 6 h 25′.

http://dx.doi.org/10.5772/intechopen.68766

Accordingly to the aim of the proposed procedure, all the distributed resources in the hybrid μG guarantee the continuity of supplying of sensitive loads.

5. Conclusions

A hybrid μG including two interconnected sections operating in DC and AC was considered starting from an existing Italian industrial facility. The DC section included non-dispatchable generation systems and controllable loads; the AC section included a BESS, a micro-turbine, non-controllable loads and non-linear loads.

A control strategy based on a real-time control, at both the fundamental frequency and harmonic frequencies, was applied. The control strategy required the solution of optimization problems and was aimed at minimizing the operation costs and improving the power quality levels.

The results of a numerical application relative to a typical working day are shown and demonstrated the effectiveness of the procedure in reducing the daily operating costs while guaranteeing higher values of the quality of the power supply.

Acknowledgements

This work was supported by Italian Ministero dell'Istruzione dell'Università e della Ricerca (MIUR) in the frame of PON03PE_00178_1: Microgrid Ibride in Corrente Continua e Corrente Alternata (MICCA) and by Università of Napoli Parthenope in the framework of "Bando per il Sostegno alla Ricerca individuale triennio 2015–2017".

Appendix

In the following, the data of the lines are provided. In particular, for each line, Tables A. I, A. II, A. III, and A. IV show the starting and ending buses, the length, resistance, and reactance. Tables A. V and A. VI report the transformers and loads data. Tables A. VII and A. VIII report the main data of the new components installed in the proposed hybrid μG.

Buses		ℓ [m]	R [mΩ/m]	X [mΩ/m]
From	To			
3	4	8	0.041	0.014
4	5	24	0.163	0.130
5	6	4	0.473	0.101
5	7	0.5	0.163	0.130
7	8	6	0.163	0.130
8	9	10	1.410	0.112

Buses		ℓ [m]	R [mΩ/m]	X [mΩ/m]
From	To			
8	10	9.3	0.163	0.130
10	11	3	0.163	0.130
11	12	2.8	0.163	0.130
12	13	3.5	0.163	0.130
13	14	3.5	0.163	0.130
14	15	11	0.163	0.130
15	16	19.1	0.163	0.130
16	17	4	1.410	0.112
16	18	1.9	0.163	0.130
18	19	4	1.410	0.112
18	20	3	0.163	0.130
20	21	10	0.236	0.097
20	22	42	1.410	0.112
20	23	61	2.240	0.119
20	24	61	1.410	0.112

Table A.I. Line parameters for the feeder "Tanks and boxes manufactory".

Buses		ℓ [m]	R [mΩ/m]	X [mΩ/m]
From	To			
4	25	31	0.041	0.014
25	26	3.5	0.163	0.130
26	27	16	2.240	0.119
26	28	9	0.163	0.130
28	29	12.5	0.163	0.130
29	30	10.5	0.163	0.130
30	31	8	0.641	0.101
30	32	1.5	0.163	0.130
32	33	10	0.641	0.101
32	34	12.5	0.163	0.130
34	35	8	0.328	0.096
34	36	13	0.163	0.130
36	37	1	0.163	0.130
37	38	35	0.665	0.260

Buses		ℓ [m]	R [mΩ/m]	X [mΩ/m]
From	To			
38	39	0.5	0.665	0.260
39	40	0.5	0.665	0.260
40	41	20	1.410	0.112

Table A. II. Line parameters for the feeder "Assembly".

Buses		ℓ [m]	R [mΩ/m]	X [mΩ/m]
From	To			
25	42	34	0.041	0.014
42	43	3.3	0.070	0.096
43	44	12	2.240	0.119
43	45	9.2	0.070	0.096
45	46	5.5	0.070	0.096
46	47	5.5	1.410	0.112
46	48	6.5	0.070	0.096
48	49	5.5	1.410	0.112
48	50	7.8	0.070	0.096
50	51	5.5	1.410	0.112
50	52	5.2	0.070	0.096
52	53	1.8	0.070	0.096
53	54	5.5	1.410	0.112
53	55	8.1	0.070	0.096
55	56	5.5	1.410	0.112
55	57	5.6	0.070	0.096
57	58	4.5	0.070	0.096
58	59	5.5	1.410	0.112
58	60	3	0.070	0.096
60	61	8	0.473	0.101
60	62	2.5	0.070	0.096
62	63	5.5	1.410	0.112
62	64	3.7	0.070	0.096
64	65	38	1.410	0.112

Table A. III Line parameters for the feeder "Winding and coils".

Buses		ℓ [m]	R [mΩ/m]	X [mΩ/m]
From	To			
2	66	30	0.094	0.090
66	67	87	0.163	0.130
67	68	7	1.410	0.112
67	69	0.5	0.163	0.130
69	70	7	0.473	0.101

Table A. IV Line parameters for the feeder "Test".

Nominal voltage	20 kV/400 V
Nominal power	630 kVA
Connection	Dyn
Group	11
Short circuit voltage	6.0%
Load losses	5721 W
No load losses	925 W
No load current	5%

Table A. V. Transformer data.

Bus	Type	Rated power [kVA]	Power factor
3	Heating and cooling system	200	0.8
6	Painting machine	75	0.8
7	Box overturning machine	4	0.99
9*	Sandblasting machine	55	0.75
11	Welder aspirators	11	0.99
15	Manual bender	8	0.99
17*	Folding walls island robot	24	0.99
19*	Wave welding machine	30	0.65
21	Automated bending robot for corrugated panels	122	0.65
22*	PLC + computer	3	0.62
23*	Plasma cutting machine	15	0.8
27	Crane	5.5	0.8
31*, 33*	Core cutting machine n. 1	60	0.8
35	Autoclaves	86	0.8
44	Furnace	5	0.99

http://dx.doi.org/10.5772/intechopen.68766

Bus	Type	Rated power [kVA]	Power factor
47*, 49*, 51*	MT winder machine	37	0.99
54*, 56*	Tuboly winder machine	37	0.99
59*, 63*	BT winder machine	37	0.99
61	Offices	36	0.99
68*	Automated bending robot for metal plates	20	0.9
70	Testing bench room	50	0.7

Table A. VI Load data.

(a)

	PV generator
Power output	60 kWp
Chopper efficiency	0.95
DC voltage output	800 V

(b)

DCµG buses		ℓ [m]	*R* [Ω/km]	*L* [H/km]
From	To			
1	2	20	1.450	0.00032
1	3	30	0.660	0.00029
1	4	10	3.950	0.00038
1	5	35	0.660	0.00029

Table A. VII DCµG components: PV generator (a) and lines data (b).

	BESS	Micro-turbine	AC/DC grids interfacing converter
Power output	120 kVA	40 kVA	100 kVA
Nominal AC voltage	400 V	400 V	400 V
Efficiency	0.95	0.95	0.95
Nominal DC voltage	800 V	–	800 V
Maximum capacity	510 kWh	–	–
SOC_{min}	20%	–	–
Maximum power rate charge/discharge	102 kW	–	–

Table A. VIII ACµG components data.

Author details

Luisa Alfieri[1], Antonio Bracale[1*], Pierluigi Caramia[1] and Guido Carpinelli[2]

*Address all correspondence to: antonio.bracale@uniparthenope.it

1 University of Napoli Parthenope, Naples, Italy

2 University of Napoli Federico II, Naples, Italy

References

[1] CIGRÉ. Working Group C6.22 Microgrids Evolution Roadmap, Microgrids 1: Engineer, Economics, & Experience. International Symposium on Smart Electric Distribution Systems and Technologies (EDST15), September 2015

[2] Hatziargyriou N. Microgrids: Architectures and Control. Wiley-IEEE Press; United Kingdom, 2014

[3] Farhangi H. The path of the smart grid. IEEE Power and Energy Magazine. 2010;**8**:18-28

[4] Hatziargyriou N, Asano H, Iravani R, Marnay C. Microgrids. IEEE Power and Energy Magazine. 2007;**5**:78-94

[5] Eghtedarpour N, Farjah E. Power control and management in a hybrid AC/DC microgrid. IEEE Transactions on Smart Grid. 2014;**5**:1494-1505

[6] Guerrero JM, Loh PC, Tzung-Lin L, Chandorkar M. Advanced control architectures for intelligent microgrids—Part II: Power quality, energy storage, and AC/DC microgrids. IEEE Transactions on Industrial Electronics. 2013;**60**:1263-1270

[7] Xiong L, Peng Wang P, Loh C. A hybrid AC/DC microgrid and its coordination control. IEEE Transactions on Smart Grid. 2011;**2**:278-286

[8] Loh PC, Ding L, Yi Kang C, Blaabjerg F. Autonomous operation of hybrid microgrid with AC and DC subgrids. IEEE Transactions on Power Electronics. 2013;**28**:2214-2223

[9] Long B, Jeong TW, Lee JD, Jung YC, Chong KT. Energy management of a hybrid AC–DC micro-grid based on a battery testing system. Energies. 2015;**8**:1181-1194

[10] Baboli PT, Shahparasti M, Moghaddam MP, Haghifam MR, Mohamadian M. Energy management and operation modeling of hybrid AC-DC microgrid. IET Generation, Transmission & Distribution. 2014;**8**:1700-1711

[11] Bracale A, Caramia P, Carpinelli G, Mancini E, Mottola F. Optimal control strategy of a DC micro grid. International Journal of Electrical Power & Energy Systems. 2015;**67**:25-38

[12] Alfieri L, Carpinelli G, Proto D, Russo G. Day-ahead optimal scheduling of loads and dispatchable resources in a hybrid AC/DC microgrid of an industrial system. In: Proceedings

of the International Conference on Renewable Energies and Power Quality (ICREPQ15); 25-27 March 2015; La Coruña, Spain

[13] Hong T, Pinson P, Fan S, Zareipour H, Troccoli A, Hyndman RJ. Probabilistic energy forecasting: Global energy forecasting competition 2014 and beyond. International Journal of Forecasting. 2016;**32**:896-913

[14] Bracale A, Carpinelli G, De Falco P. A probabilistic competitive ensemble method for short-term photovoltaic power forecasting. IEEE Transactions on Sustainable Energy. 2017; **8**(2): 551-560. DOI: 10.1109/TSTE.2016.2610523.

[15] Bracale A, Carpinelli G, De Falco P. A bayesian-based approach for the short-term forecasting of electrical loads in smart grids. Part I: Theoretical aspects. In: IEEE International Symposium on Power Electronics, Electrical Drives, Automation and Motion (SPEEDAM); 22-24 June 2016; Capri, Italy

[16] Binyan Z, Yi S, Xiaodai D, Wenpeng L, Bornemann J. Short-term operation scheduling in renewable-powered microgrids: A duality-based approach. IEEE Transactions on Sustainable Energy. 2014;**5**:209-217

[17] Fossati JP, Galarza A, Martín-Villate A, Font L. A method for optimal sizing energy storage systems for microgrids. Renewable and Sustainable Energy Reviews. 2015; **77**:539-549

[18] Divya C, Østergaard J. Battery energy storage technology for power systems: An overview. Electric Power System Research. 2009;**79**:511-520

[19] Carpinelli G, Russo A, Russo M, Verde P. Inherent structure theory of networks and power system harmonics. IEE Proceedings—Generation, Transmission and Distribution. 1998;**145**:123-132

[20] Jolly MR. On the calculus of complex matrices. International Journal Control. 1995;**61**:749-755

[21] Griffo A, Carpinelli G, Lauria D, Russo A. An optimal control strategy for power quality enhancement in a competitive environment. International Journal of Electrical Power & Energy Systems. 2007;**29**:514-525

[22] Yu KK C, Watson NR, Arrillaga J. An adaptive Kalman filter for dynamic harmonic state estimation and harmonic injection tracking. IEEE Transactions on Power Delivery. 2005;**20**:1577-1584

[23] Beides HM, Heydt GT. Dynamic state estimation of power system harmonics using Kalman filtering methodology. IEEE Transaction on Power Delivery. 1991;**6**:1663-1669

[24] Pecas Lopes JA, Moreira CL, Madureira AG. Defining control strategies for analyzing microgrids islanded operation. In: IEEE PES PowerTech Conference; 27-30 June 2005; St. Petersburg, Russia

[25] http://www.pge.com/tariffs/

[26] Standard CEI EN 50160: Voltage characteristics of electricity supplied by public distribution systems. 2015.

Chapter 3

Future of Microgrids with Distributed Generation and Electric Vehicles

Syed Abid Ali Shah Bukhari, Wen-Ping Cao,
Toufique Ahmed Soomro and Du Guanhao

Additional information is available at the end of the chapter

http://dx.doi.org/10.5772/intechopen.68819

Abstract

This chapter examines the current energy scenario for microgrids over the world and discusses the challenges and opportunities due to the increasing penetration of distributed power generation systems and electric vehicles (EVs) into the microgrids. Wind power and solar power can be generated by wind turbines and photovoltaics, respectively, while these are intermittent in nature. EVs and hybrid EVs use a battery energy storage system and charging facilities while the latter also include an Internal Combustion Engine (ICE) to provide an extra energy source. The features of these systems in the context of microgrids are studied in detail, in terms of their components, efficiency, reliability, charging and discharging arrangements, active and reactive power control. The chapter provides a reference to the development of microgrid systems especially for developing countries.

Keywords: Batteries, Carbon Free Generation, Charging and Discharging, CO_2 Emissions, Electric Vehicles, Microgrid, Renewable Energy

1. Introduction

Currently, air pollution is a serious issue, especially in heavily populated cities, such as London, Paris, Beijing, and Tokyo. Approximately 25% of global CO_2 emissions are due to passenger road and air travel, and the transport of goods. In addition to CO_2, Sox and Nox are also generated. Although, cars and trucks are responsible for the bulk of these of emissions, about 75% worldwide emissions due to aviation and shipping are growing rapidly too, and the energy used in transport could double by 2050 [1]. The chart of projected overall CO_2 emissions by different countries is depicted as shown in **Figure 1** [2], and particularly CO_2 emissions by electric cars by different countries are shown in **Figure 2** [3, 4].

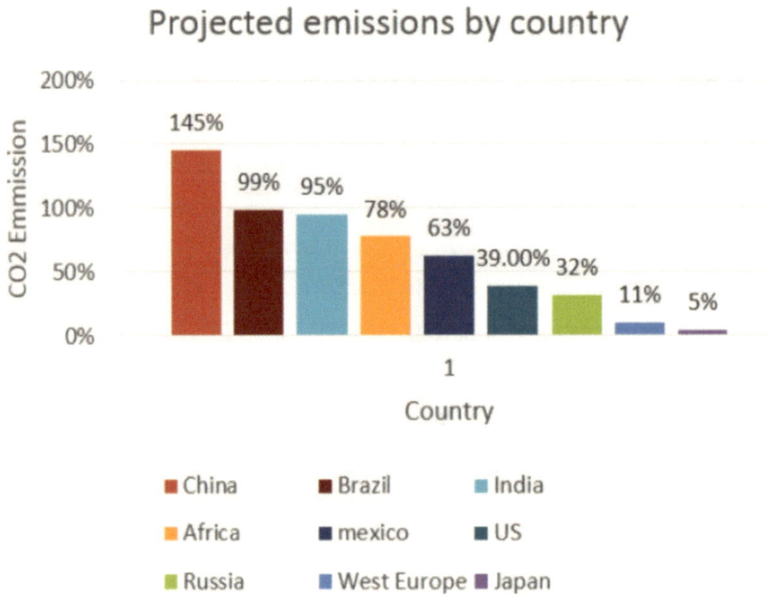

Figure 1. Projected emissions by country.

Therefore, CO_2 emissions need to be cut by approximately 50%. Priority should be given first to research technologies and practices that are mostly cost effective. Microgrid and E-propulsion systems have the capability to enable this; thus, bringing about a revolution in

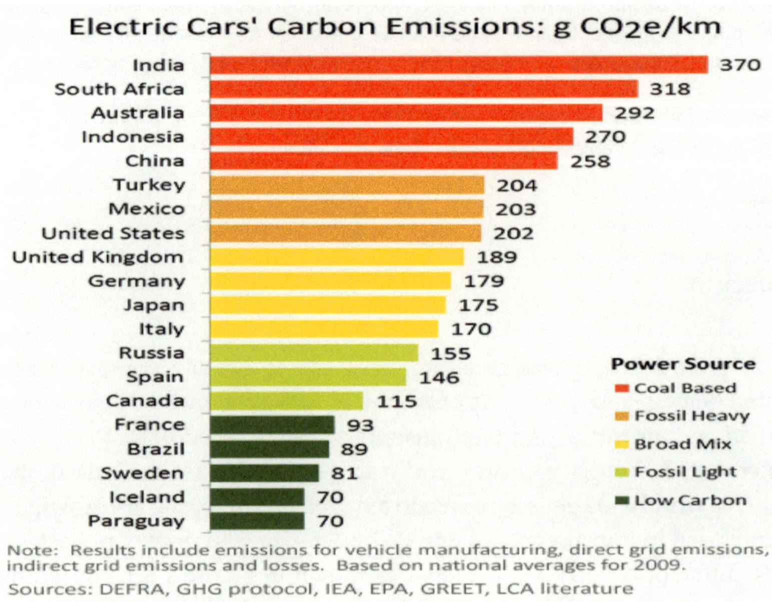

Figure 2. Electric cars' carbon emissions.

low CO_2 emission technology [5]. Thus, for the developing zero-emission, electric vehicles have become the key logical and scientific research project in many countries around the world in the twenty-first century [6]. The dynamism and energy saving motor drive technology have developed one of the key points to electrical vehicles commercialization [7].

As EVs are suitable, progressively more essential not only because they decrease noise and greenhouse gases but also they are able to diminish the dependency of carrying oil, as the power is generated from fuels other than oil [8]. Electric vehicles can contribute to decrease carbon emissions. The zero release of CO_2 production necessitates that the EV energy should be produced from alternate sources like nonfossil fuel, for example, nuclear and unconventional sources of energy. Actually, the most dangerous situation is that there is merely 40 years of oil supply is left [9, 10] with the rates of current usage. In fact, increasing inadequacy may bring the consequence in the huge amount of rises in price and ultimately the practice of continuous oil usage and extra fossil fuels will never be lavishly practicable, hence sources of oil might be well-maintained, as the procedure of consumption will decline. The production of oil can also be achieved from other fossil fuel sources such as coal [11]. Normally in this way, oil production is deliberated to be round about 10% more costlier.

This chapter provides a brief overview of the recent energy crisis in all over the world and the comparison between an electric vehicle and gasoline vehicle, and the methods by which driving range of electrical vehicle can be enhanced with the help of microgrid. The proposal which actually necessitates the development of microgrid technologies on the basis of different reasons and their implementation with respect to electrical vehicle and issues associated with the energy crisis is also given. The main purpose of this chapter is to lump together plug-in vehicles and electrical vehicles with microgrid.

1.1. Electrical energy scenario and background

The most complex and largest machine which has been ever built by the human brain is actually the electric grid. We found electricity 100 years ago. But still very much a dark planet, still a lot of places in the world there have not had an access of electricity in the last 100 years [12]. Even people living in some places do not know what electricity is [13]. So, just like many others, engineers and a group of some technical persons, if we could create an open source movement, an elaboration to build something and to start thinking about how would we electrify placed out there that governments have forgotten about. The first business should be to plan together and to find the ways to build the future. First of all, one block of location or something nearer and something that would benefit from what we are doing. So if we look at the statistics which is shown as below.

1.2. Energy poverty statistic

One in five people in this world do not have access to electricity today and it is approximately 1.2 billion people globally. If we start to look in depth that where are these people? Three of five in Africa, primarily in the rural population, have no access to electricity today [14]. If we further look at the statistics, where is the growth going to be? Where are all those new children going to be in the year 2100? The statistic of world's energy poverty is shown in **Figure 3**, and the statistic of Africa's energy poverty is shown in **Figure 4** [15].

It will be approximately 10 billion people in the year 2100 [16] and in Africa, it will be 3/2 billion people. Africa's land mass is three times bigger than the United States and Africa's population is 750 million people today and it is 4 times more [15]. The area is full of poverty and full of orphans because of HIV [17]. Now, if we try to figure out how can this community be upgraded? Let's see if we can create an open source collaborative approach to electrify this rural population, and there must be an impact on all of the community. Just for example as the first thing which is needed is an affordable and reachable electrical power, a nice thing in Africa. Everybody has cell phone and people mostly walk 5–10 miles to charge the mobile battery for their cell phone [18], so what if they could do that at home with the help of microgrid and in this way, business of microgrid will be growing. The second point is we have to be sustainable and have to keep the environment green. No greenhouse gases, no nitrous, no methane, and no carbon dioxide. So what do we have to do for alternative sources? We have lot of wind as well as lot of sunshine for photovoltaic (PV). Finally to reduce usage kerosene in homes. As kerosene is the primary source of cooking and lighting in the developing world. It is very dangerous and very polluting.

1.3. Electric vehicles

A vehicle that uses one or more electric motors or traction motors for propulsion.

As the worldwide economy originates to strain under the burden of increasing petroleum costs and environmental distresses, studies have encouraged the growth of numerous categories

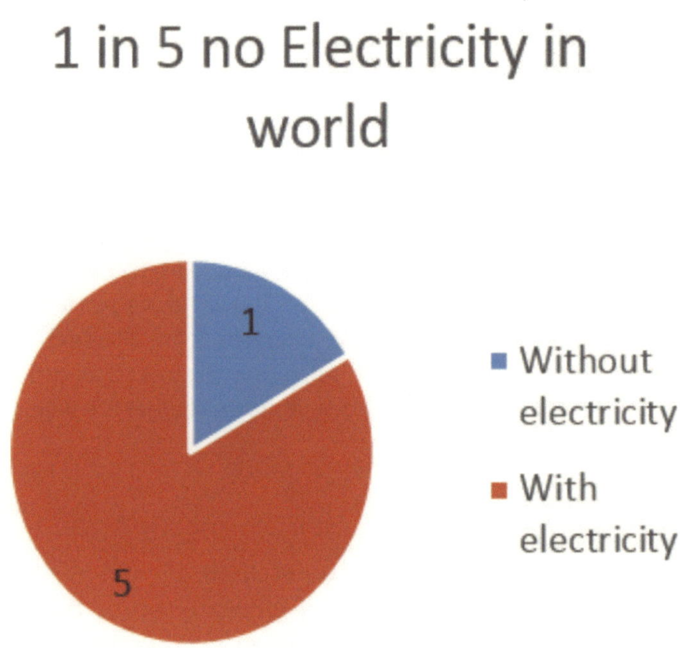

Figure 3. World's energy poverty density.

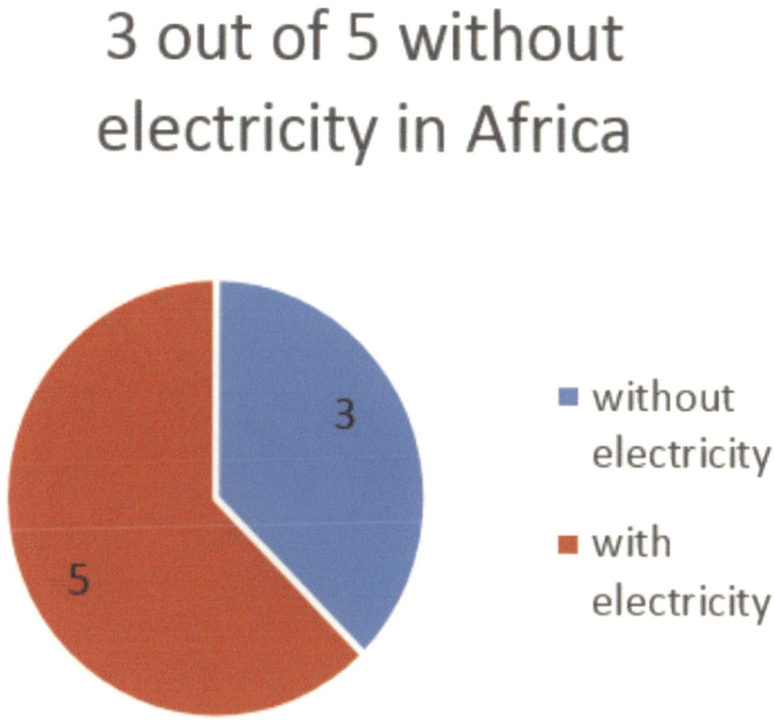

Figure 4. Africa's energy poverty density.

of clean energy transportation systems, for example, hybrid electric vehicles (HEVs), battery electric vehicles (BEVs), and plug-in hybrid electric vehicles (PHEVs).

Plug-in hybrid electric vehicles are already playing an important role in minimizing greenhouse gas emissions, as these emissions are very hazardous and are the main cause of global warming [19]. Therefore, electrical vehicle technology, electric vehicles (EV) and plug-in hybrid electrical vehicles (PHEVs), have gained the momentum in recent years and have become the focal point of research and development in finding suitable replacements for the internal combustion engine [8]. As internal combustion engine vehicles are being converted into EV, increasing amounts of oil will be saved. However, the energy for the innovative EVs will have to be generated in addition to the electricity used for surviving purposes [20]. Furthermore, there is also a need to upgrade the existing power grid if large number of EVs/HEVs are connected to it. To tackle this problem, a current trend is to develop microgrid [21]. This will not only necessitate a considerable enhancement in electricity generation, but also the microgrid distribution will have to be planned for [22]. The electric vehicles by Range Rover Evoque and Jaguar F type are shown in **Figures 5** and **6** [23–25].

Particularly, PHEVs obtain the most consideration because of combining electrical source and conventional engine. This kind of vehicle makes available the buyer a significant pure electrical choice and also an extended range, which is able to perform by means of a conventional internal combustion engine (ICE). The literature reveals that sales in electric cars have been

Figure 5. Range Rover Evoque electric vehicle.

tremendously increased in recent years. The UK growth in electric cars data taken by SMMT from 2011 to 2016 is shown in **Figure 7** [26], and the overall sales in electric cars are shown in **Table 1** [27].

Figure 6. Jaguar F type.

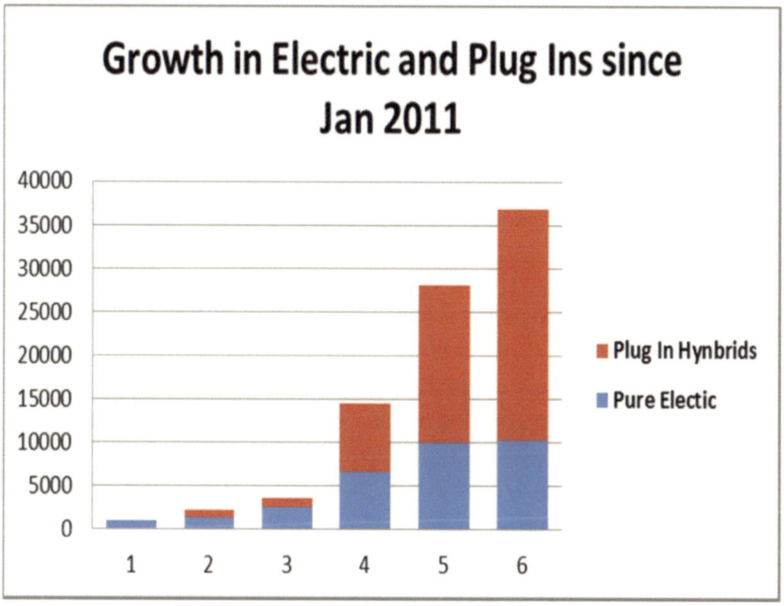

Figure 7. Growth in electric cars in UK.

Year	Number of electric cars on the road
2010	25,000
2011	80,000
2012	200,000
2013	405,000

Table 1. Electric car sale accelerates.

Due to above-rising trend in sales scenario of the electrical vehicle on the road, the demand of power is growing at an approximate rate of 100% per annum for the last 4–5 years and it is obviously anticipated that the growth in average demand over the upcoming decades will be bigger. This is an indication of load demand anxiety and stress on the distribution grid station in the years ahead, and much would be expected from electricity service provider so as to operate in such a way that it can satisfy customer needs and also enable it to stay in effective business.

The establishment of a rechargeable energy storage system (RESS) that has a provision of the output power during acceleration, proficiently use the regenerative energy to achieve a considerable cycle life is the serious aspect to be met by battery technologies.

1.4. Gasoline vs electric vehicles

The comparison between electric and gasoline vehicle is shown in **Table 2** [28].

Electric vehicle	Gasoline vehicle
Requires utility company for charging	Organization of petroleum exporting countries (OPEC)
200+/− mile range	300+ mile range
Zero emission	Huge greenhouse gases/pollution
Hours to charge	Minutes to refuel
2 cents per mile	12 cents + per mile

Table 2. Comparison between electric vs gasoline.

By considering **Table 2**, in contrast to gasoline/patrol vehicle, the electric vehicle has 6× lower cost for a 1-mile drive. However, its driving range is only 1/3 of gasoline vehicle per full charge. And this is the main drawback [29].

1.5. Improvement in driving range

The storage of electricity and resultant mass is shown in **Table 3**. We can enhance the range of driving by the accumulative capacity of the battery. But then, again at what expense? In order to have heavier and bigger battery, there is a need that it must have the greater energy density to be higher, so more costly battery/car hence lesser price charge of a storage battery is required. The trend of battery energy density is shown in **Figure 8** and the cost of electric vehicle battery storage to consumers is shown in **Figure 9** [30].

In **Figure 9**, from the year 2015 onwards, the improvement in rates is further delicate at 5% (as with the case of extraordinary technology battery). Satisfactory for electric vehicle to compete the worth and range of driving of gasoline vehicle, battery cost ($/kWh) necessities to drop by 4 times, which is unachievable even in 2035 [31].

1.6. Future of Batteries

From the above chart of cost of battery storage per kWh, we can conclude that even up to 2035, price of EV may not be able to match price of gasoline vehicle. Primarily due to cost of

Battery technology	Energy/Weight	Energy/Volume	Power/Weight	Relative	Self-discharge	Recharge time
	Watt h/KG	Watt h/L	Watt/kg	Cost	Rates	Hours
Lead acid	30–40	60–75	180	1	5%	8–16
Lithium/Ion	160	270	1800	4	10%	2–4
Li-Ion Polymer	130–200	300	2800	4	10%	2–4

Table 3. Storage of electricity and resultant mass.

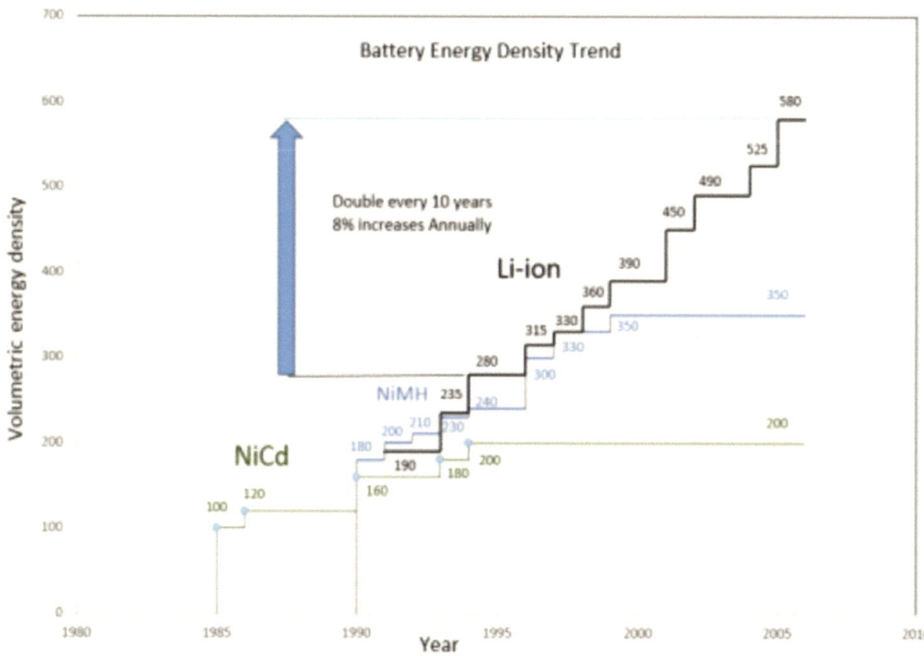

Figure 8. Energy density trend for battery.

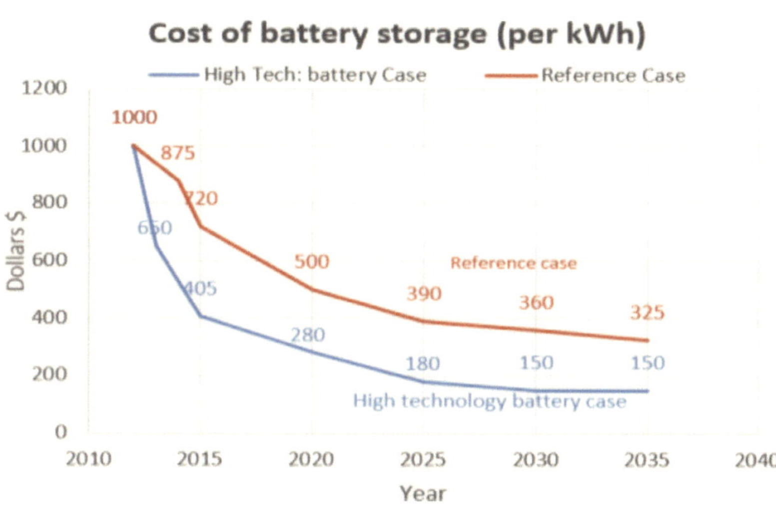

Figure 9. Cost of electric vehicle battery storage to consumers in two annual energy outlook (AEO).

battery only drop by 5% annually. Only in 2047, battery is capable to get closer with gasoline in terms of energy density. Therefore, other choices are needed to drive the penetration of EV to the consumer market.

1.7. Proposed solution

In order to cope with the battery issue, six solutions can be proposed as follows,

Community microgrid–sustainable

Feasibility of charging stations

Provide local residents access to affordable power

Reduced usage of kerosene for lighting

Future for charging stations

Sustainable greenhouse gas free environment

2. Introduction to microgrid

A microgrid is a small-scale power supply setup that is aimed to offer power for a small or private community. The definition of a small community could range from a distinctive housing estate, inaccessible or remote rural communities and societies such as universities or schools, to commercial areas, industrial, and business sites and trading estates or municipal regions. In fact, the Technical and economic viability of the distributed energy resource (DER) technologies for distribution voltage class applications has resulted in the development of the microgrid concept [32]. In other words, the microgrid is a small-scale power grid that is able to work independently or in combination with the area's central electrical grid. Whichever small-scale localized station with its own power resources, generation and loads and definable boundaries be nominated as a microgrid [33], or it is a cluster of interconnected loads and circulated energy assets surrounded by clearly defined electrical boundaries that act as a solitary well-regulated entity with respect to the grid [34, 35]. "The main important point is, it can attach and detach from the grid to allow it to operate in both grid-connected or island mode in a coordinated, controllable way" [36]. The microgrid can also be planned to meet its special needs [37, 38] such as,

Improve local reliability.

Decrease feeder losses.

Support indigenous voltages.

Offer improved efficiency through use waste heat.

Voltage sag adjustment or afford uninterruptible power supply functions.

Microgrids have a provision and flexibility of efficient electric grid by facilitating the integration of growing supplies of distributed energy resources such as renewables like solar and wind [39]. In addition, the practice of local sources of energy to assist local loads helps reduce energy losses in transmission and distribution, further increasing an efficiency of the electric delivery system, choices will have to be made whether electricity for transport will come from carbon-free generation or whether the standard combination of energy currently used will be followed to. An assessment of the amount of electrical energy which would be essential throughout the world. If power for transportation from fossil fuels is to be swapped, extensive investment in new methods of power for transport exist in nuclear or unconventional energy will be necessary; therefore, is fairly conceivable to generate electrical power used for transportation and further uses by approaches which should not cause carbon releases or which reduce any further fossil fuels [22]. Now, approximately all such procedures will comprise a mixture of nuclear power generation units and a choice of substitute energy resources. Solar energy from the sun is expected to be the principle substitute of energy [40]; nonetheless, the combination of solar with wind energy, tidal and underwater currents surrounded by others are about to feature intensely. The microgrid architecture and its applications are shown in **Figure 10** [39].

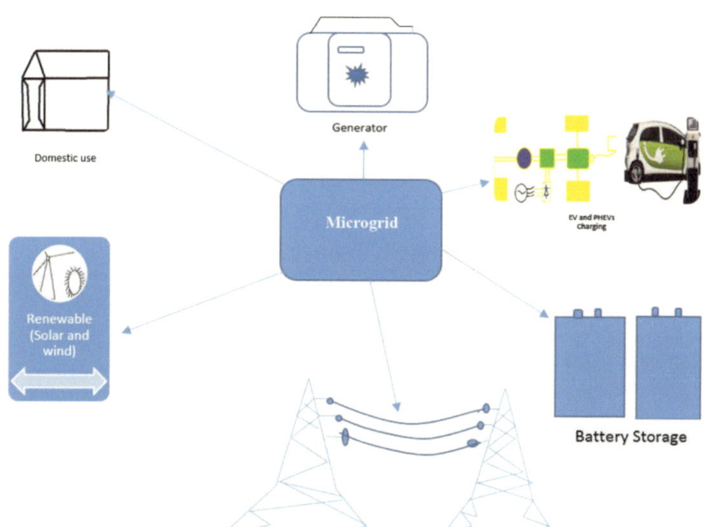

Figure 10. Microgrid architecture and applications.

2.1. Domestic Use

As depicted above, the applications of microgrid are very vast, which can be applied for domestic use, battery storage, electric vehicle charging, utility grid, and renewable energy generation. As most of the appliances are AC power, such as television set, air conditioner, washing machine, refrigerator, microwave oven, and with the implementation of microgrid technologies, these mentioned appliances can easily be operated and can decrease the impact on distribution grid station. Recently, DC power fans are also emerging, which consume very low electricity and have a better performance.

2.2. Industrial Use

With the bi-directional information and power flow provided by the microgrid, several novel planning is being promoted to expand and advance the energy usage in diverse settings. One such extension is the concept of microgrids in different industries such as oil and gas exploration industries. These kinds of industries are usually situated at site area where the actual reservoirs are discovered through different seismic surveys and central facility (Plant) is built for gas and oil processing. The actual place from where the oil and gas is explored is called wellhead and is located at far distance (40–100 km away) from the central facility and has no access of electricity, in order to control the process variable like pressure, temperature, flow, and level. There is a need for proper instrumentation for process flow lines and its control variables. So, there are potential benefits of using microgrid especially oil and gas industries which is shown in **Figure 11** [41, 42].

2.3. Benefits of Microgrids

The main benefits of microgrid are listed below.

A more suitable climate

Electric vehicle integration

Better public health

Lower electric bills for consumers growing revenues for utilities

Energy savings–direct DC charging

Renewable energy integration

Figure 11. Flow control valve, pressure gauges, and process flow lines.

http://dx.doi.org/10.5772/intechopen.68819

Improved control and monitoring

Improved system reliability

Facilitation of entrepreneurial opportunities

Energy storage systems

Wireless charging

2.4. Locations of microgrid

Residential and commercial buildings consume about 32% of the worldwide energy use, as they are liable for about 30% of the total end use energy-related radiations if the indirect upstream emissions are considered [43]. As the proper renewable energy generation, including photovoltaic arrays or wind turbines, is the main concern in the development of microgrid, in this connection parking facilities are the ideal locations available for systems linking electric vehicles with solar and wind energy sources. Thus, these tactics permit a high number of EVs to be charged at an office building, even with a limited number of charging points, due to the large standstill times [44]. The microgrid with a solar array and windmill is shown in **Figure 12** [45].

A microgrid that mixes renewable generation and vehicle energy storage offers many advantages such as (1) the security of energy, (2) savings in cost, and (3) reliability benefits. The charging stations will work as energy management portals, enabling unidirectional and bidirectional power flow with vehicle energy storage. In the world market 2013–2020, the

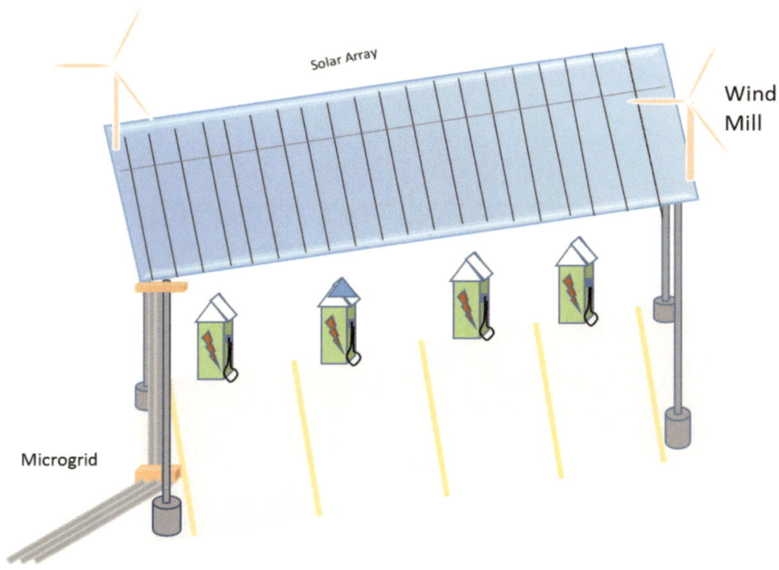

Figure 12. Microgrid with solar array and wind mill.

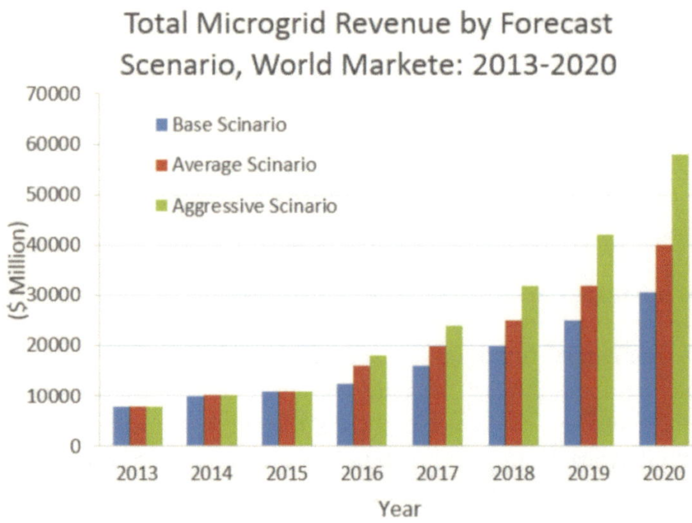

Figure 13. Microgrid revenue by forecast scenario, world market: 2013–2020 [38].

average scenario of microgrid revenue is ($40 billion) and from the aggressive forecast scenario, the revenue is approximately 58,000 $ million, from the world forecast scenario of microgrid revenue in the world market 2013–2020 is shown in **Figure 13** [46], and the charging standard and cost are shown in **Table 4** [47].

Table 4 shows the types of chargers according to different levels; however, the rate of one direct current fast-charge (level 2) is the identical as the cost of one 24 kWh battery, which is used by Nissan LEAF; therefore, constructing additional charging stations will accelerate occasion of opportunities to install lesser capacity of battery, consequently proposing an inexpensive electric vehicle. The domestic electrical vehicle charger is shown (**Figure 14**) [48].

Charging level	Power supply	Charging power	Miles of range for 1 h of charge	Charging times BEV	Charging times PHEV
Level 1	120 VAC Single phase	1.4 kW @ 12 amp (on-board charger)	~3–4 miles	~ 17 h	~7 h
Level 2	240 VAC Single Phase up to 19.2 kW (upto 80 amp)	3.3 kW (on-board)	~8–10 miles	~7 h	3 h
		6.6+ kW (on-board)	~17–20 miles	~3.5 h	~1.4 h
DC fast charger level 2	200–450 VDC upto 90 kW (approx.: 200 amp)	45 kW (off-board)	~ 50–60 miles (80% per 0.5 h charge)	~ 30–45 min (to ~80%)	~ 10 min (to~80%)

Table 4. Charging standards and costs.

Figure 14. Domestic electrical vehicle charger.

2.5. Suggestions for further improvement in microgrid

- Building the future for EVs
- Licensing of third party activities
- Microgrids for local power generation–peak shaving and commercial prospects for trade of surplus energy back into grid
- EV battery charging, rental, replacement services
- Advertisement on charging stations
- Software and mobile applications to find the nearest charging station

2.6. Future of charging stations

For the future of charging stations, the private companies are already developing their electric vehicles charging solutions [49]. However, building a strong infrastructure for EV charging can overcome the problem of low mile range. Still, there are many efforts and initiatives, which need to be taken for building the case for wireless electrical vehicle charging (WEVC) up to mainstream. The government should distribute its available resources in the favor of research in electrical vehicle demands. Like: (1) low price and availability of power supply to drive electrical vehicle demand, or vice versa. (2) The wireless charging lines on major expressways which are shown below in **Figure 15**, and building the smart metering system with the secure payment for the wired and wireless electrical vehicle charging. Beyond the EV-Extension of Wireless Transmission Applications. Infrastructure availability for EV charging (lowering power supply costs). Directly lowering EV costs (increasing EV demand). With EVs capable to simply and easily locate charging facilities/achieve and get charging on the move when required. Have need of complementary enhancements in smart powering and metering systems for calculation and payment of charging fee.

2.7. Room for Improvements

Rates of enhancements enabling development in use of microgrid, electric vehicle, and expertise in transfer of power without wire: such as

- ICs
- MOSFETs
- Roll printing for thin film substrates

The improvement rates in the above skills exceed the rate of extension in (car) battery skills. Consequently, the task in short mile range of electric vehicle will be overcome more rapidly while smoothing the advance of economical wireless facilities for charging, through powered microgrid. The implementation of the above-mentioned suggestions will lead to a healthier and sustainable life, which is depicted in building the dedicated WEVC on major expressways as shown in **Figure 15** [50].

Figure 15. Building the dedicated WEVC on major express ways.

3. Conclusion

This chapter has presented the development of microgrids incorporating renewable energy and electric vehicles while the associated energy crisis and their challenges have been extensively discussed. Addressing these challenges would contribute to simplifying the execution of novel renewable technologies in a microgrid. As the microgrid research involves many areas of power systems by considering the latest trend, charging infrastructure can be developed to better cater for EVs in the future. Also, the worldwide anxieties about diminishing fossil energy assets and ever-increasing pollution have augmented significantly in recent years. Distributed generation and EVs have acknowledged enormous consideration because of their environmentally friendly nature as a power source.

The impact on the grid can be reasonably eased by means of developing smart onboard or offboard charging facilities in residential or office areas. These approaches permit a high quantity of EVs to be fast charged with a minor impact on the main grid and an increased DER self-consumption. With the help of microgrids, EVs can reduce the reliance on crude oil fossil fuel energy so as to reduce CO2 emissions. The record numbers of EV chargers based on microgrids and charging infrastructures are anticipated to rapidly increase surrounded by domestic areas in the near future.

The future microgrids will have a huge quantity of distributed generation systems and adaptable power generation from renewable energy resources. As microgrid systems can provide much promise for integrating large numbers of distributed energy resource systems and minimise the impact on the main grids, their future are bright. In addition to the electric vehicle application, microgrid technologies can also be applied to aircraft, rails, ships, submarines or other forms of electrified transportation.

Author details

Syed Abid Ali Shah Bukhari[1]*, Wen-Ping Cao[1], Toufique Ahmed Soomro[2] and Du Guanhao[1]

*Address all correspondence to: bukhars2@aston.ac.uk

1 Aston University, Birmingham, United Kingdom

2 Charles Sturt University, Australia

References

[1] Lomonova E, Paulides J, Wilkins S, Tegenbosch J. ADEPT: "ADvanced electric powertrain technology"-Virtual and hardware platforms. In Ecological Vehicles and Renewable Energies (EVER), 2015 Tenth International Conference on; 2015. pp. 1-10.

[2] Speight JG. The Chemistry and Technology of Coal. CRC Press, Taylor & Francis Group 6000 Broken Sound Parkway NW, Suit 300 Boca Raton, FL 33487-27; 2012.

[3] Wu Y, Zhang L. Can the development of electric vehicles reduce the emission of air pollutants and greenhouse gases in developing countries?. Transportation Research Part D: Transport and Environment. 2017;**51**:129-145.

[4] Casals LC, Martinez-Laserna E, García BA, Nieto N. Sustainability analysis of the electric vehicle use in Europe for CO_2 emissions reduction. Journal of Cleaner Production. 2016;**127**:425-437.

[5] Zhang T, Su W, Duan Q, Meng X, Yu J, Chen X. A simulation study of electric vehicle charging in microgrids. In Power and Energy Engineering Conference (APPEEC), 2013 IEEE PES Asia-Pacific; 2013. pp. 1-5.

[6] Van Roy J, Leemput N, Geth F, Büscher J, Salenbien R, Driesen J. Electric vehicle charging in an office building microgrid with distributed energy resources. IEEE Transactions on Sustainable Energy. 2014;**5**:1389-1396.

[7] Yan W, Haochun Y, Tianming Y. Applications of SR Drive Systems on Electric Vehicles: INTECH Open Access Publisher; Beijing Tongdahuaquan Ltd. Company 2011.

[8] Situ L. Electric vehicle development: the past, present & future In Power Electronics Systems and Applications, 2009. PESA 2009. 3rd International Conference on, 2009; pp. 1-3.

[9] Rutledge D. Estimating long-term world coal production with logit and probit transforms. International Journal of Coal Geology. 2011;**85**:23-33.

[10] Devezas T, LePoire D, Matias JC, Silva AM. Energy scenarios: Toward a new energy paradigm. Futures. 2008;**40**:1-16.

[11] Weller SW. Catalysis and catalyst dispersion in coal liquefaction. Energy & Fuels. 1994; **8**:415-420.

[12] Panos E, Densing M, Volkart K. Access to electricity in the World Energy Council's global energy scenarios: An outlook for developing regions until 2030. Energy Strategy Reviews. 2016;**9**:28-49.

[13] Gronewold, Nathanial. "One-quarter of world's population lacks electricity." *Scientific American* **24** (2009).

[14] Louie H, Grady EO, Acker VV, Szablya S, Kumar NP, Podmore R. Rural Off-grid Electricity Service in Sub-Saharan Africa [Technology Leaders]. IEEE Electrification Magazine. 2015;**3**:7-15.

[15] Rosenberg NA, Pritchard JK, Weber JL, Cann HM, Kidd KK, Zhivotovsky LA, et al. Genetic structure of human populations. Science. 2002;**298**:2381-2385.

[16] Lutz W, Sanderson W, Scherbov S. The end of world population growth. Nature. 2001;**412**:543-545.

[17] Timaeus IM. Impact of the HIV epidemic on mortality in sub-Saharan Africa: Evidence from national surveys and censuses. AIDS (London, England). 1997;**12**:S15-S27.

[18] Adkins E, Oppelstrup K, Modi V. Rural household energy consumption in the millennium villages in Sub-Saharan Africa. Energy for Sustainable Development. 2012;**16**:249-259.

[19] Crowley TJ. Causes of climate change over the past 1000 years. Science. 2000;**289**:270-277.

[20] Dresselhaus M, Thomas I. Alternative energy technologies. Nature. 2001;**414**:332-337.

[21] Kempton W, Tomić J. Vehicle-to-grid power implementation: From stabilizing the grid to supporting large-scale renewable energy. Journal of Power Sources. 6/1/2005;**144**:280-294.

[22] Whitefoot JW. Optimal Co-design of Microgrids and Electric Vehicles: Synergies, Simplifications and the Effects of Uncertainty. The University of Michigan, Horace H. Rackham School of Graduate Studies; 2012.

[23] Ivanov V, Savitski D, Augsburg K, Barber P. Electric vehicles with individually controlled on-board motors: Revisiting the ABS design. In Mechatronics (ICM), 2015 IEEE International Conference on. 2015, pp. 323-328.

[24] Raynauld O., Fath S. Magneride suspension for the range rover evoque. ATZ Worldwide. 2012;**114**:16-21.

[25] Lyons W, Walmsley W, Mistry C. P, Callum I, Popham P, Motors T, *et al.*, "Jaguar Cars." Pediaview.com. open source Encyclopedia.

[26] Slowik P, Pavlenko N, Lutsey N. Assessment of Next-Generation Electric Vehicle Technologies. International Council on Clean Transportation 1225 I Street NW Suite 900 Washington, DC 20005 USA; 2016.

[27] Nykvist B, Nilsson M. Rapidly falling costs of battery packs for electric vehicles. Nature Climate Change. 2015;**5**:329-332.

[28] Offer G, Howey D, Contestabile M, Clague R, Brandon N. Comparative analysis of battery electric, hydrogen fuel cell and hybrid vehicles in a future sustainable road transport system. Energy Policy. 2010;**38**:24-29.

[29] Granovskii M, Dincer I, Rosen MA. Economic and environmental comparison of conventional, hybrid, electric and hydrogen fuel cell vehicles. Journal of Power Sources. 2006;**159**:1186-1193.

[30] Thomas C. Fuel cell and battery electric vehicles compared. International Journal of Hydrogen Energy. 2009;**34**:6005-6020.

[31] Administration EI. Annual Energy Outlook 2012: With Projections to 2035: Government Printing Office;2012.

[32] Jiayi H, Chuanwen J, Rong X. A review on distributed energy resources and MicroGrid. Renewable and Sustainable Energy Reviews. 2008;**12**:2472-2483.

[33] Hatziargyriou N, Asano H, Iravani R, Marnay C. Microgrids. IEEE Power and Energy Magazine. 2007;**5**:78-94.

[34] Lasseter RH. Smart distribution: Coupled microgrids. Proceedings of the IEEE. 2011;**99**: 1074-1082.

[35] Katiraei F, Iravani R, Hatziargyriou N, Dimeas A. Microgrids management. IEEE Power and Energy Magazine. 2008;**6**.

[36] Jiang Z, Yu X. Hybrid DC-and AC-linked microgrids: Towards integration of distributed energy resources. In Energy 2030 Conference, 2008. ENERGY 2008. IEEE, 2008. 1-8.

[37] Li Z, Yuan Y, Li F. Evaluating the reliability of islanded microgrid in an emergency mode. In Universities Power Engineering Conference (UPEC), 2010 45th International IEEE, Cardiff, Wales, UK; 2010, pp. 1-5.

[38] Colson CM, Nehrir MH, Sharma RK, Asghari B. Improving sustainability of hybrid energy systems part II: Managing multiple objectives with a multiagent system. IEEE Transactions on Sustainable Energy. 2014;**5**:46-54.

[39] Gopalakrishnan A, Biswal AC. Applications of emerging communication trends in automation. In 2016 IEEE 6th International Conference on Power Systems (ICPS); 2016, pp. 1-6.

[40] Samith MV, Rashmi MR. Controller for integrating small scale power generation to hybrid AC/DC grid. In: 2016 International Conference on Inventive Computation Technologies (ICICT); 2016, pp. 1-5.

[41] Lipták BG, Venczel K. Measurement and Safety: Instrument and Automation Engineers' Handbook: CRC Press, Taylor & Francis Group Boca Raton London New York. 2016.

[42] Nazeri M, Maroto-Valer MM, Jukes E. Performance of Coriolis flowmeters in CO_2 pipelines with pre-combustion, post-combustion and oxyfuel gas mixtures in carbon capture and storage. International Journal of Greenhouse Gas Control. 2016;**54**:297-308.

[43] Bozell JJ. National Renewable Energy Laboratory 1617 Cole Boulevard Golden, CO 80401.

[44] Liu N, Chen Q, Liu J, Lu X, Li P, Lei J, et al. A heuristic operation strategy for commercial building microgrids containing EVs and PV system. IEEE Transactions on Industrial Electronics. 2015;**62**:2560-2570.

[45] Nunes P, Figueiredo R, Brito MC. The use of parking lots to solar-charge electric vehicles. Renewable and Sustainable Energy Reviews. 2016;**66**:679-693.

[46] Lasseter RH, Paigi P. Microgrid: A conceptual solution. In Power Electronics Specialists Conference, 2004. PESC 04. 2004 IEEE 35th Annual; 2004, pp. 4285-4290.

[47] Yilmaz M, Krein PT. Review of battery charger topologies, charging power levels, and infrastructure for plug-in electric and hybrid vehicles. IEEE Transactions on Power Electronics. 2013;**28**:2151-2169.

[48] Ibrahim M, Bernard L, Pichon L, Labouré E, Razek A, Cayol O, et al. Inductive charger for electric vehicle: Advanced modeling and interoperability analysis. IEEE Transactions on Power Electronics. 2016;**31**:8096-8114.

[49] Bayless SH, Neelakantan R, Guan A. Connected Vehicle Assessment: Vehicle Electrification and the Smart Grid: The Supporting Role of Safety and Mobility Services. 2012. Aug 16 2013

[50] Imura T, Yasuda T, Oshima K, Nayuki T, Sato M, Oshima A. Wireless power transfer for electric vehicle at the kilohertz band. IEEJ Transactions on Electrical and Electronic Engineering. 2016;**11**.

Section 2

Operational and Technical Issues of Microgrids

Chapter 4

A Generalized Voltage Control Algorithm for Smooth Transition Operation of Microgrids

Jing Wang and Bouna Mohamed Cisse

Additional information is available at the end of the chapter

http://dx.doi.org/10.5772/intechopen.69402

Abstract

The chapter proposes a generalized control algorithm that can reject the disturbances associated with microgrid transition operation to facilitate smooth microgrid transition operation. Firstly, the literature review of the state-of-the-art gives a deep analysis of the disturbances associated with microgrid transition operation and it reveals that the same controller should be adopted in the inverter control layer to prevent some harmful transients during transition. Then, a generalized voltage control algorithm in inverter control layer that can achieve smooth transition of microgrid is developed including the formulation of the problem, description of the design methodology and design procedures, and analytical study. The salient feature of the developed generalized voltage control algorithm is that the disturbances associated with microgrid transition are fully cancelled by using inverse dynamic model, and the inverter control layer can be seen as a bypass for the application layer. The practical feasibility of the proposed control algorithm is demonstrated by implementing and testing in a signal level hardware-in-the-loop (HIL) platform.

Keywords: microgrid transition operation, inverter control, inverse dynamic model, voltage control

1. Introduction

According to a survey performed by Microgrid Knowledge, electric reliability is the number one reason customers install microgrids, thanks to their ability to provide uninterrupted power supply (in particular, for critical loads) when the utility is lost. For this reason, a microgrid should be controlled to operate both in grid-connected and in islanded mode, as well as to transit seamlessly between the two [1–3]. In the transition between grid-connected and islanded mode, two types of disturbances are expected to occur and therefore the controller of the inverter may have to deal with (1) frequency disturbances related to a sudden change of the angle/frequency reference for the inverter control, and (2) current and voltage disturbances associated with switching between different operating modes [4]. Therefore, the inverter controller should reduce the impact from those disturbances to acceptable limits, or, at best case, eliminate them completely.

For the first type of disturbance, the seamless transfer techniques focus on the application layer and the essential effort is to improve power angle/voltage transients during transition. Thanks to the smooth modifications of the references (i.e. frequency, voltage and current) for the inverter layer (voltage and current controller), smooth transition operation can be achieved. As for the second type of disturbance, extensive research works have been undertaken in the inverter controller to improve the disturbance rejection performance. However, the impact from the disturbances can only be reduced and the robustness of these controllers is not guaranteed during the transition of different operating modes.

Motivated by the research gap, a novel inverter control algorithm is developed based on the inverse dynamic model of the *LC* filter and the inverter, transforming the closed loop transfer function of the inverter control level into the 'unitary gain'. The inverter controller with the unitary gain property automatically eliminates the second type of disturbance during the microgrid transition operation; therefore, smooth transition operation is achieved.

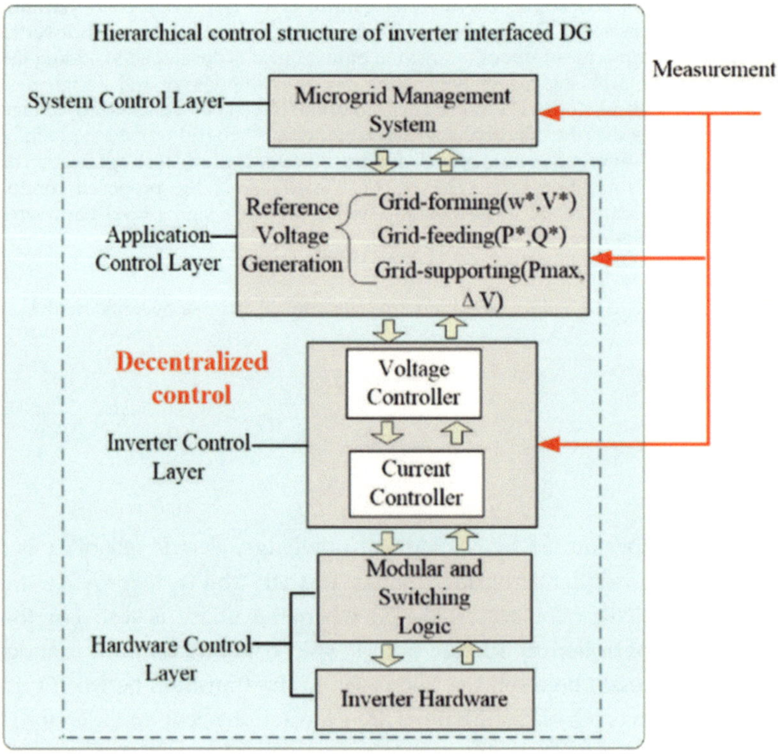

Figure 1. Modified control structure for DGs in microgrid operating in different modes.

2. A generalized control architecture for DGs in microgrid

The control structure for DG interface in microgrids is mapped in **Figure 1** according to the IEEE Std. 1676, compatible with all the microgrid operation modes [5]. The highest layer, the system control layer, is implemented in microgrid central controller (energy management system (EMS)). The lower layers are locally applied in decentralized controller of each DG unit.

The main task of the system control layer is to manage the operation modes (grid-forming, grid-feeding and grid-supporting mode) of DGs in microgrid based on the network characteristic and distribution network operator (DNO)'s request and then to send the corresponding reference signals and control commands to the application and the inverter control layers. The application control layer generates a specific voltage reference for the inverter layer according to the chosen operation. The inverter control layer executes the commands to fulfil the task set by the system control layer, and it hosts the proposed generalized voltage controller, designed to reject the current/voltage disturbances associated with microgrid transition operation. The functions of the hardware control layer of inverter can reference the IEEE Std. 1676 for PEBB system.

3. Key challenge and a promising solution: an intelligent control algorithm in inverter control layer

As stated in Section 1, the key challenge in designing a controller to reject the harmful voltage and current transients caused by the shift of microgrid operation mode is in the design of the inverter layer control algorithm. Microgrids in islanded mode and grid-connected mode applications have been considered in the past separately from the point of view of control design in the inverter control layer. This control strategy causes unnecessary harmful transients in the control system [5, 6]. Moreover, the commonly adopted control algorithm in inverter control layer has a non-negligible drawback: if the voltage is controlled, the system performance is sensitive to the output current of the *LC*/*LCL* filter and the output impedance (gain of the output current) of the inverter, which act as disturbances to the output voltage [7]. The same applies to the current control if the control variable is the output current of the *LC*/*LCL* filter. Therefore, a promising solution to solve the issues in the existing control methods is direct voltage control method adopted in the inverter control layer, and a good disturbance rejection strategy should be considered to fully cancel the disturbances in the inverter control layer. As the disturbances in the inverter control system can be measured, full feedforward compensation through inverse dynamic model can be applied to totally cancel the disturbances: a detailed methodology and principles are explained in the following section.

3.1. Principle of inverse dynamic model

The basic structure of a single-input/single-output plant using the inverse dynamic model feedforward is shown in **Figure 2**. The plant input u is composed of two parts: the feedback control input u_d and the inverse dynamic model u_f. The inverse dynamic model $G_i(s)$ is used in the feedforward path of the controller to compute the desired actuator inputs u_f to the plant.

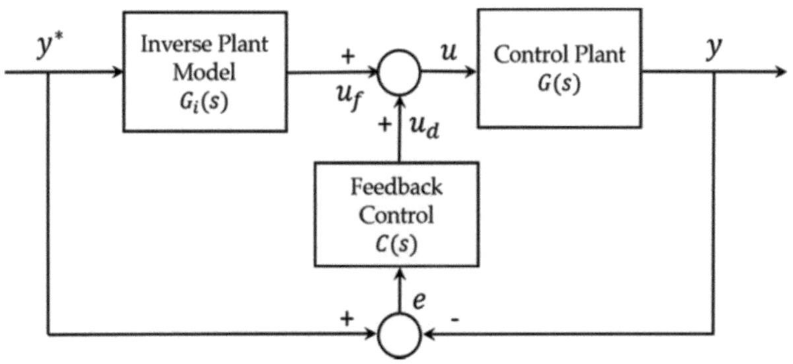

Figure 2. Control structure with inverse plant model in the feedforward path.

The feedback control $C(s)$ eliminates the tracking error. The feedforward control and feedback control can be designed separately, which follows the design concept of the two-degree-of-freedom control.

The transfer function of the above system is given by

$$\frac{y}{y^*} = \frac{G(s)G_i(s) + G(s)C(s)}{1 + G(s)C(s)} \tag{1}$$

If an accurate inverse model of the plant is obtained ($G_i(s) \equiv G^{-1}(s)$) and in a proper form, then the transfer function of the controlled system has a unitary gain for all frequencies. Thus, all the internal disturbances imposed upon the controlled output are fully cancelled.

3.2. Control strategy in the inverter control layer

Different control strategies in the application layer regulate different outputs of the DG (e.g. desirable active and reactive power generation, voltage and frequency regulation, and maximum active power injection). Essentially, these control strategies translate the various outputs into references for the injected current or the terminal voltage in the inverter control layer. To achieve universality, a new inverter control algorithm is required to control either one of these variables in a flexible way. As the impedance between the DG and the grid is a known parameter and the main grid voltage is an external measurable variable, the injected DG current can be indirectly regulated by controlling the terminal voltage [5]. Thus, universality is attained by introducing a voltage-based control algorithm into the inverter control layer.

By integrating disturbance rejection and universality of the control algorithm in the inverter control layer, a control algorithm based on the inverse dynamic model method is developed [5]. The control structure of this algorithm is presented in **Figure 3**. As seen in the figure, the control algorithm includes double loops: outer voltage and inner current loop. The inverse dynamic model 1 shows the analytical relationship between the inverter output voltage v_o and the inverter current i_i and also illustrates the input-output relationship between them. To compel the control system to achieve the target output v_o*, the corresponding control input i_{iff}* needs to be brutally imposed on the system. Incorporating the additional feedback control

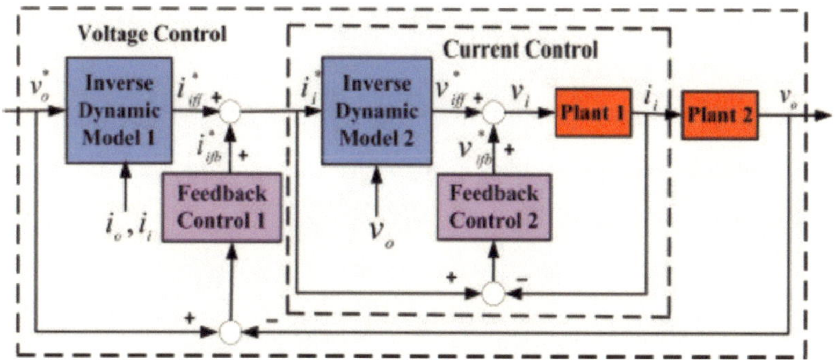

Figure 3. Control structure of the developed intelligent control algorithm using inverse dynamic model.

block 1, the closed loop transfer function of the outer loop is 'unitary gain'. The inner current loop respects the same design strategy and achieves also the closed loop transfer function with 'unitary gain'.

4. Design of the inverter layer control algorithm

The design of the generalized voltage control algorithm in the inverter control layer is based on the model of the *LC* filter and the inverter (the inverter is treated as a gain '1'). **Figure 4** shows the circuit diagram of the DG interface and *LC* filter together with the simplified structure of the control algorithm in the inverter control layer. All the variables are represented in *synchronous frame* and the dynamics of the *LC* filter are formulated in Eqs. (2)–(5):

$$\frac{di_{id}}{dt} = -\frac{R_f}{L_f} i_{id} + \frac{1}{L_f}(v_{id} - v_{od}) + \omega_o i_{iq} \tag{2}$$

$$\frac{di_{iq}}{dt} = -\frac{R_f}{L_f} i_{iq} + \frac{1}{L_f}(v_{iq} - v_{oq}) - \omega_o i_{id} \tag{3}$$

$$\frac{dv_{od}}{dt} = -\frac{1}{C_f}(i_{id} - i_{od}) + R_c\left(\frac{di_{id}}{dt} - \omega_o i_{iq}\right) - R_c\left(\frac{di_{od}}{dt} - \omega_o i_{oq}\right) + \omega_o v_{oq} \tag{4}$$

$$\frac{dv_{oq}}{dt} = -\frac{1}{C_f}(i_{iq} - i_{oq}) + R_c\left(\frac{di_{iq}}{dt} + \omega_o i_{id}\right) - R_c\left(\frac{di_{oq}}{dt} + \omega_o i_{od}\right) - \omega_o v_{od} \tag{5}$$

where i_{id}, i_{iq} ,i_{od}, i_{oq} are the inverter currents and inverter output currents in the *dq* frame, respectively, and v_{id}, v_{iq}, v_{od} ,v_{oq} are the inverter voltages and inverter terminal voltages in the *dq* frame, respectively. R_f, L_f, C_f and R_c are the per-phase resistance, inductance and capacitance of the *LC* filter.

By using Laplace transformation, we obtain the transfer functions for the inverter current and inverter terminal voltage:

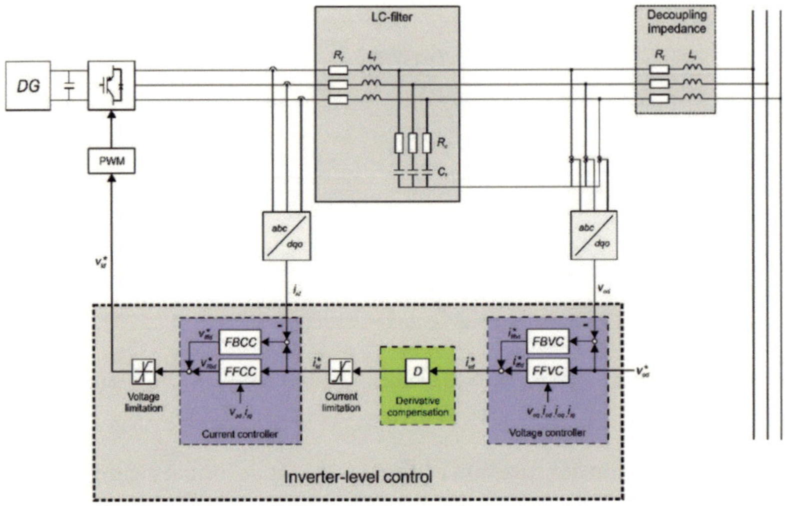

Figure 4. Circuit diagram and control structure of the inverter control layer.

$$i_{id} = \frac{1}{sL_f + R_f}(v_{id} - v_{od} + \omega_o L_f i_{iq}) \tag{6}$$

$$i_{iq} = \frac{1}{sL_f + R_f}(v_{iq} - v_{oq} - \omega_o L_f i_{id}) \tag{7}$$

$$v_{od} = \left(\frac{1}{sL_f} + R_c\right)(i_{id} - i_{od}) - \omega_o \frac{R_c}{s}(i_{iq} - i_{oq}) + \frac{\omega_o}{s} v_{oq} \tag{8}$$

$$v_{oq} = \left(\frac{1}{sL_f} + R_c\right)(i_{iq} - i_{oq}) + \omega_o \frac{R_c}{s}(i_{id} - i_{od}) - \frac{\omega_o}{s} v_{od} \tag{9}$$

It becomes evident that the system described above is highly coupled. For instance, the currents are functions of both voltages and the coupling terms of voltage, the latter of which interferes with voltage as well. The block in **Figure 5** shows the coupled system and resistance of the LC filter.

The dynamics of the *LC* filter can be expressed as one equation for the inverter current and one for the terminal voltage for each component. This structure suggests a cascaded control structure for the inverter control containing one inner current loop and one outer voltage loop. In the controller design, the inverter current i_i and the output voltage v_o are the controlled variables for current controller and voltage controller, respectively. The choice of the voltage loop as outer loop is a natural consequence of the fact that the inverter output voltage is the filter's outermost variable. The general idea is to force the controlled variables to quickly follow the reference signal and to be robust against disturbances and coupling terms. At the same time, in order to achieve maximum transparency to higher control levels, the proposed controller is designed in such a way that both the current closed loop and voltage closed loop have a unitary transfer function. Therefore, all effects of disturbances are removed and the inverter is

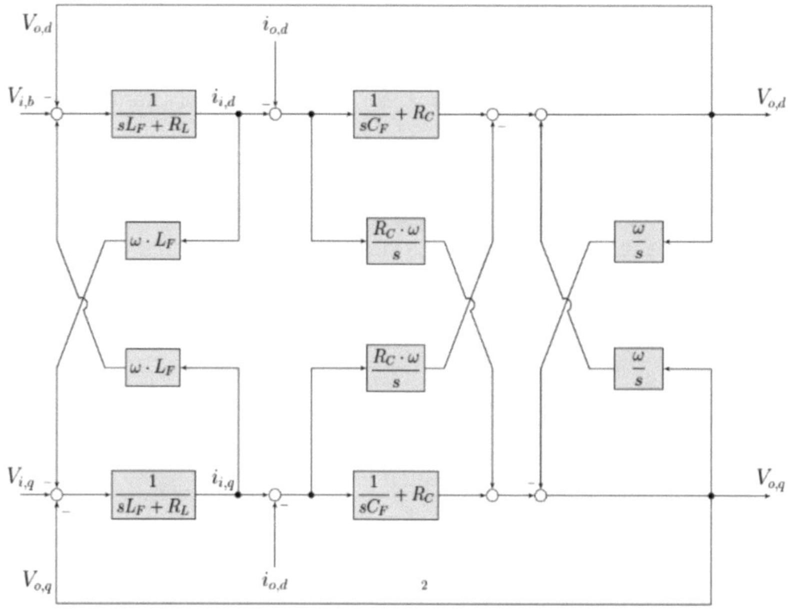

Figure 5. Block diagram of dynamics of the *LC* filter.

theoretically transformed into a virtual bypass to the current or voltage reference signal. Once this is achieved, the whole inverter can be operated as a perfectly controllable voltage source.

Figure 6 illustrates the abstract control structure of a controlled variable in *d* axis (i_{id} and v_{od}) containing all the blocks: the inverse dynamics feedforward control, feedforward control of disturbance, feedback control and decoupling effects. There are two different implementations as shown in **Figure 6**: the parallel and the series connection of the FF and FB blocks. As the series implementation results in more complex internal dynamics and it is more vulnerable to measurement noises, both current and voltage control have been implemented using the parallel connection. An important assumption made is that we neglected the dynamics of the inverter, thus making the inverter act as bypass ($v_i^* = v_i$) as well. This approximation is valid since advanced synchronized sampling techniques can reduce the time-delay of inverter digital implementation to $0.25T_{sampling}$ [8]; thus, the inverter can be approximated as a unitary gain without delay. If the control for inverter could be successfully implemented according to the strategy shown in **Figure 6** (top), the inverter's control algorithm would yield the closed loop system displayed in the block diagram shown in **Figure 4**.

Figure 4 shows the simplified control block of the cascaded voltage and current controller in inverter control layer, which is in line with the structure illustrated in **Figure 3**. The outer voltage loop has two components: FBVC and FFVC denote feedback voltage control and feedforward voltage control, respectively. The inner current loop consists of two components: FBCC and FFCC, standing for feedback current control and feedforward current control, respectively. The feedforward control is a crucial element in this control system, which contains the inverse dynamic model of the *LC*-filter shown in Eqs. (2)–(5) and its main effect is

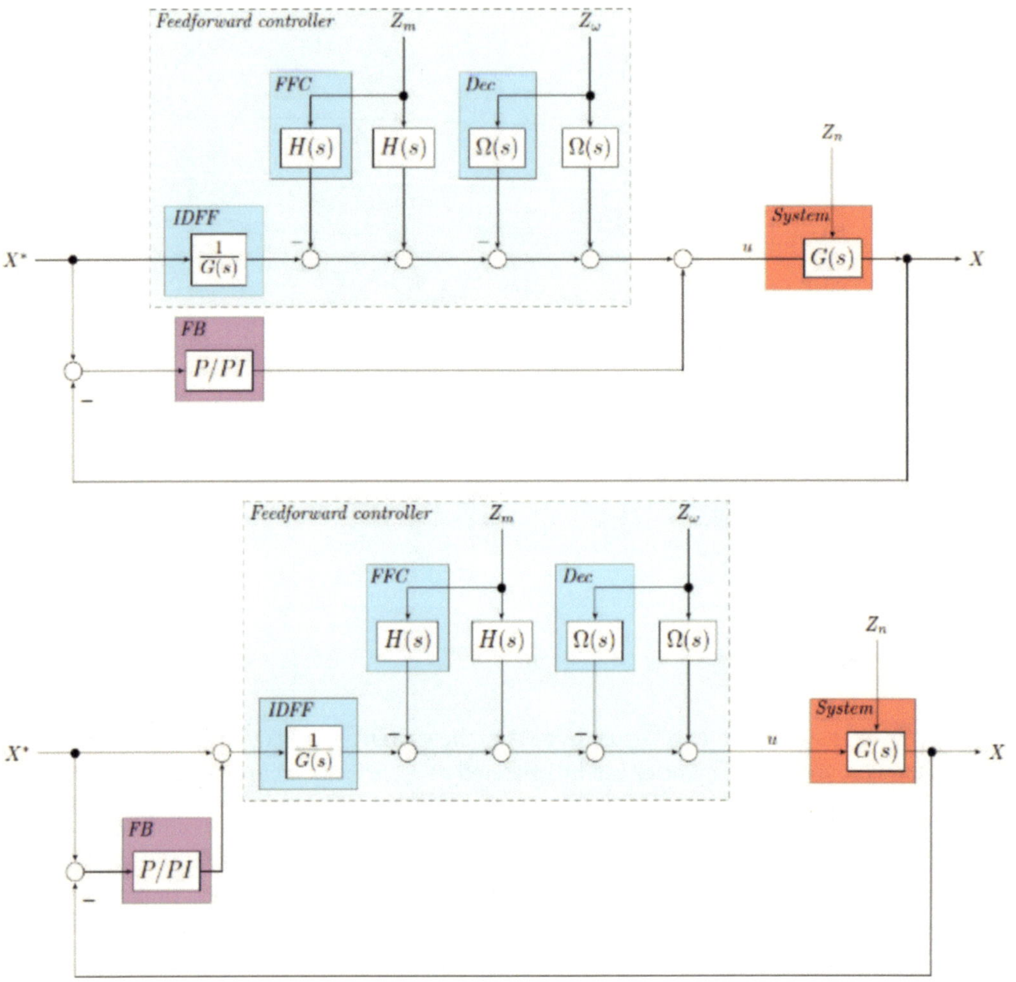

Figure 6. Abstract structure of the proposed control algorithm and its components. Parallel (top) and serial (bottom) connection of feedback and feedforward control.

to perform ideal compensation of the filter's dynamics within each loop. A deeper insight into the feedforward control indicates that there are three components in each loop. The first component consists of the decoupling elements; these remove the coupling between the *d* and *q* variables. The second is the disturbance compensation; this eliminates the effect of measurable variables acting as disturbances to each loop (including an active damping function based on back electromotive force (EMF)-decoupling). Finally, the third component counteracts the dynamics of the control path, transforming the dynamics of the controlled variable into a virtual bypass for the reference value [9, 10].

4.1. Inner current loop controller

The inner current control loop is seen as a bypass in the perspective of the outer voltage control loop. Accordingly, the following relationship should hold:

$$i_{id} = i_{id}^*,\ i_{iq} = i_{iq}^* \tag{10}$$

In order to achieve this, the feedforward component of the current control simply inverts the dynamics described in Eqs. (6) and (7) and then we obtain the following control law for the feedforward component v_{iff}^*:

$$v_{iffd}^* = (sL_f + R_f)i_{id}^* - \omega_o L_f i_{iq} + v_{od} \tag{11}$$

$$v_{iffq}^* = (sL_f + R_f)i_{iq}^* + \omega_o L_f i_{id} + v_{oq} \tag{12}$$

These equations could already be used directly as the control law for current control. They include the inverse dynamic feedforward term for the reference current signal. This feedforward term compensates the inverter disturbances including the terminal voltage and coupling elements.

Nonetheless, the first term in Eq. (11) and (12) contains a component with differential behaviour (denoted by the Laplace operator). It is not recommended to employ this control law in the inner control loops which could directly amplify and feed the high-order harmonics in the inverter system which in the end results in undesirable low-order harmonics. In the worst case, this derivative term could lead to unacceptable THD or even instabilities to the control system. Furthermore, the equations must include a feedback term to cancel the deviation between the actual output and the reference current. A P-control is parallel connected to the feedforward control. The use of a P-control instead of a PI-control can be explained by the fact that the steady-state error in the inner loop is automatically sensed and compensated by the outer control loop with the cascaded structure. The feedback controls of the inner current loop are given by

$$v_{ifbd}^* = K_{CP}(i_{id}^* - i_{id})\ ,\ v_{ifbq}^* = K_{CP}(i_{iq}^* - i_{iq}) \tag{13}$$

Therefore, after deleting the derivative element in the feedforward original control law, the updated control law for the current control including the P-control is given by

$$v_{id}^* = v_{iffd}^* + v_{ifbd}^*\ ,\ v_{iq}^* = v_{iffq}^* + v_{ifbq}^* \tag{14}$$

where K_{cp} is the proportional gain of the feedback control.

The transfer function for the inner loop can be obtained by substituting Eq. (14) into Eq. (6) and (7), and then we obtain

$$i_{id} = \frac{R_f + K_{cp}}{s \cdot L_f + R_f + K_{cp}} i_{id}^*,\ i_{iq} = \frac{R_f + K_{cp}}{s \cdot L_f + R_f + K_{cp}} i_{iq}^* \tag{15}$$

4.2. Outer voltage loop controller

In an analogous manner, the closed loop of the voltage controller can be designed as a bypass for the application layer controller as well. Therefore, the closed loop transfer function of the voltage controller is expressed in the form of 'unitary gain' shown as follows:

$$v_{od} = v^*_{od}, \; v_{oq} = v^*_{oq} \tag{16}$$

Therefore, the same strategy will be applied in the design of the voltage control law. The control algorithm is obtained by substituting Eqs. (4) and (5) into Eq. (16) and rearranging in terms of the reference for the inverter current. As shown in **Figures 3** and **4**, the voltage controller includes both feedforward control and feedback control. The control laws of the outer voltage loop are formulated as follows:

$$i^*_{iffd} = \frac{s \cdot C_f}{1 + s \cdot C_f \cdot R_c} v^*_{od} - \frac{C_f}{1 + s \cdot C_f \cdot R_c} v_{oq} + \omega_o \frac{C_f R_c}{1 + s \cdot C_f \cdot R_c} i_{iq} - \omega_o \frac{C_f R_c}{1 + s \cdot C_f \cdot R_c} i_{od} + i_{od} \tag{17}$$

$$i^*_{iffq} = \frac{s \cdot C_f}{1 + s \cdot C_f \cdot R_c} v^*_{oq} + \omega_o \frac{C_f}{1 + s \cdot C_f \cdot R_c} v_{od} - \omega_o \frac{C_f R_c}{1 + s \cdot C_f \cdot R_c} i_{id} + \omega_o \frac{C_f R_c}{1 + s \cdot C_f \cdot R_c} i_{od} + i_{oq} \tag{18}$$

$$i^*_{ifbd} = \frac{s \cdot K_{vp} + K_{vi}}{s} (v^*_{od} - v_{od}), \; i^*_{ifbq} = \frac{s \cdot K_{vp} + K_{vi}}{s} (v^*_{oq} - v_{oq}) \tag{19}$$

where K_{vp} and K_{vi} are the proportional and integral gains of the voltage feedback control, respectively. The reference current generated by the voltage controller is given by

$$i^*_{vcd} = i^*_{iffd} + i^*_{ifbd}, \; i^*_{vcq} = i^*_{iffq} + i^*_{ifbq} \tag{20}$$

4.3. Unitary transfer function compensation

Based on Eqs. (16)–(20), the transfer function of the voltage controller is unitary, whereas the transfer functions of the current controller are given by Eq. (15).

In order to improve the system's stability and dynamic performance, it is necessary to make the inner current loop appear as a unitary gain from the perspective of the outer voltage loop. A derivative compensation is used between the voltage controller and the current controller to make the current transfer function also equal to unitary gain. The transfer functions of the derivative compensation are obtained by inverting the inner loop transfer functions represented in Eq. (15)

$$\frac{i^*_{id}}{i^*_{ivd}} = \frac{i^*_{iq}}{i^*_{ivq}} = \frac{s \cdot L_f + R_f + K_{cp}}{R_f + K_{cp}} = D(s) \tag{21}$$

By substituting Eq. (15) into Eq. (21), the transfer function of the current controller is equal to the ***unitary gain***. In digital implementation, the derivative compensation should operate at the same bandwidth as the voltage control. As we have $R_f + K_{cp} \gg L_f$, we have i^*_{id} i^*_{ivd} and i^*_{iq} i^*_{ivq}. The derivative term would not cause significant distortion in case there is nonlinear load connected.

5. Analysis of the voltage control algorithm in the inverter control layer

5.1. Analytical verification of the closed loop transfer function

Analytical verification is provided here to derive the closed loop transfer function of the inverter control layer, and to prove that the design target is satisfied. A simplified structure of the double-loop controller is presented in **Figure 7**. W_1 represents the control plant of the inner loop and its formula is shown in Eq. (6) and (7). W_2 represents the control plant of the outer loop and its formula is shown in Eq. (8) and (9). Only derivation process in *d*-coordinate is presented and it is shown in **Figure 8**. Following the steps listed in **Figure 8**, we can obtain the closed loop transfer as designed, $v_{od}(s) = v^*_{od}(s)$.

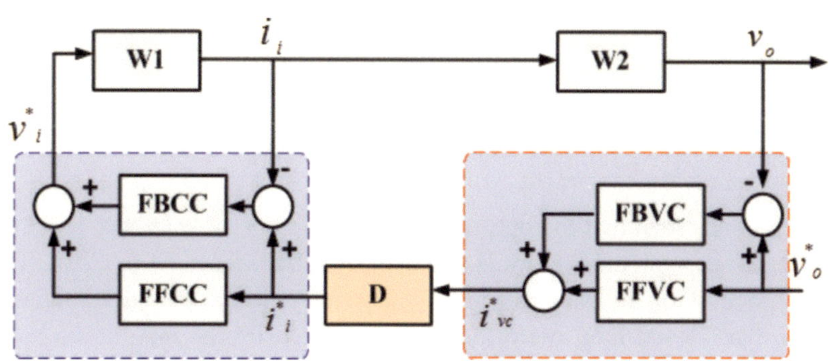

Figure 7. Simplified structure of the control algorithm in the inverter layer.

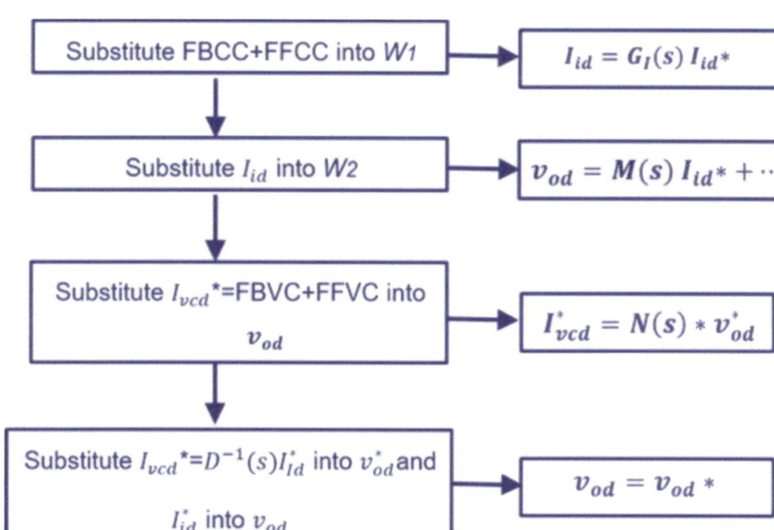

Figure 8. The derivation process of the closed loop transfer function.

5.2. Sensitivity analysis

The developed voltage control algorithm in the inverter layer is model-based control method, which requires very accurate modelling of the control plant together with correct parameters. If there is mismatch between the estimated plant parameter and the real one, the reverse model would not perfectly compensate for the plant's dynamics and transform the open loop into a unitary gain. As a result, disturbance terms would appear in the closed loop transfer function, which degrades the steady-state and transient performance. In this sensitivity analysis, how the stability and control performance of the proposed controller is affected by the variations of the control plant parameters (e.g. C_f and L_f) is investigated.

Assuming that the capacitance of the physical plant is C_f and the estimated capacitance of the inverse dynamic model for the control system is C'_f, we can derive the following closed loop transfer function:

$$v_{od} = v_{od}^{*} + \frac{\left(\frac{C'_f}{C_f} - 1\right)s}{(C'_f + K_{vp}C'_f R_c)s^2 + (K_{vp} + K_{vi}C'_f R_c)s + K_{vi}}(i_{id} - i_{od}) \tag{22}$$

From Eq. (22), it can be seen that the output v_{od} is a function of the reference input v^*_{od} and the disturbance i_{id}-i_{od}. The system stability is not degraded by variations of C_f as that the poles of the transfer function for the disturbance i_{id}-i_{od} are always in the left-half-plane regardless of variations of C_f. The closed loop performance is evaluated by the magnitude bode diagram of the transfer function for the disturbance for different C_f values, as shown in **Figure 9**. It can

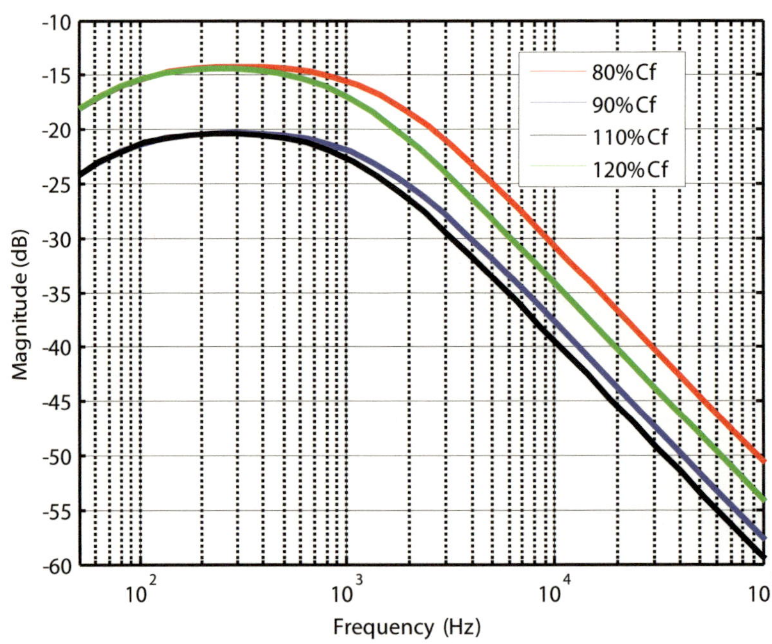

Figure 9. Gain magnitude curve of disturbance (i_{id}-i_{od}) caused by C_f variation.

http://dx.doi.org/10.5772/intechopen.69402

be observed that the effect of modelling mismatch caused by variations of C_f is very small since the gain magnitude is below -15 dB for all frequencies.

If the inductance of the real plant is L_f and the inductance of the inverse dynamic model is L'_f, the transfer function of the inner current loop is given by

$$i_{id} = \frac{L'_f s + R_f + K_{cp}}{L_f s + R_f + K_{cp}} i^*_{id} + \frac{\omega_o (L_f - L'_f)}{L_f s + R_f + K_{cp}} i_{iq} \tag{23}$$

and the derivative compensation term is

$$D'(s) = \frac{i^*_i}{i^*_{vc}} = \frac{sL'_f + R_f + K_{cp}}{R_f + K_{cp}} \tag{24}$$

Therefore, the closed loop transfer function of the whole system is given by

$$v_{od} = \frac{A(s) + B(s)}{1 + B(s)} v^*_{od} - \frac{(1 + sC_f R_c)(1 - A(s))}{sC_f (1 + B(s))} i_{od} + \frac{\omega_o (1 + sC_f R_c)(L_f - L'_f)}{sC_f (1 + B(s))(L_f s + R_f + K_{cp})} i_{iq} \tag{25}$$

where $A(s) = \frac{(R_f + K_{cp} + sL_f)^2}{(R_f + K_{cp} + sL_f)(R_f + K_{cp})}$, $B(s) = \frac{1 + sC_f R_c}{sC_f} A(s) \frac{K_{vp} s + K_{vi}}{s}$.

The stability of the system is studied by checking the position of the poles of each input expressed in Eq. (25) (v^*_{od}, i_{od}, i_{iq}) when L_f varies from 80 to 120% of L'_f. The results show that all the components have the same dominant poles (-409) with variations of L_f. Consequently, the stability of system is not degraded with mismatches between L_f and L'_f. The closed loop performance of the system is analysed with the bode plots for each individual input in Eq. (25). It can be observed in **Figure 10(a)** that the gain for the reference tracking at the fundamental frequency is equal to the unitary gain. The magnitude of the disturbance caused by the output current i_{od} is very low over the frequency range of interest, so that it has little influence on the system's performance. The magnitude of the disturbance caused by the coupling component i_{iq} is also quite low and almost negligible over the frequency range of interest.

The sensitivity study to variations of C_f and L_f indicates that the stability of the system will not be degraded. However, variations of plant parameter make the system unable to suppress

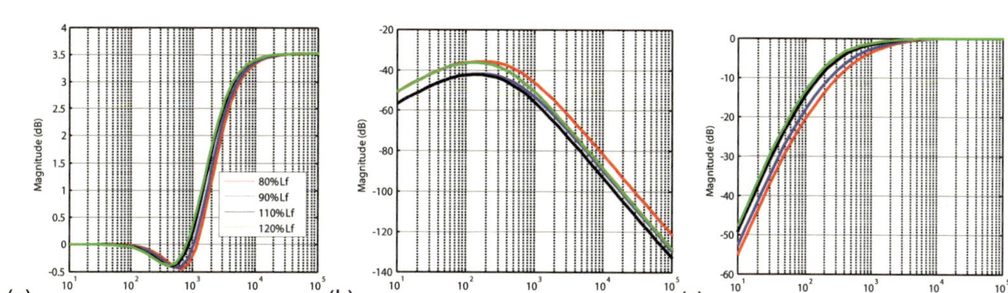

Figure 10. Gain magnitude curves of components of v_{od} caused by L_f variation. (a) Gain of v^*_{od}. (b) Gain of i_{od}. (c) Gain of i_{iq}.

completely all disturbances and degrade the system's reference tracking and disturbance rejection performance. The performance loss, however, is small and acceptable for practical applications.

5.3. Comparison with the conventional voltage controller

For comparison purpose, analysis is also carried out with the cascaded voltage controller proposed in Refs. [2, 6], which is a popularly employed controller in academic and industry. The structure of this conventional voltage controller is shown in [4]. This conventional controller requires the same number of sensors as the proposed controller. The analysis starts from the inner loop current controller. Based on $v_i = v_i^*$, we obtain the following equation (only equation in d-axis is shown here):

$$(sL_f + R_f)i_{id} - \omega_o L_f i_{iq} + v_{od} = K_{cp}(i_{id}^* - i_{id}) - \omega_o L_f i_{iq} + v_{od} \tag{26}$$

After rearranging the above equation, we obtain the closed loop transfer function of the inner loop

$$\frac{i_{id}}{i_{id}^*} = \frac{K_{cp}}{sL_f + R_f + K_{cp}} \rightarrow i_{id} = i_{id}^* \frac{K_{cp}}{sL_f + R_f + K_{cp}} \tag{27}$$

According to the control law for the outer voltage loop, we have

$$i_{id}^* = \frac{K_{vp} + sK_{vi}}{s}(v_{od}^* - v_{od}) - \omega_o C_f v_{oq} + i_{od} \tag{28}$$

Substituting Eq. (28) into Eq. (27), we obtain the following equation:

$$i_{id} = \left(\frac{K_{vp} + sK_{vi}}{s}(v_{od}^* - v_{od}) - \omega_o C_f v_{oq} + i_{od}\right)\frac{K_{cp}}{sL_f + R_f + K_{cp}} \tag{29}$$

Based on the dynamic model of the outer voltage loop, we have the following equation:

$$i_{id} = sC_f v_{od} + i_{od} - \omega_o C_f v_{oq} \tag{30}$$

Following the same logic as the inner loop, we have the following equation:

$$\left(\frac{K_{vp} + sK_{vi}}{s}(v_{od}^* - v_{od}) - \omega_o C_f v_{oq} + i_{od}\right)\frac{K_{cp}}{sL_f + R_f + K_{cp}} = sC_f v_{od} + i_{od} - \omega_o C_f v_{oq} \tag{31}$$

By rearranging the equation, we obtain the closed loop transfer function for the entire system

$$v_{od} = \frac{B_1}{A_1}v_{od}^* + \frac{C_1}{A_1}i_{od} + \frac{D_1}{A_1}v_{oq} \tag{32}$$

http://dx.doi.org/10.5772/intechopen.69402

where $A_1 = C_f L_f s^3 + C_f(R_f + K_{cp})s^2 + K_{cp}K_{vp}s + K_{cp}K_{vi}$, $B_1 = K_{cp}K_{vp}s + K_{cp}Kv_i$, $C_1 = -L_f s^2 + \left(\frac{K_{cp}}{sL_f+R_f+K_{cp}}K_{cp} - R_f - K_{cp}\right)s$ and $D_1 = \omega_o L_f C_f s^3 + \omega_o R_f C_f s^2$.

As seen from Eq. (32), the output voltage v_{od} is a function of the reference input v_{od}^* and two disturbance inputs i_{od} and v_{oq}. The two disturbance inputs result in static error and harmonics distortion in the output voltage. **Figure 11** shows the bode plot of the gain of each input. **Figure 11(a)** displays the bode diagram of output voltage to voltage reference closed loop transfer function. It demonstrates that the output voltage can track the reference very well at the fundamental frequency. However, it may amplify the harmonics with frequency around 550 Hz if there are harmonics in the reference voltage. This may happen due to the voltage reference in some applications generated based on some measured variables and power references rather than a fixed value. If there are distortions in the grid, the measured variables can be distorted and then the generated voltage reference can be distorted as well. The gain of i_{od} represents the equivalent harmonic impedance which indicates the main reason of the steady-state error in tracking the target reference. The bode diagram shown in **Figure 11(b)** indicates that the harmonics with frequency around 550 Hz (resonance peak) in the output current will be significantly amplified and results in large distortion in the output voltage. The decoupling

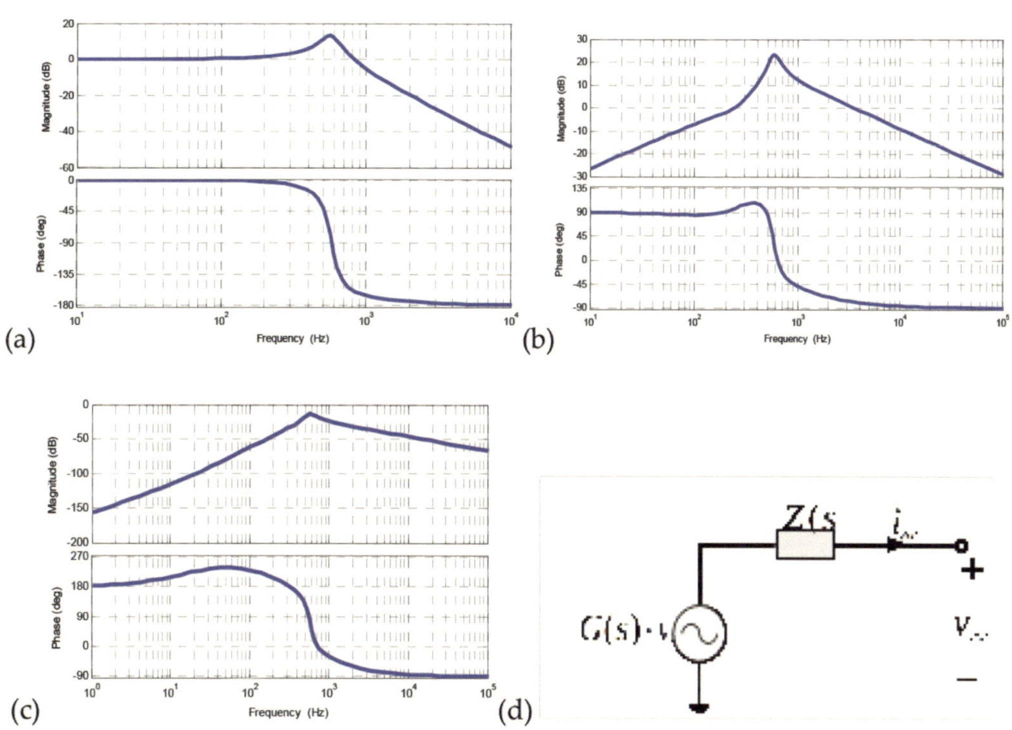

Figure 11. Analytical study of the conventional double-loop voltage controller. (a) Bode plot of the gain of v_{od}. (b) Bode plot of the gain of i_{od}. (c) Bode plot of the gain of v_{oq}. (d) Equivalent circuit of the closed loop system.

term of the output voltage v_{oq} has negligible impact on the output voltage as seen from **Figure 11(c)**. Hence, the output voltage of inverter can be treated as a voltage source series connected with output impedance, which is represented in Figure 5.9(d). $G(s)$ is equal to $\frac{B_1}{A_1}$ and $Z(s)$ is equal to $\frac{-C_1}{A_1}$. In order to achieve fast dynamic response and eliminate steady-state error and voltage distortions, the output impedance should be as small as possible [2, 11].

For the proposed voltage controller, the output impedance is equal to 'zero' which therefore predicts superior control performance compared to the conventional voltage controller. For instance, the tracking error for steady-state and transient moment is forced to be 'zero' in a symmetric and non-distorted grid. However, the proposed control method contains derivative part in the current reference generation which affects not only the feedforward term of the current controller but also the feedback controller. In fact, the analysis shown in Ref. [5] proves that the proposed method has good robustness against harmonic distortion.

6. Results and discussion

The variation of the developed control algorithm was conducted by simulating a microgrid with two inverter-interfaced DGs whose specifications are described in Ref. [5]. **Figure 12** shows the topology of the microgrid for the case study. In the application layer, the voltage reference generation algorithms for the three operating modes are running parallel to prevent the latter to start from scratch after every transition. Before connection/disconnection microgrid to the main grid, the angle difference between grid-forming and grid-supporting/grid feeding (θ_{inv}-θ_g) is compared to guarantee small phase/frequency deviations.

The European Standard EN 50160 defines standard operating conditions of frequency and voltage for islanded and interconnected power systems [12] . Assuming that the nominal root-mean-square (RMS) voltage is 230 V, the RMS value of voltage for interconnected systems should maintain between 207 and 253 V (±10%), whereas for islanded systems, it should be between 195.5 and 253 V (±15%). In terms of frequency, for interconnected systems, it should remain in the range of 49.5 and 50.5 Hz (±1%), and in the range of 49 and 51 Hz for islanded systems (±2%). These standard operating limits are used to evaluate the work in this chapter.

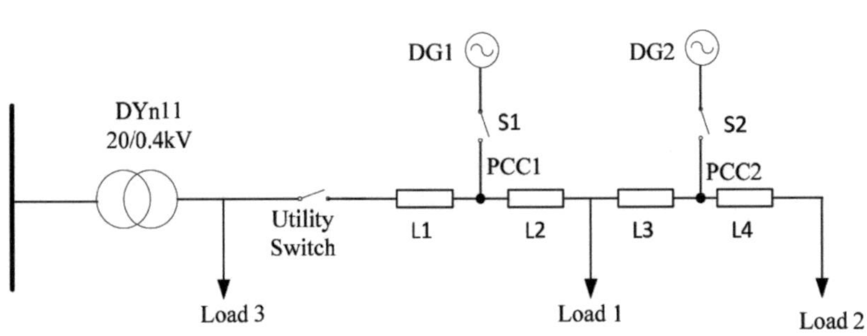

Figure 12. Microgrid configuration used as a case study [5].

6.1. Simulation results

6.1.1. Grid-supporting to grid-forming mode

When a fault is detected in the main grid during the grid-supporting operation, it requests the microgrid to disconnect from the main grid. In this scenario, the transient behaviour of the DGs controlled with the developed algorithm is investigated. The maximum amounts of active power injections of the DGs running in grid-supporting mode are 20 kW for DG#1 and 15 kW for DG#2. The two DGs are connected to the main grid at 0.6 s to prevent the transients from the black start and phase-lock-loop (PLL) is used to synchronize each DG with the grid before connection. From 0 to 0.6 s, the controllers of DGs are pending to run in grid-supporting mode while the references P_{max} are set to be 'zero'. The DGs work in grid-supporting mode from 0.6 to 3.8 s, then they switch to grid-forming mode after the microgrid disconnects from the main grid. The transient behaviours of the DGs are presented in **Figure 13**.

Figure 13(a) indicates that the two DGs can inject maximum amount of active power (20 and 15 kW, respectively) in the grid-supporting mode, and achieve accurate power sharing (15 kW) in the grid-forming mode. The output voltages of both DGs during operating mode transition, as illustrated in **Figure 13(b)**, indicate that the two DGs exhibit neither overshooting nor harmful transient in voltage, and that both rapidly arrive at the steady state. **Figure 13(c)**

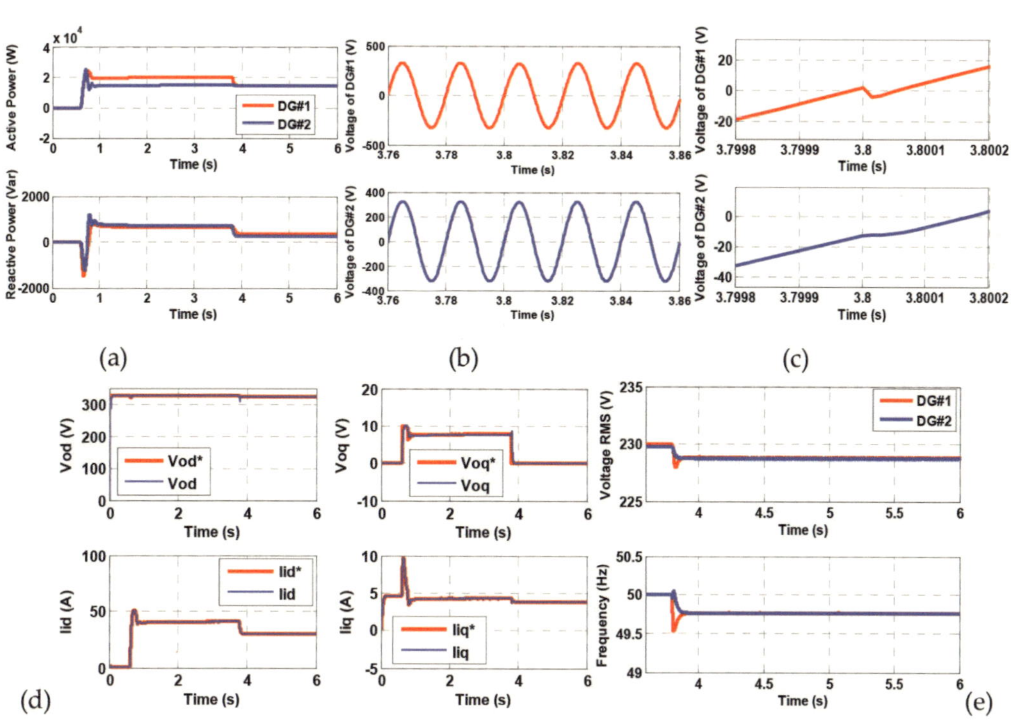

Figure 13. Transient performance from grid-supporting mode to grid-forming mode. (a) Active and reactive power of DGs. (b) Output voltage of DGs during mode change. (c) Zoomed-in plot of output voltage of DGs at transient moment. (d) DG#1: reference signal (red), actual output (blue) of voltage controller and current controller in d -q frame. I RMS voltage and frequency of DGs.

shows the zoomed-in plot of output voltage of DGs at the transition moment (3.8 s); negligible transients are observed for both DGs at the transition moment. Thus, very smooth transition performances are achieved as expected. The frequency of DGs shown in **Figure 13 (e)** also indicates the smooth transition process. The tracking performance of the inner and outer loop of DG#1 is presented in **Figure 13(d)**. The upper graph of **Figure 13(e)** illustrates that the voltages of both DGs lie inside the satisfactory limits in both grid-supporting and grid-forming mode. Moreover, the transient voltages remain within the required operating limits of the islanded microgrid. The lower graph of **Figure 13(e)** illustrates the frequency at the moment of transition. In less than 0.2 s, the two DGs accomplish accurate power sharing and arrive at the identical frequency of 49.76 Hz, which lies within the required operating limits. Furthermore, the largest frequency variation of the two DGs at the moment of transition is 49.52 Hz, which does not violate the operating range of EN 50160. The results show that the controlled variables can perfectly track their respective reference with zero steady-state error during steady state and transient. This is consistent with the specification set in Section 4 that the transfer functions of the inner and outer loops are both equivalent to the unitary gain.

6.1.2. Grid-supporting to grid-forming mode

In this test scenario, the two DGs are switched intentionally from grid-forming mode to grid-feeding mode at 2 s, when the phase differences of voltages between the main grid and the DGs are lower than 0.5 rad. In grid-feeding mode, the received active and reactive power references from the system control layer are 15 kW and 200 Var for DG#1, and 15 kW and 300 Var for DG#2. **Figure 14** shows the simulation results. **Figure 14(a)** illustrates the process of the inverters synchronizing with the grid. As it can be seen, the voltages of DG#1 and DG#2 can be synchronized with the grid voltage very rapidly and with small negligible transient after the switch is reconnected, thanks to fast dynamics and good disturbance rejection performance of the proposed method. **Figure 14(c)** presents the performance of the developed outer voltage and inner current controller. We can observe that each of the actual output variables follows the corresponding reference target perfectly; this confirms the validity of the design laid out in Section 4. The output current of DG#1 presented in **Figure 14(e)** shows the smooth transient performance of the microgrid.

The same test is investigated with the conventional method shown in Ref. [5] for comparing the controller performances (performance in steady state and transient state) with the developed control method. **Figure 14** shows the simulation results. **Figure 14(b)** illustrates that though DG#1 is able to synchronize with the main grid very rapidly, larger transient is exhibited; **Figure 14(d)** demonstrates that there are some tracking errors with respect to the reference and real output, and large transient currents shown at the transition moment, and **Figure 14(f)** further shows the undesirable transient current as well. Compared with the results in **Figure 14(b)**, **(d)** and **(f)**, the proposed method has smaller transient current during transition and better tracking and disturbance rejection performances.

The DG's output voltages in RMS values are shown at the top and the frequency is shown at the bottom of **Figure 15**. As seen from **Figure 15**, the RMS values of voltage and the frequencies

http://dx.doi.org/10.5772/intechopen.69402

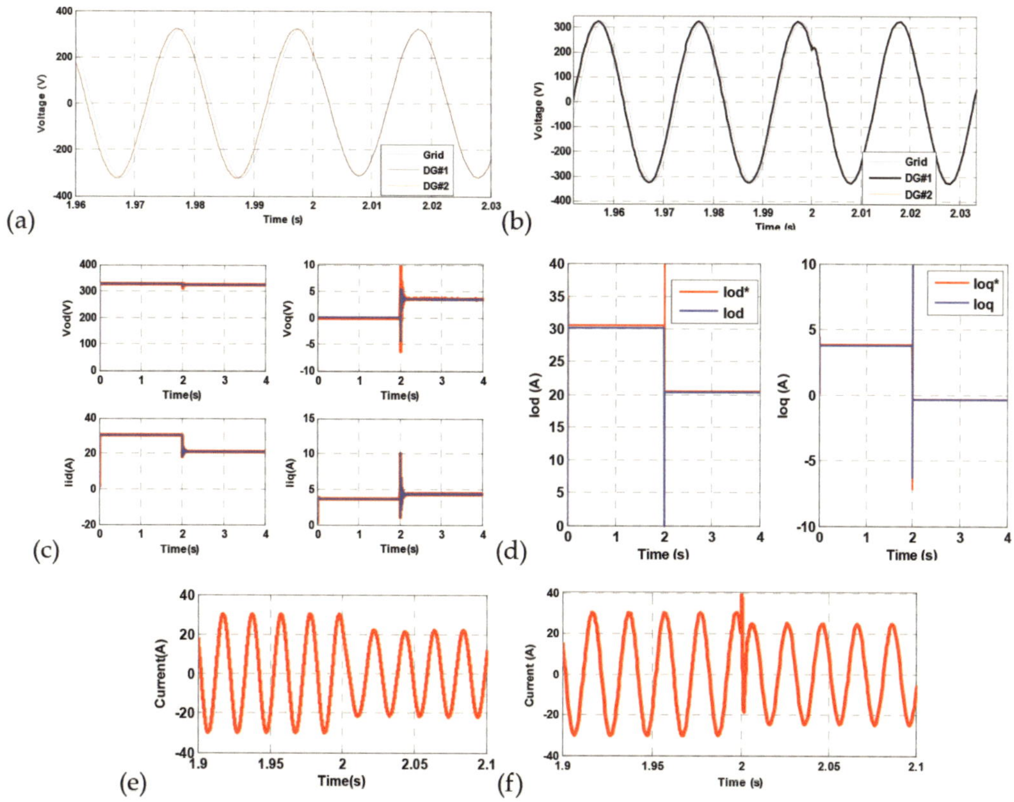

Figure 14. Transient performance from grid-forming mode to grid-feeding mode: proposed method (left column) and conventional method (right column). (a, b) Up: Grid voltage (blue) and inverter voltage (red) Down: Error between grid and inverter voltage. (c, d) DG#1: Reference signal (red), actual output (blue) of voltage controller and current controller in d-q frame. (e, f) Output current of DG#1.

of the DGs are near their nominal values pre- and post transition. As per the standard EN 50160, the DGs′ voltage and frequency both lie within the satisfactory operating ranges.

6.2. Experimental results

All simulations are performed in a hardware-in-the-loop (HIL) platform, and practical implementations such as delays caused by digital sampling, computation time and inverter switching are included. The structure including system layer, application layer, inverter control layer and switching layer is presented in **Figure 16**.

The microgrid with two inverter-interfaced DGs and the main grid model is built in a real-time digital simulator (RTDS). The application and inverter layer control algorithms have been programmed using Texas Instrument TMS28335 DSP. Each inverter is controlled using one DSP. There is a conditioning interface between RTDS and DSP to scale the output voltage level of RTDS (±5 V) to the output voltage level of DSP (0–3.3 V), and vice versa. The system layer controller is implemented in a Xilinx ML507 board; the board directly communicates with a RTDS gigahertz processor card through fibre optics to obtain measurements and transmit the

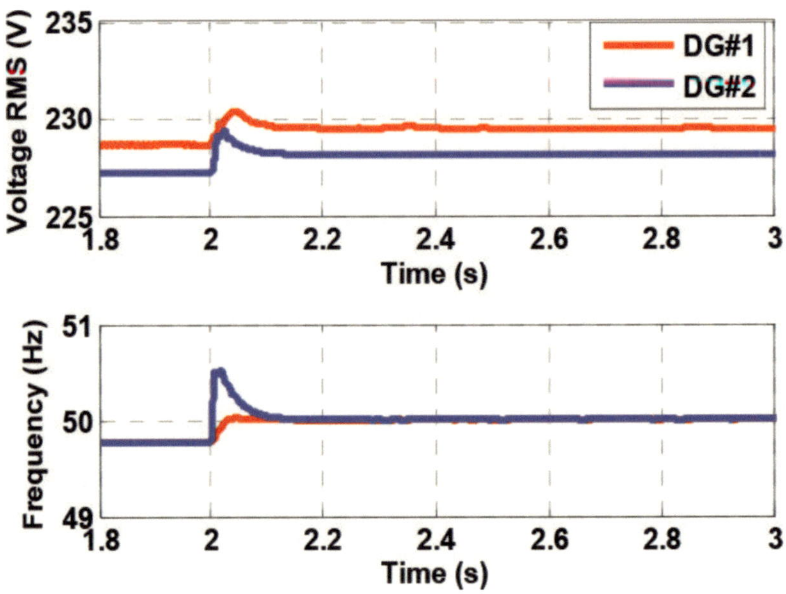

Figure 15. DGs' output voltage RMS values and frequencies.

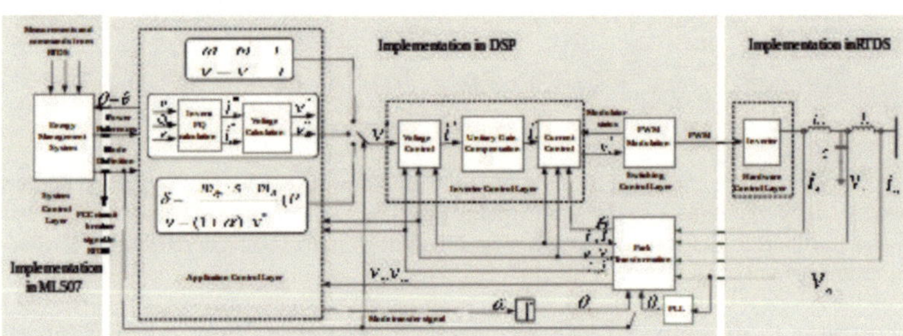

Figure 16. The structure of the universal control algorithm for flexible microgrid operation.

PCC circuit breaker control signal to RTDS. A control algorithm in the system layer, consisting of operating mode management and control reference generation, is implemented in ML507, and the operation mode signal and power references are sent to DSPs via serial peripheral interface. The algorithms implemented in the application layer are grid-forming (top), grid-feeding (middle) and grid-supporting (bottom), respectively, and the output is the voltage reference for the inverter layer. The schematic diagram and setup of HIL platform are presented in **Figure 17**. In this signal HIL platform, both the ML507 board and DSP board are controlled under test. The oscilloscope panels related show 5 V for 400-V voltage and 5 V for 50-A current.

6.2.1. Transition from grid-supporting to grid-forming mode

The two DGs begin in grid-supporting mode (the command signal sent from ML507 to RTDS is '1' and from ML507 to DSPs is '10'). The algorithm of grid-supporting mode in application

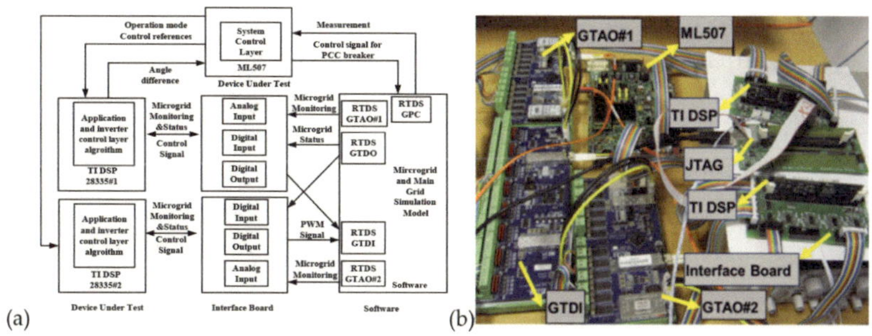

Figure 17. (a) Schematic diagram of the signal level HIL platform. (b) Set-up of HIL platform.

layer is selected and the voltage reference is passed to the inverter control layer. After letting the system run for a pre-defined time, the microgrid is intentionally switched to islanding mode by transmitting '00' to DSPs and '0' to RTDS from ML507. Then, the grid-forming mode in the application layer algorithm is selected to produce the voltage reference for the inverter control layer, and a new angle θ_{inv} for Park transformation is also transmitted to the inverter control layer. **Figure 18** presents details of the DGs' performance, showing the responses of the voltage and current during the transition between the two operating modes [5].

No transient change is shown in the voltage waveform and only very slight transients are exhibited in the current waveform. The calculated frequencies of both DGs are approximately 50 Hz in grid-supporting mode and 49.7 Hz in grid-forming mode, with the largest deviation during transition being 0.35 Hz. The calculated RMS values of voltage remain within the acceptable range around the rated value of 230 V during the transition process, with negligible deviations. Therefore, the expected smooth transition from grid-supporting mode to grid-forming mode is achieved with the proposed control algorithm, and the operating range of DGs′ frequencies and voltages is within the range as per EN 50160.

6.2.2. *Transition between grid-forming and grid-feeding mode*

Two scenarios are studied in this case: the transition from grid-forming to grid-feeding mode requested by the DNO, and the opposite situation, the transition from grid-feeding mode to grid-forming mode caused by intentional islanding of the microgrid. During the first transition, the system controller sends operating mode signal '0' to RTDS and '00' to DSP, after which the voltage reference from grid-forming mode and angle θ_{inv} is calculated and sent to the inverter

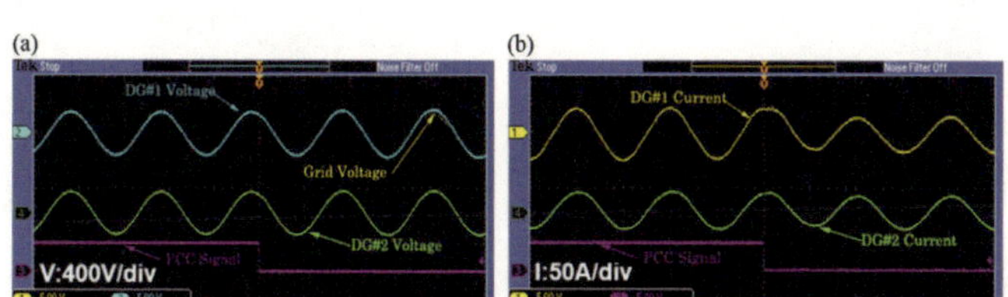

Figure 18. Transition from grid-supporting to grid-forming mode. (a) Voltage response during transition. (b) Current response during transition.

control layer. In addition, the difference in angle between the main grid voltage and DG#1 output voltage is compared. The microgrid will only be connected (reconnected) to the main grid when an 'enable' signal ('1') is transmitted from the system controller to the main grid (modelled and controlled in RTDS) and also the difference in angle is equal to or less than 0.5 rad. When these conditions are met, the system layer controller transmits '1' to RTDS, and '01' and power references to DSPs, which commands the inverters to operate in grid-feeding mode. Then, after a pre-defined time, the system layer controller transfers intentionally the microgrid to islanding mode, and the corresponding command signals are transmitted to DSPs and RTDS.

The control performance of the developed control algorithm is compared to the popularly used control algorithm presented in Ref. [6]. To attain valid simulation results, the operating conditions for the two algorithms are made identical. The microgrid is connected to the main grid when the phase difference between the grid voltage and the voltage at the terminal of DG#1 is equal to or less than 0.5 rad. Then, the two DGs transfer from grid-forming to grid-feeding mode [5]. After certain time (235 s), the microgrid is disconnected from the main grid, switching the two DGs back to grid-forming mode. The waveforms of voltage and current for both scenarios are different during transition moment because the DSP boards don't start to run at the same time after RTDS is running. However, the switching conditions (e.g. phase difference between main grid voltage and DG#1 output voltage) are maintained the same for both scenarios.

Figure 19(a) and **(b)** denotes that both control methods cause negligible distortion of the grid voltage at the moment of transition. This is expected, since the main grid is strong. **Figure 19(c)** and **(d)** shows a less distorted current waveform after transition when the proposed control algorithm is applied.

Figure 20 illustrates that at the moment of disconnection, both control algorithms swiftly generate a clean voltage waveform, with a marginally better performance than the conventional

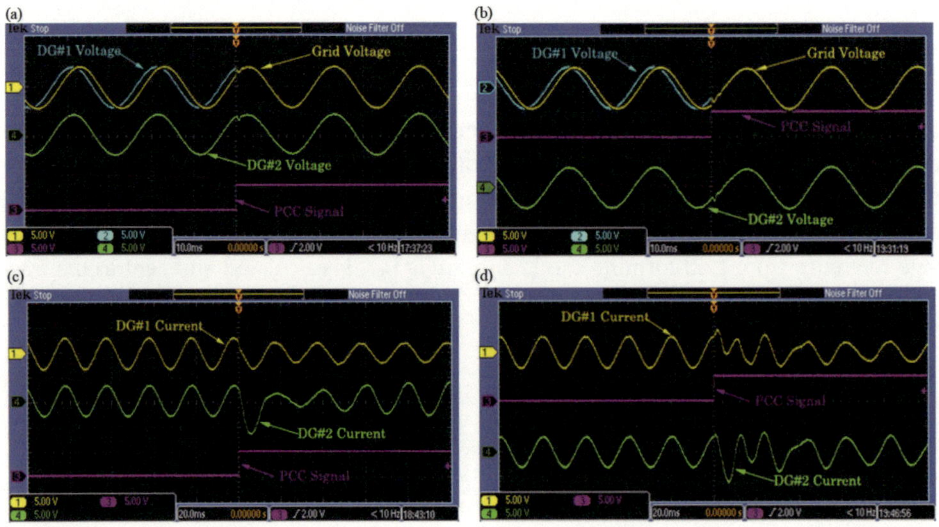

Figure 19. Transition from grid-forming to grid-feeding mode. (a) Voltage response with the proposed method. (b) Voltage response with the conventional method. (c) Current response with the proposed method. (d) Current response with conventional method.

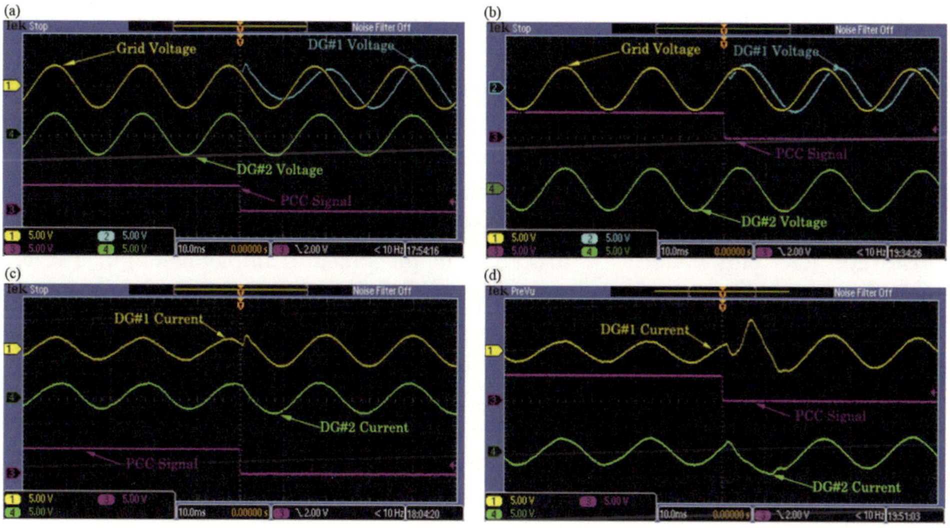

Figure 20. Transition from grid-feeding to grid-forming mode. (a) Voltage response with the proposed method. (b) Voltage response with the conventional method. (c) Current response with the proposed method. (d) Current response with conventional method.

method. In addition, the output currents of the developed control algorithm exhibit considerably superior qualities with fewer overshoots and quicker dynamic responses.

The calculated RMS values of voltage and frequencies of the two DGs with the developed control method during the transition process between grid-forming and grid-feeding modes are summarized as follows: in transition from grid-forming to grid-feeding mode, the calculated frequencies of the two DGs are approximately 49.7 Hz in grid-forming mode and 50 Hz in grid-supporting mode, with the largest deviation during transition being 0.45 Hz, from DG#2 [5]. The calculated RMS values of voltage remain within the acceptable range around the rated value of 230 V during the transition process, with the largest deviation being less than 5 V, from DG#2; in the transition from grid-feeding to grid-forming mode, the calculated RMS values of voltage and frequencies of the two DGs are within the acceptable operating range. Note that the largest deviations of frequency and voltage are 0.6 Hz and 7.5 V, respectively, from DG#1. The results indicate that the standard EN 50160 is respected and smooth transition behaviours of DGs are achieved. A small test is performed to compare the computation time of the proposed method and conventional method. The results show that the total sampling and calculation time is 48 μs for the proposed controller and 37 μs for the conventional controller. This indicates that the control complexity of the proposed controller is comparable with that of the conventional controller. The sampling interval is 300 μs which is long enough for both controllers to accomplish the sampling and computation.

7. Summary

The developed generalized control algorithm in the inverter control layer of DGs facilitates the seamless transition of microgrids. This is obtained by designing the multi-loop controller in the inverter control layer in such a way that the closed loop dynamics of the inverter together with the *LC* filter present unitary gain. Thus, the current/voltage disturbances associated with the mode transition are fully cancelled. The proposed voltage control algorithm and the conventional double-loop voltage controller are compared with analytical study and experimental implementation. The output (harmonic) impedance is the cause of the tracking error and distortion caused by the output current. With the proposed voltage controller, the output impedance is ***'zero'*** which theoretically eliminates the tracking error and reduces/eliminates the distortion (can only reduce the distortion in the practical implementation due to the measurement noise, digital quantization errors, etc.). The work has shown that the developed voltage control algorithm has superior control performance than the conventional controller, and it is a high-performance controller, easy to be implemented in the practical application.

Author details

Jing Wang[1,2]* and Bouna Mohamed Cisse[3]

*Address all correspondence to: jingwang1108@gmail.com

1 Institute for Automation of Complex Power System, E. ON ERC, RWTH Aachen University, Aachen, Germany

2 GE Energy Connections—Grid Solutions, Stafford, United Kingdom

3 GE Digital Energy, Philadelphia, USA

References

[1] Chowdhury S, Chowdhury SP, Crossley P. Microgrids and Active Distribution Network. Stevenage: IET; 2009

[2] Zhong QC, Hornik T. Control of Power Inverters in Renewable Energy and Smart Grid Integration. Chichester: Wiley-IEEE; 2013

[3] Rocabert J, Luna A, Blaabjerg F. Control of power converters in AC microgrids. IEEE Transactions on Power Electronics. 2012;**27**(11):4734–4749

[4] Mohamed YAR. New control algorithms for the distributed generation interface in grid-connected and microgrid system [dissertation]. Ontario: University of Waterloo; 2008

[5] Wang J, Chang NP, Feng X, Monti A. Design of a generalized control algorithm for parallel inverters for smooth microgrid transition operation. IEEE Transactions on Industrial Electronics. 2015;**62**(8):4900–4914

[6] Balaguer IJ, Lei Q, Yang S, Supatti U. Control for grid-connected and intentional islanding operations of distributed power generation. IEEE Transactions on Industrial Electronics. 2011;**58**(1):47–157

[7] He J, Li Y. Hybrid voltage and current control approach for DG-grid interfacing converters with LCL filters. IEEE Transactions on Industrial Electronics. 2013;**60**(5):1797–1809

[8] Parker S, Mcgrath B, Holmes D. Regions of active damping control for LCL filters. IEEE Transactions on Industry Applications. 2014;**50**(1):424–432

[9] Teodorescu R, Liserre M, Rodr?guez P. Grid Converters for Photovoltaic and Wind Power Systems. Hoboken: John Wiley & Sons; 2011

[10] Kashem MA, Ledwich G. Multiple distributed generators for distribution feeder voltage support. IEEE Transactions on Energy Conversion. 2005;**20**(3):676–684

[11] Deng H, Oruganti R, Srinivasan D. A simple control method for high-performance UPS inverters through output impedance reduction. IEEE Transactions on Industrial Electronics. 2008;**55**(2):888–898

[12] Bollen M, et al. Performance indices and objectives for microgrids. In: Proceedings of the 20th International. CIRED; June 2009; Prague: IEEE; 2009. pp. 1–4

Chapter 5

Design and Operation of an Islanded Microgrid at Constant Frequency

Daming Zhang and John Fletcher

Additional information is available at the end of the chapter

http://dx.doi.org/10.5772/intechopen.69401

Abstract

This chapter presents a method for operating an islanded microgrid at a constant frequency. The proposed method uses de-coupled PQ control plus real power reference generation based on voltage variation to control the grid-forming generator and grid-supporting generators. Its effectiveness has been validated by a three-phase microgrid system where there is one grid-forming generator, one grid-supporting, and one grid-feeding generator. The grid-forming generator produces its own voltage reference with a constant frequency of 50 Hz, while the grid-supporting and grid-feeding generators take the voltage as a reference at their respective coupling point with the microgrid. It is found that the grid-forming and grid-supporting generators work collaboratively to keep voltages at each bus around the rated value. For a practical microgrid, it is necessary to determine the location and sizing of each grid-supporting generator in order to keep the voltage profile within specification under all operating conditions. To achieve these two purposes and also to reduce the computational demand of modeling and to shorten simulation time, a single-phase equivalent microgrid has been adopted in this research. Such approach is useful for the design of a practical microgrid.

Keywords: constant frequency, grid-forming, grid-feeding, grid-supporting, microgrid, reactive power compensator

1. Introduction

An islanded microgrid is normally composed of three groups of distributed generators (DGs), one being grid-forming, the other being grid-supporting and the grid-feeding DGs [1]. To avoid loss of synchronism, normally only one grid-forming DG is adopted in an islanded microgrid. But there could be as many grid-supporting DGs as necessary. Either conventional power sources such as hydropower plants or renewable energy sources like wind or photovoltaic can be used to power grid-feeding generators.

In recent decades, intensive research has been conducted on the operation of islanded microgrids, but it is yet to standardize their control methods.

Frequency and voltage droop are normally adopted [1–4]. Such conventional droop control methods have several disadvantages, including (1) ignoring load dynamics that can result in failure subsequent to a large or fast load change; (2) inability to narrow down frequency within certain limit independent of system loading conditions [4–8].

To overcome the drawbacks of the droop control method, a constant frequency method has been applied to operate both three-phase and single-phase microgrids and is described in this chapter.

In the case of the three-phase system, the microgrid is composed of one grid-forming DG, one grid-supporting generator and one grid-feeding DG. The grid-feeding DG is powered by a time-varying solar source. Both the grid-forming DG and grid-supporting DG are powered by fuel cell energy to manage power balance due to load dynamics and solar power variation (PV). The grid-feeding DG adopts PQ control with the capability of maximum power point tracking. The grid-forming DG produces the reference voltage by itself with a constant frequency of 50 Hz and outputs real power according to the system demand after islanding occurs. The grid-supporting DG adopts its terminal voltage as a reference and uses its terminal voltage variation to generate its real power reference while reactive power is set at either zero or a lower value. The system always operates at constant frequency 50 Hz.

The grid-forming DG acts as reactive power sensor as well, which indicates system reactive power demand change due to switch-on or switch-off of loads absorbing reactive power. Once its output reactive power exceeds its set limit, the accompanying instantaneous var compensator takes over the extra reactive power. By doing so, the grid-forming DG's output real power can follow its reference accurately and is adaptive to meet varying load demand. A multiplying factor can be adopted to ensure fast response as described in [9].

As three-phase modeling needs to use a lot of computer memory and results in long simulation times for a practical microgrid, a single-phase model of the microgrid has been developed to identify the locations of each necessary grid-supporting generator and size its necessary capacity to keep the voltage profile at each bus of the microgrid within limits.

This chapter is organized as follows: in Section 2, the overall system is introduced; Section 3 presents the terminal properties of the fuel cell and solar panel. It also presents the control method for extracting maximum power from the solar panel; Section 4 shows results and discussion for the three-phase microgrid under study; in Section 5, the results for the single-phase microgrid are presented and discussed. Section 6 concludes this chapter.

2. Overall system

2.1. System description

Figure 1 shows the three-phase microgrid under study, where DG1 acts as the grid-forming generator and is powered by fuel cell energy, DG2 acts as a grid-supporting generator and is also powered by fuel cell energy, and DG3 acts as the grid-feeding generator and is powered by solar energy. In practice, in view of the slow response of the fuel cell, at the DC-link of DG1 and DG2, extra circuits such as DC/DC converter interfaced super capacitors can be adopted to ride through transient power demands.

The power rating of each DG is 40 kW in the microgrid system as shown in **Figure 1** and their voltage ratings are 415 V (LL).

$$G_{ol}(s) = \frac{G_c(s) \cdot K}{(L_1S + R_1) \cdot (L_2s + R_2) \cdot Cs + K \cdot Cs \cdot (L_2s + R_2) + (L_1S + R_1) + (L_2s + R_2)} \quad (1)$$

In each DG, the fundamental converters are the same: DC/DC converter + DC/AC inverter with LCL filter (**Figures 2–5**). **Figures 2** and **3** show such converters for fuel cell and solar energy conversion. **Figures 4** and **5** show their control flow and AC side reference current generation.

The open-loop transfer function of the DC/AC inverter with LCL filter is given by Eq. (1). The Laplace description of the proportional resonant controller is given by Eq. (2) and the closed-loop transfer function is given by Eq. (3).

$$G_c(s) = K_p + \frac{K_i s}{s^2 + \omega_0^2} \quad (2)$$

$$G_{cl}(s) = \frac{G_{ol}(s)}{1 + G_{ol}(s)} \quad (3)$$

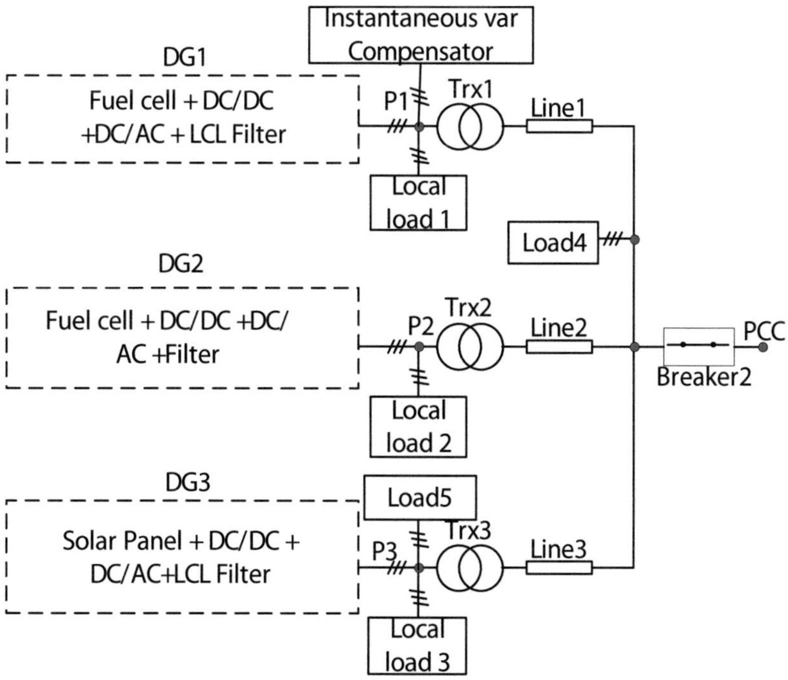

Figure 1. Overall microgrid system.

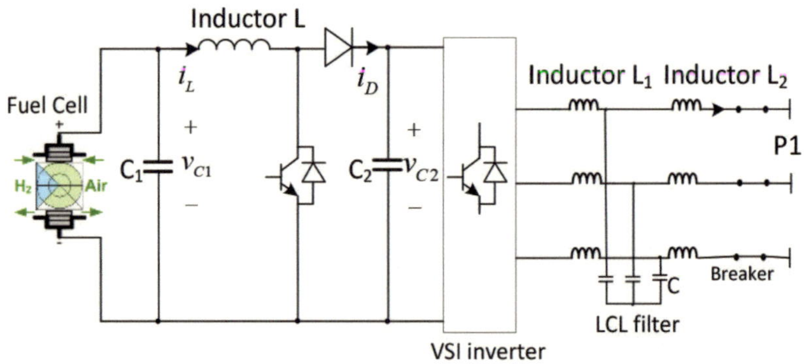

Figure 2. Fuel cell power conversion unit.

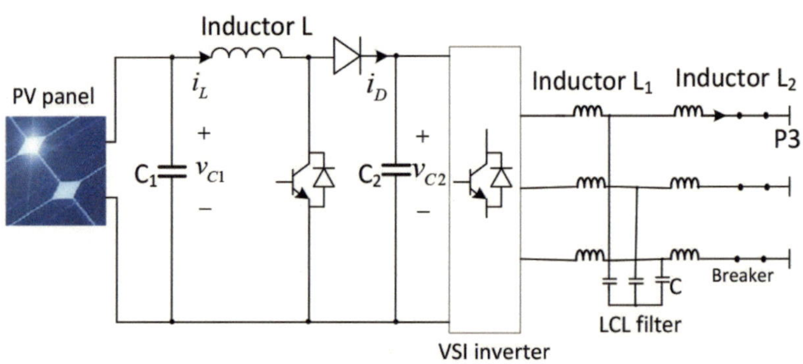

Figure 3. PV panel coupled to a dc-dc boost converter and grid feeding inverter.

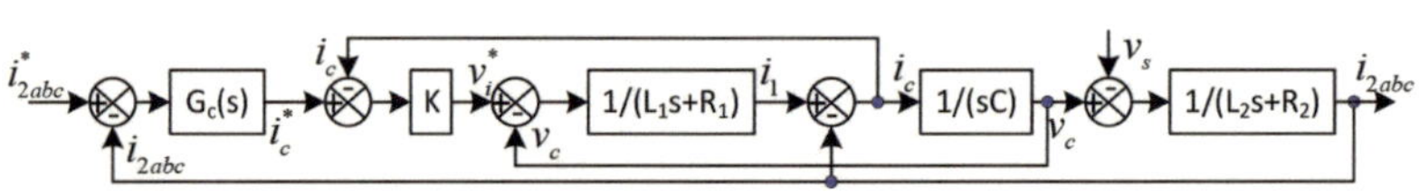

Figure 4. Control flow for the VSI with LCL filter.

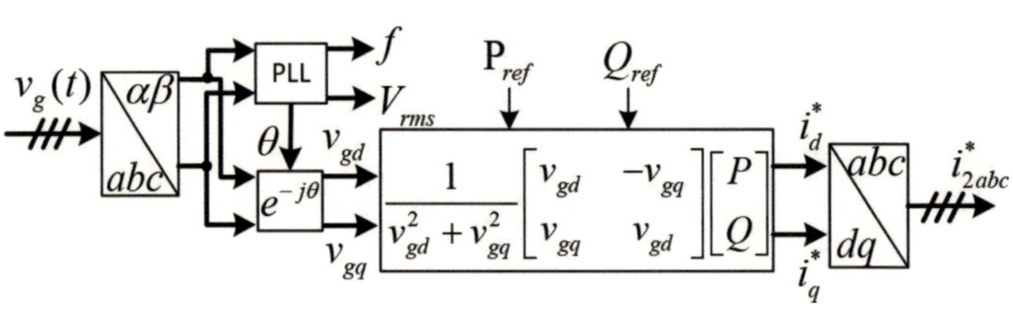

Figure 5. Generation of reference currents for the given grid voltage and real and reactive power references.

2.2. Parameter design for grid-tied inverter

The stability of each DG plays a vital role in the overall system operation. The inverter in each DG needs to be well designed to facilitate this purpose.

The method adopted in Ref. [5] allows one to choose appropriate L_1, L_2, and C parameters. It can also allow one to select appropriate K_p, K_i, and K parameters. Furthermore, the parameters determined can automatically avoid resonance. To facilitate such optimization, a one-phase circuit model with virtual resistance that consumes the rated power at rated voltage is adopted as shown in **Figure 6**.

The resonant frequency of the LCL circuit is given by the following expression

$$f_{res} = \frac{1}{2\pi}\sqrt{\frac{L_1 + L_2}{L_1 L_2 C}} \tag{4}$$

For good damping of switching frequency harmonic components, the resonant frequency needs to be carefully chosen. Normally it is set below the switching frequency divided by a factor of 1.5–2 and 10 times greater than the fundamental frequency, 50 Hz [5, 11, 14].

Furthermore, a combination of partial direct-pole-placement and differential evolution algorithm is used to determine the basic parameters of the proportional resonant controller for the inverter as described in Ref. [5].

Below are the designed parameters:

L_1 = 6.55 mH, L_2 = 0.295 mH, C = 34.4 μF, K_p = 2.278, K_i = 28.34, K = 4 2.9, $f_{sampling}$ = 100 kHz, f_{sw} = 5 kHz

The designed zeros and poles of the closed-loop transfer function are shown in **Figure 7**, from which one can see the poles closest to imaginary axis have a real part close to the target −50. One can also see that two zeros almost overlap with two of the five poles. Hence the optimization follows control theory that the lower the order of the closed-loop transfer function, the less susceptible the system is to noise.

Figures 8 and **9** show the open-loop and closed-loop transfer functions, from which one can see that at the resonant frequency f_{res} = 1.616 kHz, the attenuation is more than 10 dB in the closed loop system. Hence resonance is avoided.

In summary, by choosing appropriate real parts for the two poles of the closed-loop transfer function closest to imaginary axis and ensuring that they are far enough apart from the

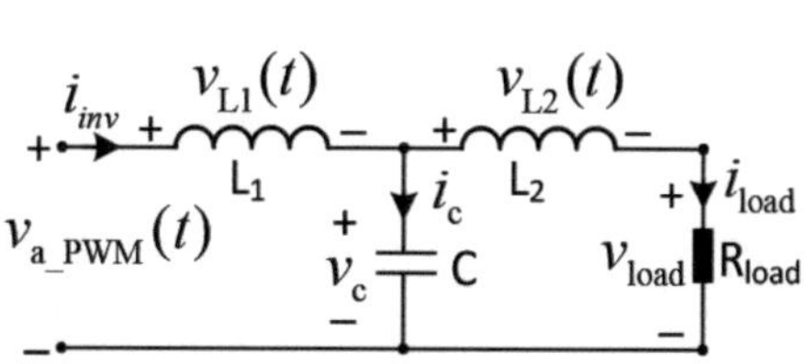

Figure 6. One-phase equivalent circuit used to choose proper LCL to contain harmonics.

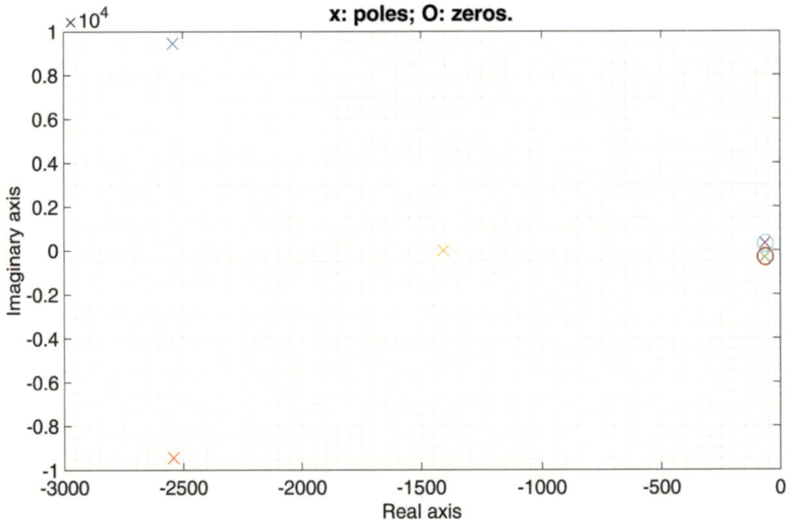

Figure 7. Designed zeros and poles of closed-loop transfer function using differential evolution optimization.

imaginary axis, resonance at the frequency range of interest can be avoided. Then either passive or active dampening as adopted in Refs. [12–14] is not necessary.

More information on DC/AC inverter design can be found in Ref. [5].

2.3. Overall control strategy

Before islanding occurs, Breaker 2 at the point of common coupling (PCC) in **Figure 1** is closed and the three DGs run in PQ control mode. As the grid voltage is almost constant, each DG can

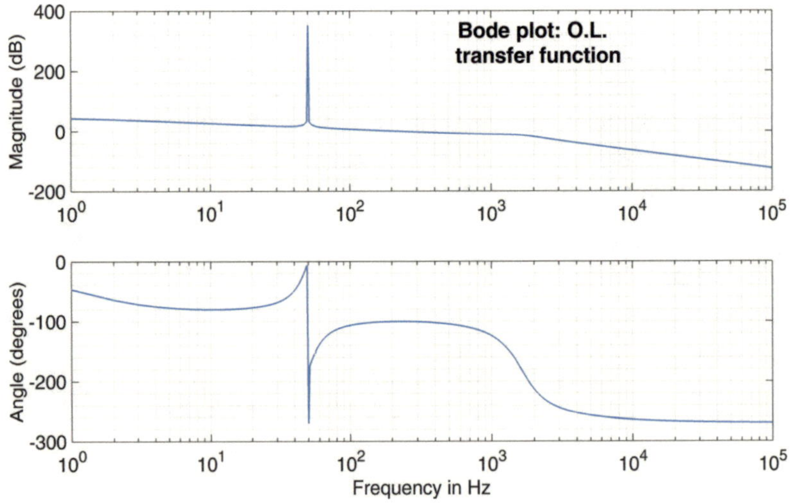

Figure 8. Bode plot of open-loop transfer function using Eq. (1).

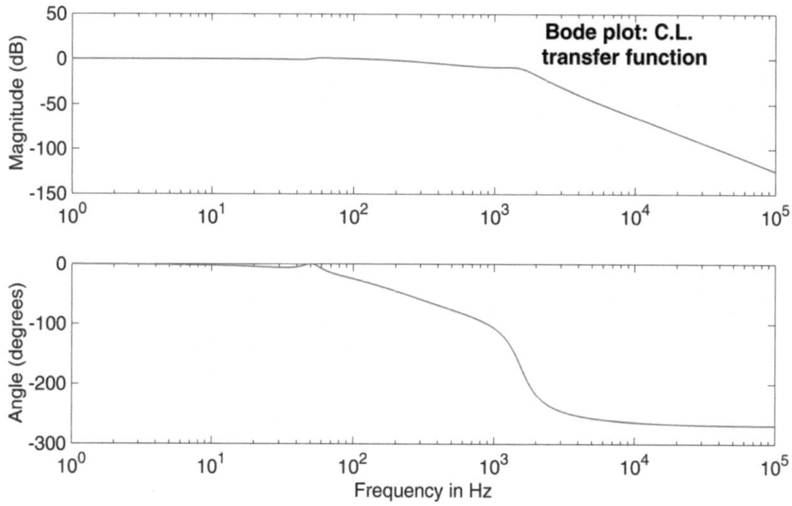

Figure 9. Bode plot of closed-loop transfer function using Eq. (3).

produce real and reactive power the same as their settings. After islanding occurs at 3.02 s, DG1 operates as a grid-forming generator. DG2 works as a grid-supporting generator, while DG3 works as the grid-feeding generator.

Totally there are five loads whose information is shown in **Table 1**. Loads 1, 2, and 3 local to each DG are always connected in the system while loads 4 and 5 are switched on and switched off to test whether the designed system can tolerate the disturbance due to such a dynamic change of loads.

The control method after islanding occurs is of paramount importance for the operation of the microgrid. After islanding occurs, the voltages in the microgrid are uncertain. Sustaining a stable voltage, both in terms of magnitude and frequency for the system becomes the main control target. In this research, the grid-forming DG1 takes de-coupled PQ control with self-generated voltage reference. **Figures 10** and **11** illustrate the method of generating reference real and reactive power for DG1 after islanding happens. A PLL method shown in **Figure 10** is adopted to obtain the magnitude and angle of both voltage and current at point $P1$ in **Figure 1**. Then such information is utilized to generate an error signal which is fed to a $P1$ controller and low-pass filter to generate reference real and reactive power as shown in **Figure 11**. In addition, a multiplying factor F2 is introduced to adjust the reference real power in **Figure 11**. More description can be found in [9, 16] on how to choose proper factor F1 and the influence of F2 on the performance of the control.

To generate i^*_{2abc} reference by the method shown in **Figure 12**, both reference power and the reference voltage are required. Instead of taking conventional voltage and frequency droop control to generate the reference, constant magnitudes of voltage and frequency are taken for the grid-forming generator. As the microgrid works at a constant frequency, there is no possibility of maloperation by frequency protection systems. The formulae to generate angular frequency and voltage are given by Eqs. (5) and (6). With the generated reference power, frequency and voltage, the reference current is obtained by the method shown in **Figure 12** to control DG1.

The method in **Figure 11** is also taken to generate a real power reference for the grid-supporting generators. Then the method shown in **Figure 5** is taken to produce reference currents, where v_g(t) is the voltage at each coupling point with the microgrid. For the accurate control of real power, the reactive power reference for the grid-supporting generators can be set at a very small value.

Load	**1 (PQ)**	**2 (PQ)**	**3 (PQ)**	**4 (RL)**	**5 (PQ)**
P (kW)	20	20	20	15	10
Q (kvar)	10	0	0	10	0
Connection	On	On	On	Initial: On Off: 6.02 s In: 7.02 s	Initial: Off In: 4.02 s Off: 5.02 s

Table 1. Load information.

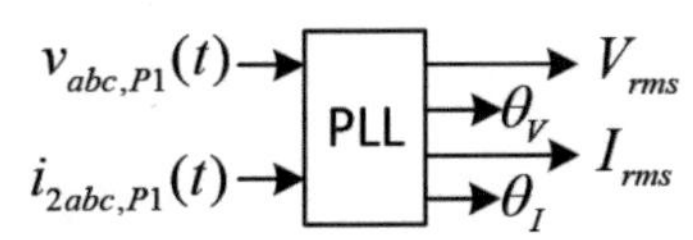

Figure 10. PLL to extract magnitude and angle.

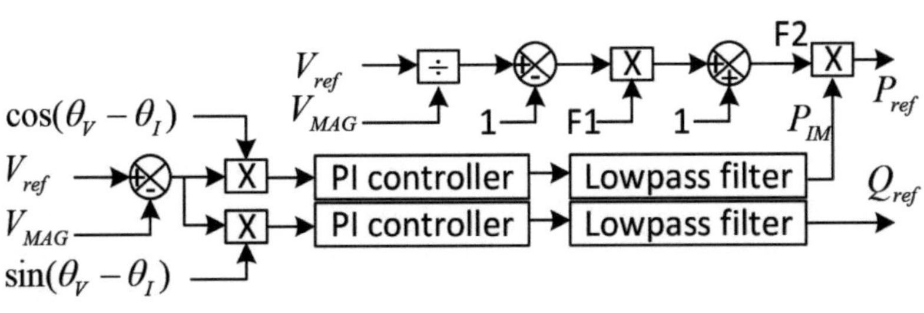

Figure 11. Generation of reference power.

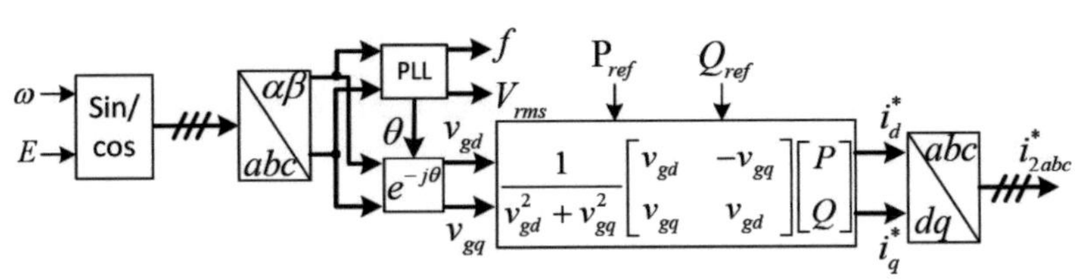

Figure 12. Generation of reference current i_{2abc}^* after islanding occurs.

$$\omega = \omega_0 \tag{5}$$

$$E = E_0 \tag{6}$$

3. Terminal properties of the fuel cell, solar panel, and wind generator

The updated fuel cell model in Matlab/Simulink has been validated by experiment [15]. So, it is an effective model to use for microgrid system level research. The adopted fuel cells have a power rating of 32 kW each. Their other parameters are the same as those in Ref. [9].

To examine the terminal properties of the fuel cell, a Matlab/Simulink circuit shown in **Figure 13** is adopted. A controlled current source (CCS) is used to control the output current of the fuel cell. At each level of fuel flow rate, the control signal to the CCS linearly increases with time to above a value which could produce maximum power extraction from the fuel cell.

By changing the fuel flow rate, one can obtain the terminal properties shown in **Figure 14**, where curves of terminal current, voltage, and power against fuel flow rate are shown.

The fuel cell model is used to power DG1 which works as a grid-forming generator and to power DG2, which works as the grid-supporting generator. No matter whether it is a grid-forming or grid-supporting generator, its reference power should be adaptive to dynamic load changes and the variation of solar power injected by DG3. Hence, the terminal properties of current, voltage, and fuel flow rate against power are neccessary for fulfilling real power management by DG1 and DG2. So, curves of terminal current, voltage, and fuel flow rate against output power have been obtained from the model as shown in **Figures 13** and **15**. With these curves, one may use a polynomial approximation to fit these curves and create a set of coefficients in each of the formulae as shown in Eqs. (7)–(9), where order 3 is adopted.

$$V(P) = A_0 + A_1P + A_2P^2 + A_3P^3 \tag{7}$$

$$I(P) = B_0 + B_1P + B_2P^2 + B_3P^3 \tag{8}$$

$$\text{fuel flow rate}(P) = C_0 + C_1P + C_2P^2 + C_3P^3 \tag{9}$$

The control for the DC/DC converter and fuel cell in **Figure 2** is shown in **Figure 16**. It is found that control of voltage across capacitor C1 is more suitable for this kind of application. For a demanded real power from a DG, Eq. (9) is used to produce fuel flow rate. Eq. (7) is used to produce the reference voltage across C1. Then the difference between the reference and measured voltage is PI-compensated and compared with a sawtooth to generate a pulse-width modulation (PWM) gating signal. As the curve in **Figure 15** has maximum power point tracking feature, for each controlled fuel flow rate, the system can operate at its maximum power output from the fuel cell.

The terminal properties of the solar panel can be found by using the method as shown in Ref. [10]. For the solar panel chosen, its terminal properties are shown in **Figure 17**.

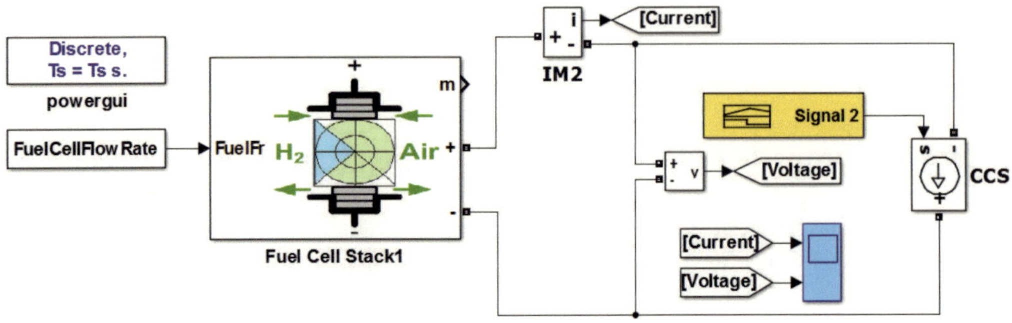

Figure 13. Circuit for studying termimal properties of the fuel cell.

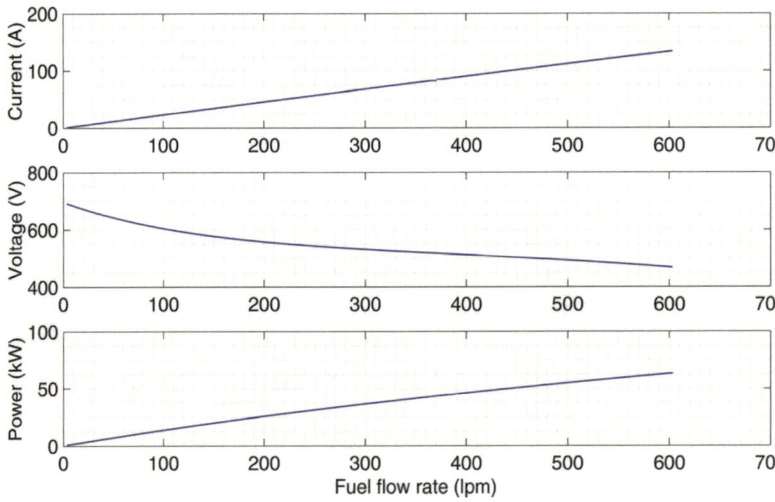

Figure 14. Terminal current, voltage and power against fuel flow rate.

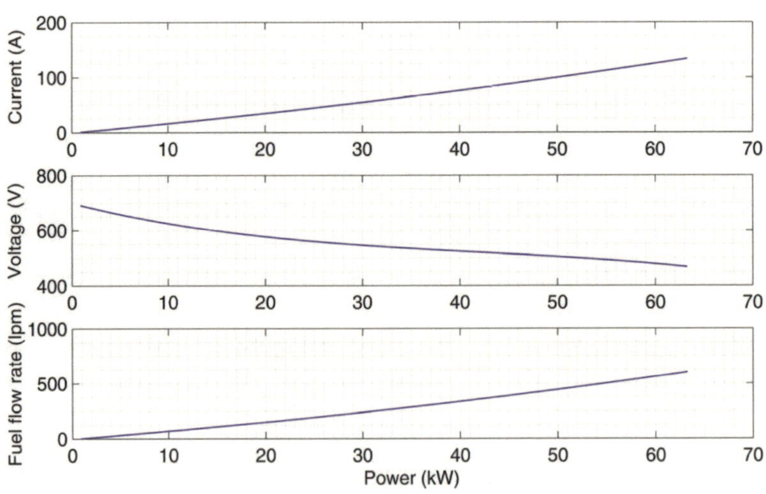

Figure 15. Terminal current, voltage, fuel flow rate against power.

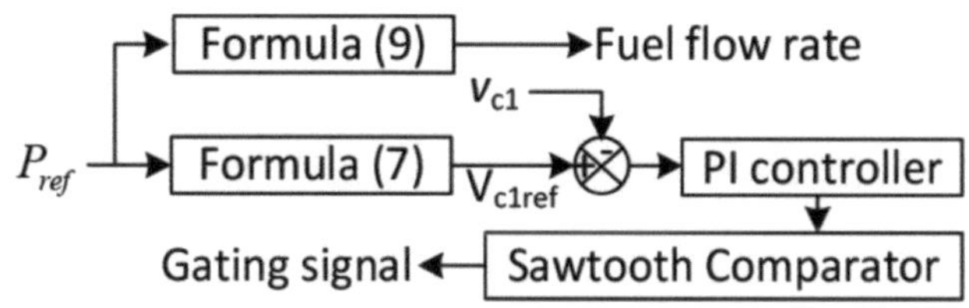

Figure 16. Controller for fuel cell and DC/DC boost converter.

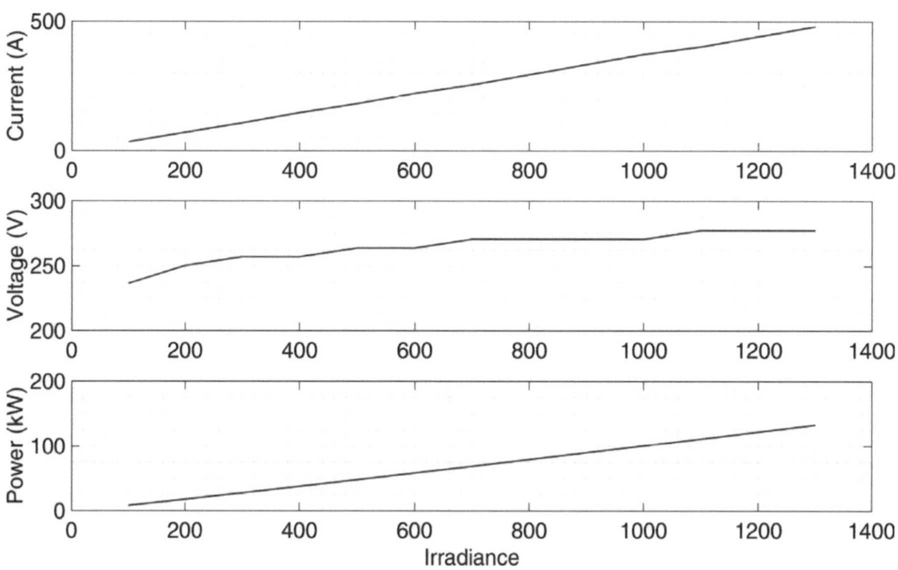

Figure 17. Operating points at different solar irradiance level (W/m^2) with MPPT.

To extract maximum power from the solar panel, one may regulate the voltage across capacitor C1 or current flowing through inductor L in **Figure 3** or use a two-loop controller to regulate both voltage and current to have better noise rejection capability.

4. Results and discussion for the three-phase microgrid

For the operation of the islanded three-phase microgrid, DG1 powered by the first set of fuel cells acts as a grid-forming generator while DG2 powered by another set of fuel cells acts as a grid-supporting generator, and DG3 powered by solar panels acts as the grid-feeding generator.

Figure 18 shows the percentage error of voltage at $P1$, $P2$, and $P3$, from which one can see that percentage error of each voltage under steady state is almost zero and each voltage is maintained almost at rated voltage after islanding occurs. This is especially true for the

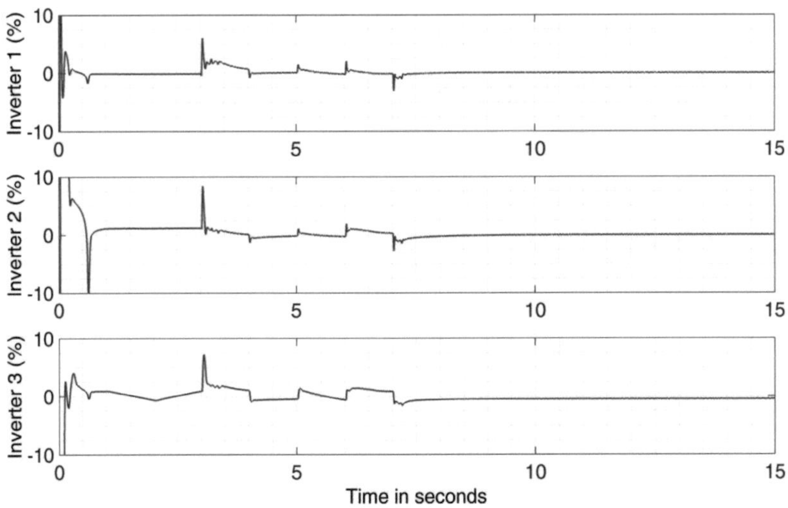

Figure 18. Percentage error of voltages at *P*1, *P*2, and *P*3.

voltages at *P*1 and *P*2 in **Figure 1** because DG1 operates as the grid-forming generator and DG2 acts as grid-supporting generator, both of which produce reference power based on the variation of their respective terminal voltages from the nominal value. As one may use grid-supporting generators as many as necessary, it is foreseeable that with the sufficient number of the grid-supporting generators installed at proper locations in the microgrid, they can work collaboratively with the grid-forming generator to keep the voltage profile within an acceptable limit. Also, the frequency of the system is kept constant at 50 Hz and power quality is ensured.

Figures 19–22 show the power from DG1, DG2, DG3, and the instantaneous var compensator. DG1 and DG2 can cooperate to produce enough power to balance load demand and dynamic changes in the solar power without communication links.

As DG2 is a grid-supporting generator with its reference voltage taken from *P*2 and operates under PQ control, its reference real power has the flexibility of varying either with the voltage at *P*2 (as adopted for this modeling) or it can be produced from a range with particular lower limit and upper limit. One can even fix it at a value when the variation of load demand and renewable energy generation does not change too much. Different from conventional droop control which demands the change of set point with system operating condition, the proposed method just demands that each grid-forming and grid-supporting DG outputs real power as they can within their limit. Each grid-supporting DG can switch from flexible operating mode which outputs real power according to its terminal voltage variation to fixed real power generation mode [16].

Figures 23 and **24** show the factor F2, pre-tuned reference real power and reference real power, which is equal to pre-tuned reference real power multiplied by F2, for DG1 and DG2. From

http://dx.doi.org/10.5772/intechopen.69401

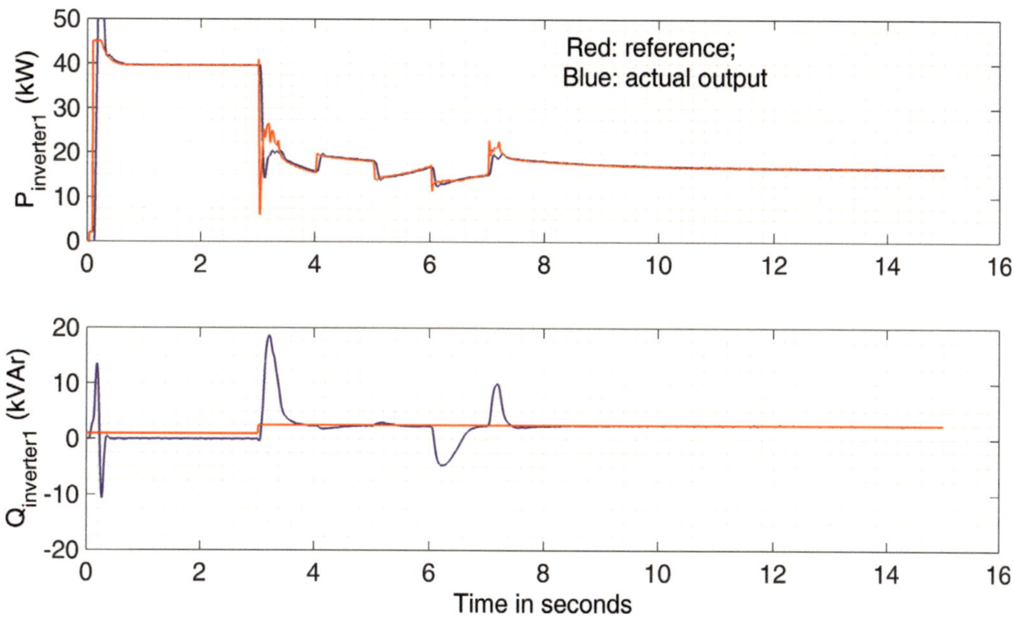

Figure 19. Power from DG1.

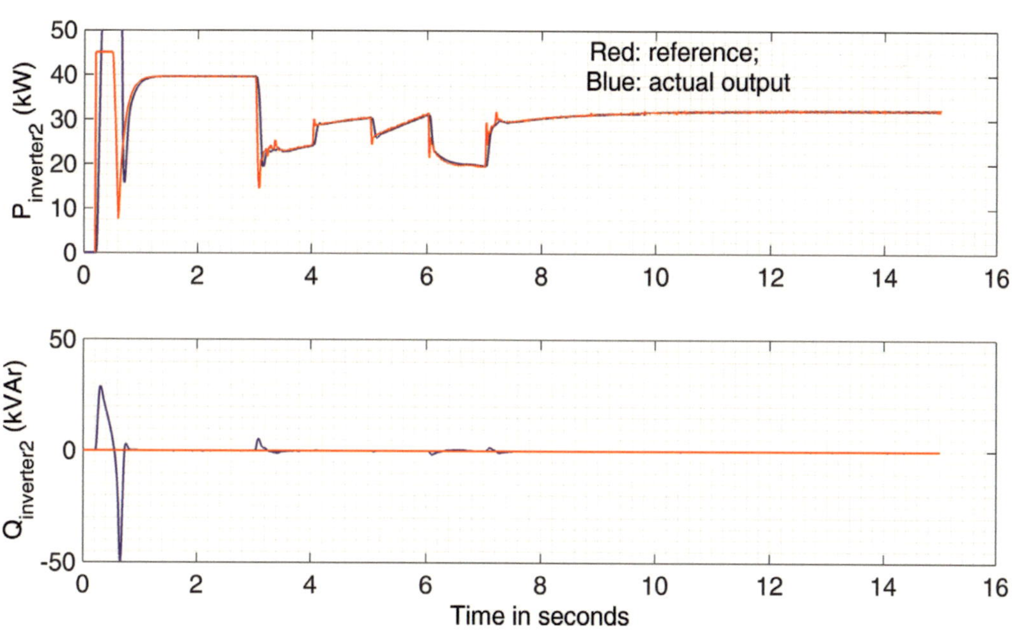

Figure 20. Power from DG2.

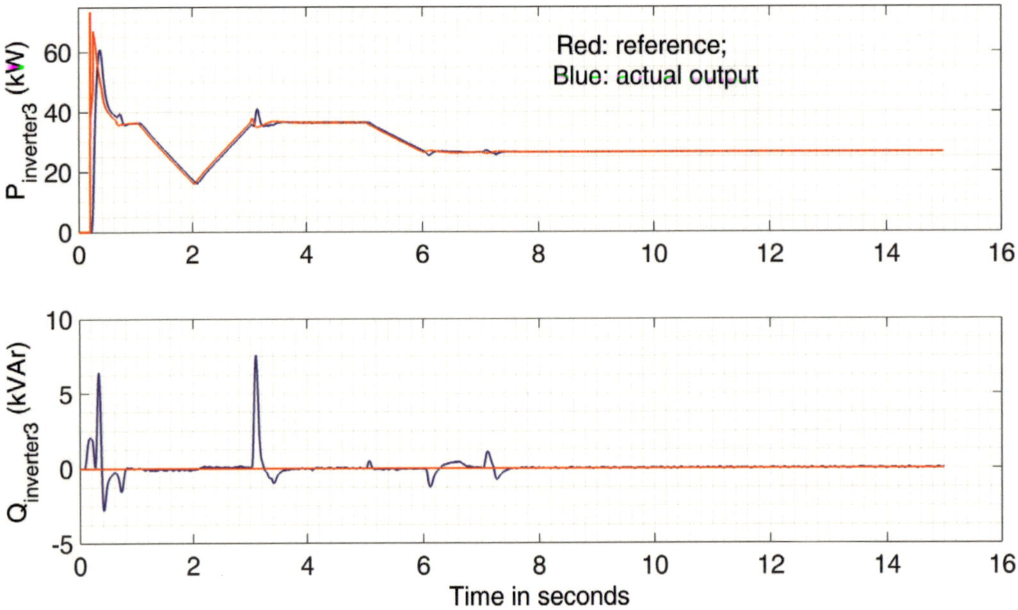

Figure 21. Power from DG3.

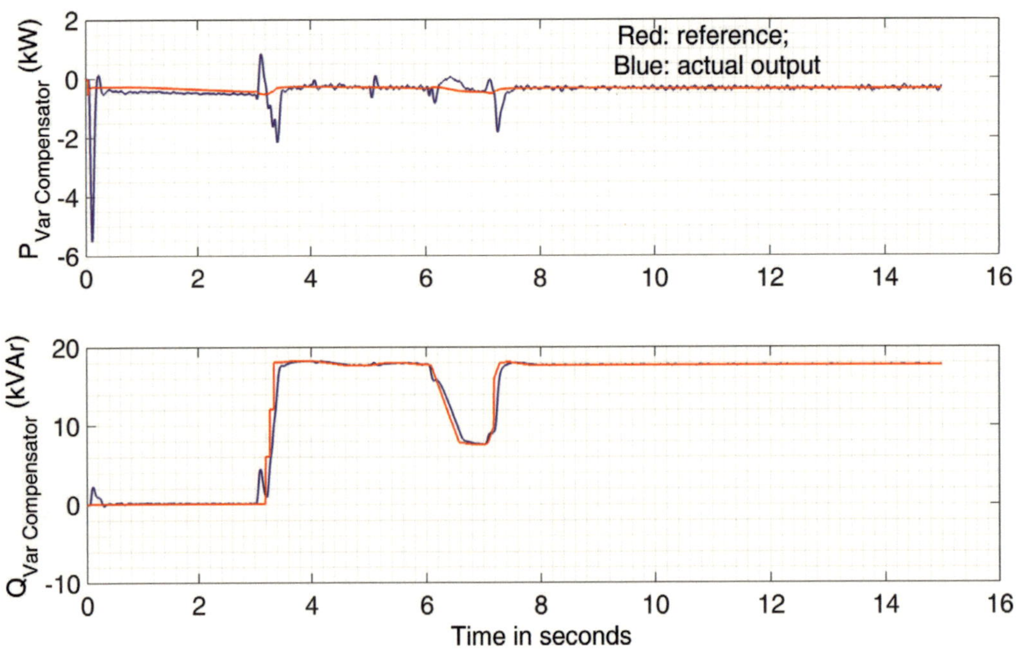

Figure 22. Power from instantaneous var compensator.

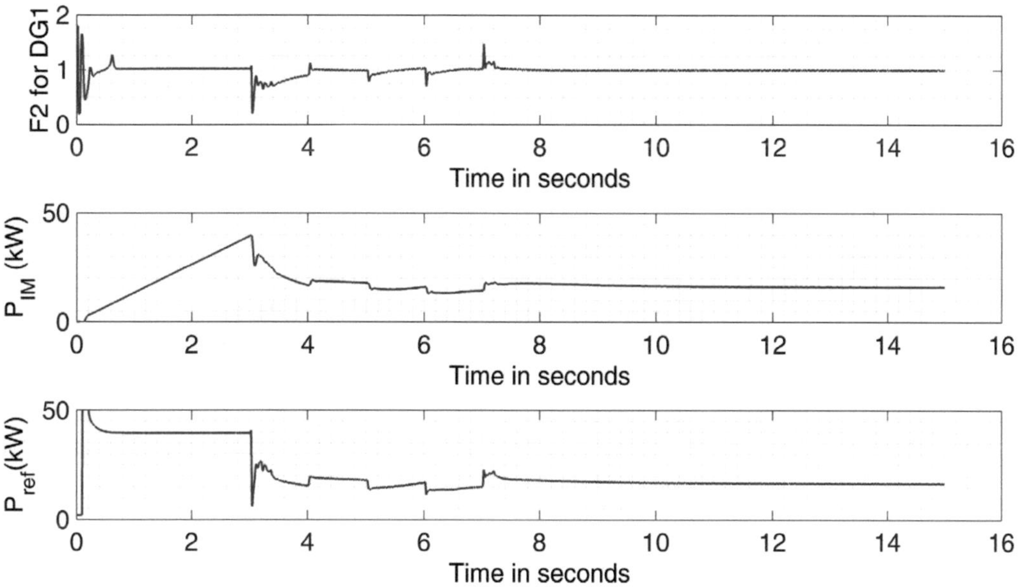

Figure 23. Factor F2, $P_{IM,}$ and P_{ref} in **Figure 11**(b) for DG1.

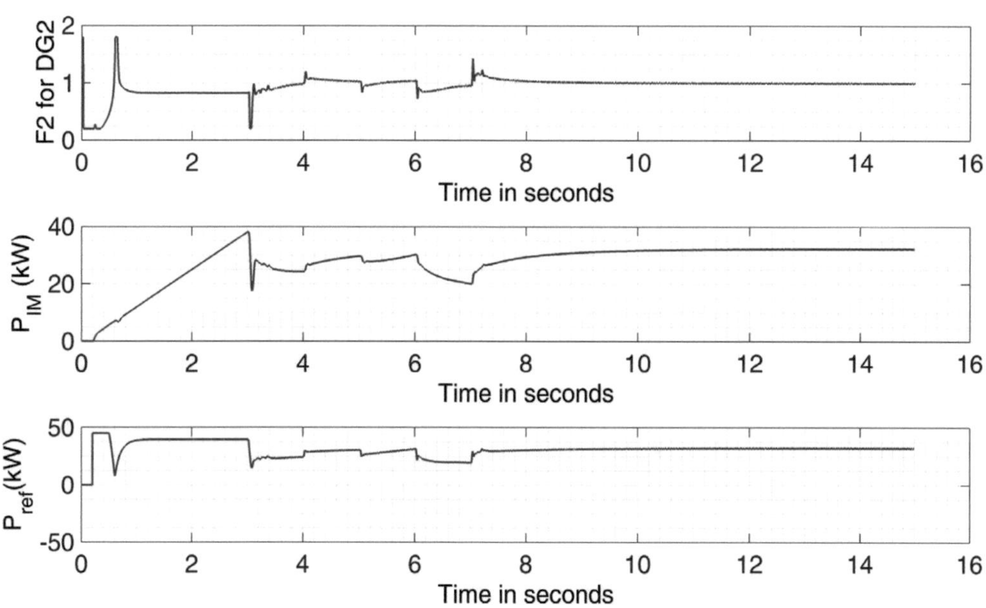

Figure 24. Factor F2, P_{IM} and P_{ref} in **Figure 11**(b) for DG2.

these two figures, one can see that introduction of F2 can quickly produce the appropriate real power reference, which helps stabilize the system voltage when there is a sudden change of load demand or renewable energy generation.

Figures 25–30 show the results of DC/DC converter for each of DG1, DG2, and DG3. From Figures 26, 28, and 30, one can see that the voltage across C1 can trace its respective reference accurately. For DG1 and DG2, such accurate tracing ensures that for a given fuel flow rate, the maximum power is output at the terminal of fuel cell and fuel wastage is avoided. To generate

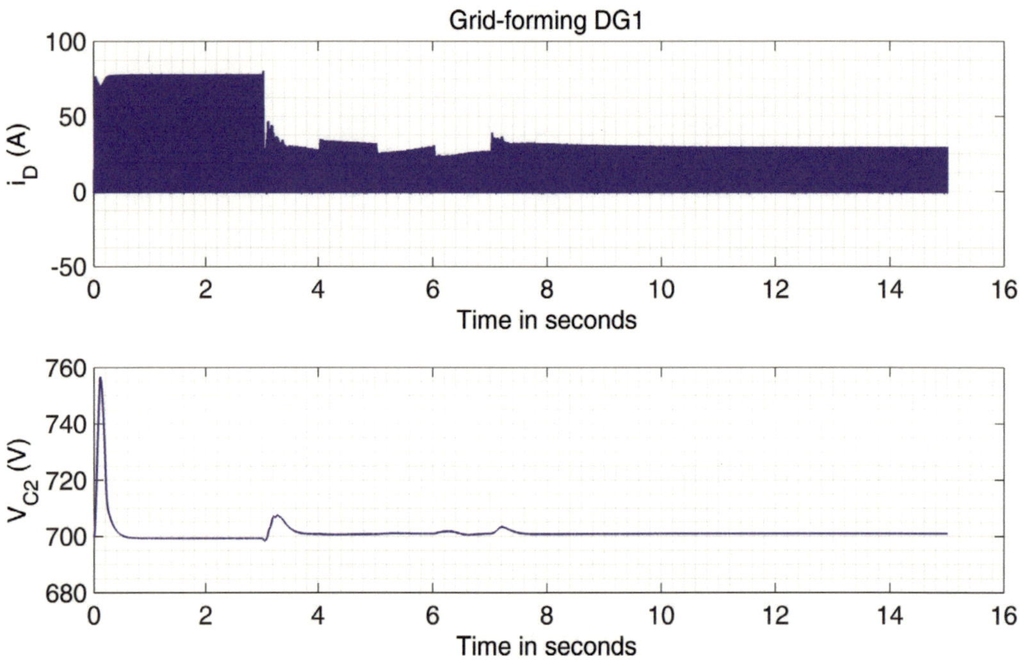

Figure 25. Current flowing through diode and voltage across C2 in the DC/DC converter for DG1.

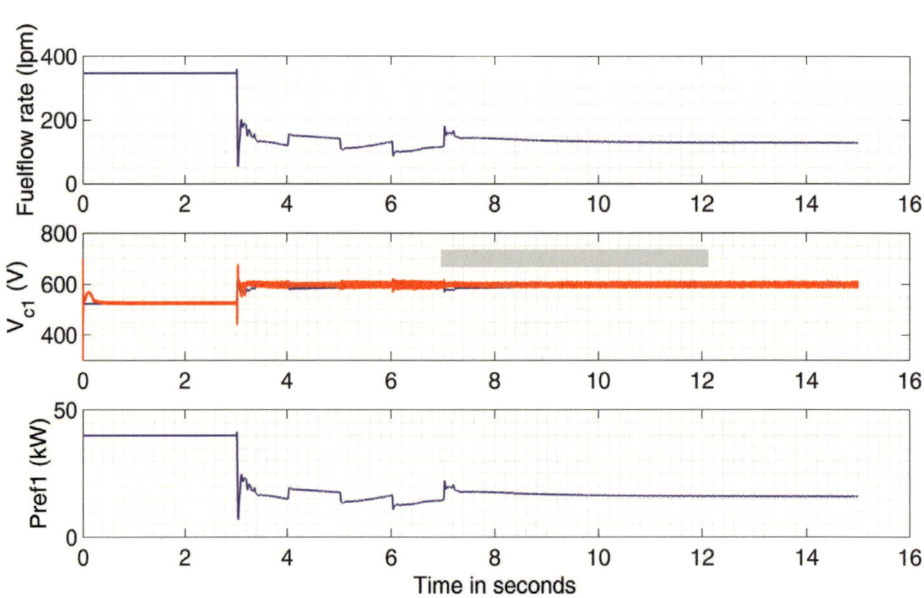

Figure 26. Fuel flow, voltage across C1, and reference power for DG1.

such reference fuel flow rate dynamically, one needs to use the curves in **Figure 15** or formula shown in Eq. (9) for a demanded reference real power. **Figures 27** and **29** show the diode current and voltage across capacitor C2 in the circuits as shown in **Figures 2** and **3**.

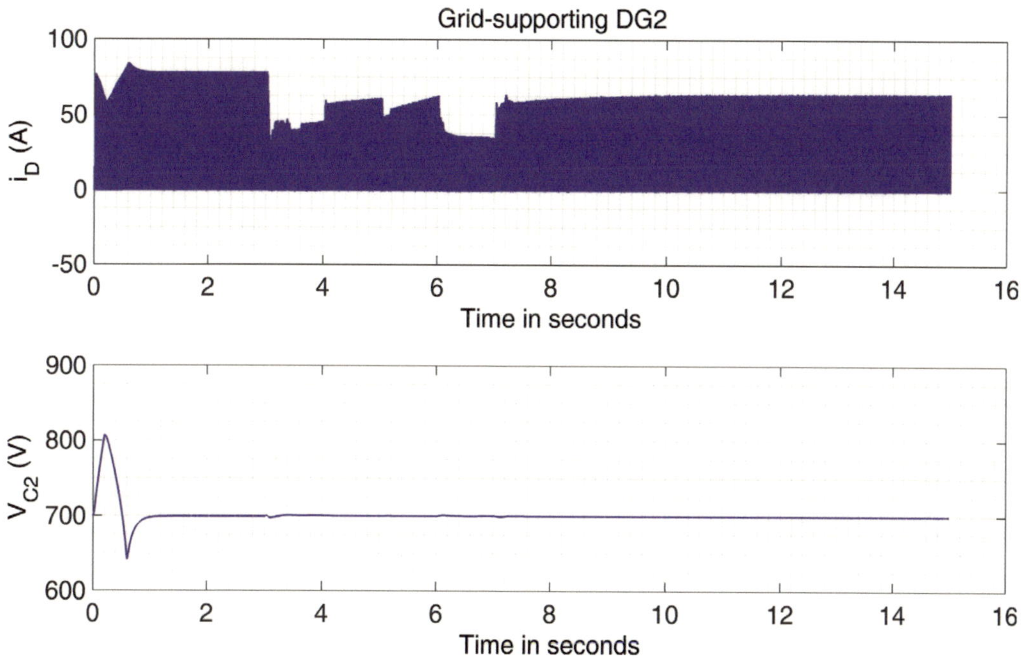

Figure 27. Current flowing through diode and voltage across *C*2 in the DC/DC converter for DG2.

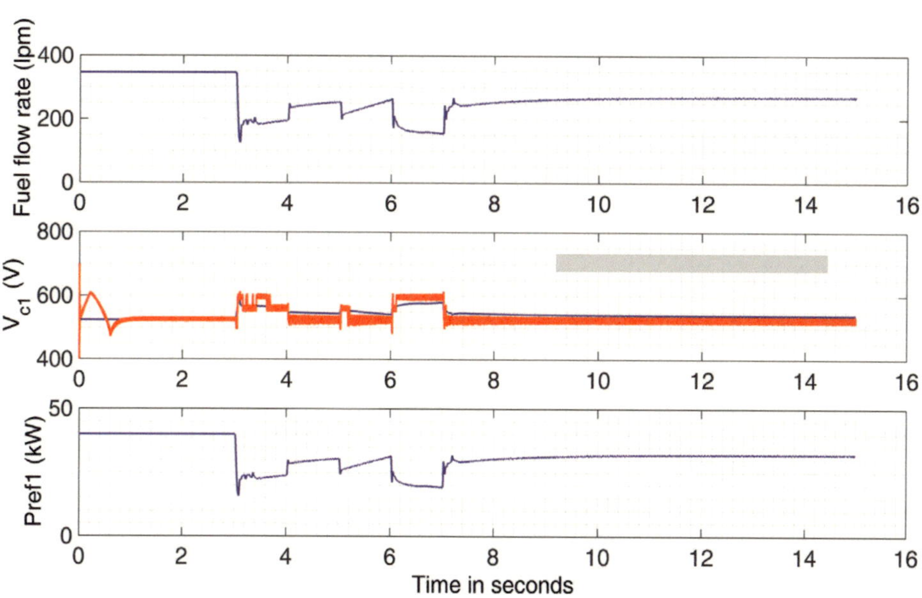

Figure 28. Fuel flow, voltage across *C*1, and reference power for DG2.

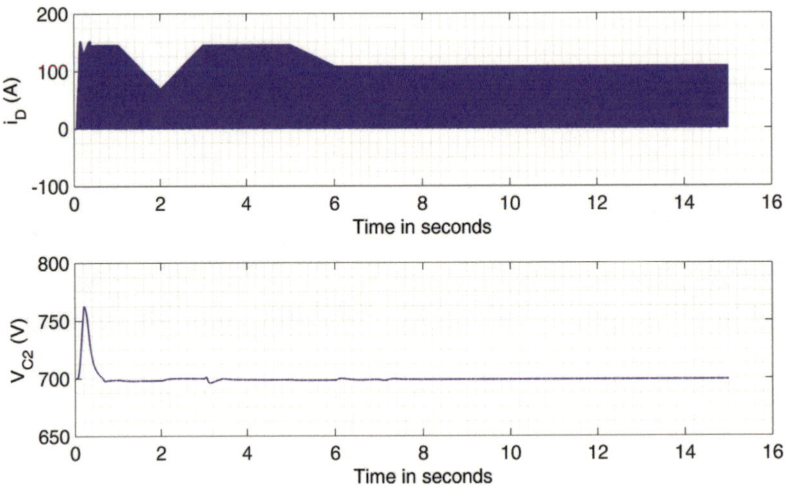

Figure 29. Current flowing through diode and voltage across *C*2 in the DC/DC converter for DG3.

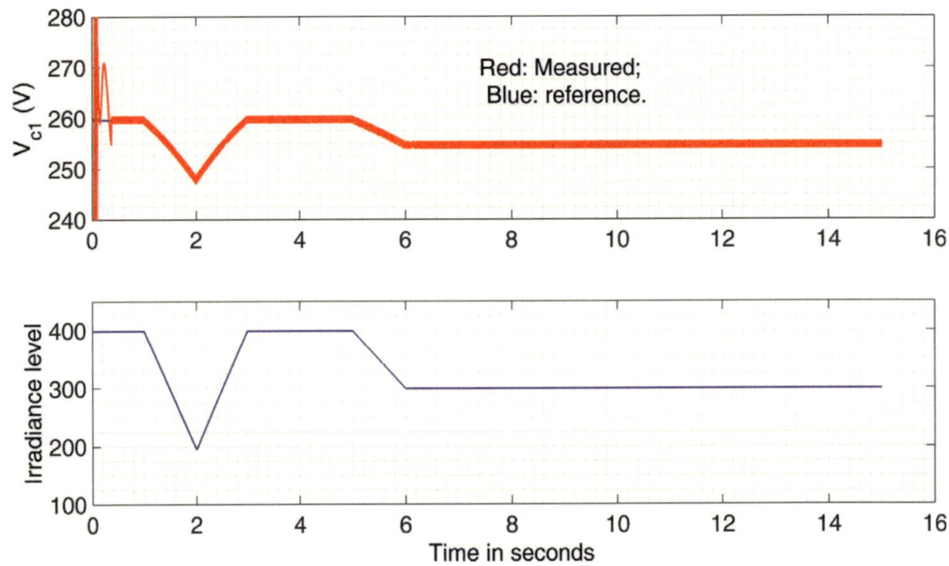

Figure 30. Voltage across *C*1 and irradiance level in **Figure 4** of DG3 for solar energy harnessing.

5. Power sharing among distributed grid-supporting generators using the single-phase model

In this section, the grid-forming generator produces a voltage reference with constant frequency for the system, and both grid-supporting and grid-feeding generators take their respective terminal voltages as a reference. Hence, the overall system operates at a constant frequency.

Then the main design target for such method is to keep voltage profile at each bus within the limits.

Time-stepped discrete code-based modeling of the three-phase power system in Matlab/Simulink is closer to its real hardware implementation. Nevertheless, it is time-consuming and takes a lot of computing resources for a practical, large microgrid system. Hence, a new approach needs are developed to solve this problem. From the point of view of designing a microgrid system, it is important to identify a suitable location for each grid-supporting generator and size each of them in order to keep the overall voltage profile of the microgrid within an acceptable limit under all conditions of possible loading and renewable generation conditions. To suit such purposes, a microgrid formed by three-phase components could be reduced to a microgrid formed by a single-phase power system. This is because within the concern of the current study, the microgrid is composed of only microgrid-tied inverter-based generators and storage and does not contain directly connected conventional synchronous generators. When a three-phase system is reduced to a single-phase system, the number of differential equations describing the system is reduced to one-third. For example, modeling the LCL filter used with a voltage source inverter could be reduced to one-third as only one-phase LCL filter instead of three phases needs be modeled.

Figure 31 shows a single-phase microgrid. This is for studying the power sharing among distributed grid-supporting generators. It can also be adopted to identify the locations of

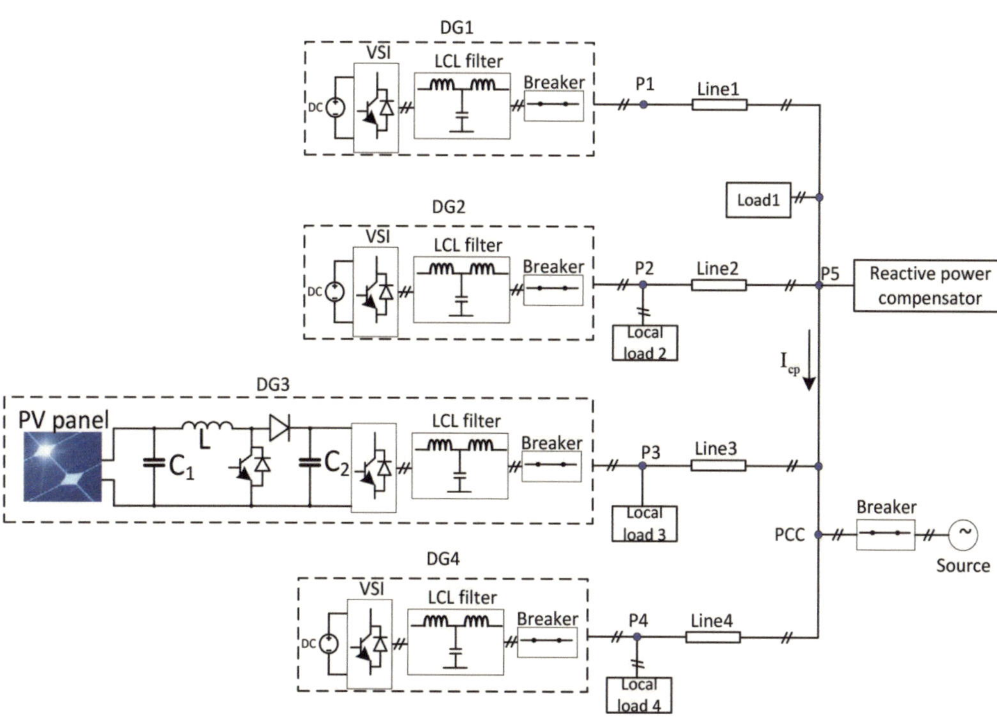

Figure 31. A single-phase microgrid system.

Load	1 (PQ)	2 (PQ)	3 (PQ)	4 (RL)
P (kW)	30	10	10	10
Q (kvar)	0	7.5	15	10
Connection	On	Initial: off On: 3.5 s Off: 4.0 s	Initial: off On: 4.5 s Off: 5.0 s	Initial: off On: 5.5 s Off: –

Table 2. Load information for single-phase microgrid.

distributed grid-supporting generators and size their capacity in order to keep the voltage profile at each bus within limits.

The information on loads in the single-phase microgrid is shown in **Table 2**. Initially, this microgrid is connected with a single-phase source at the point of common coupling (PCC). At 2.5 s, the source is disconnected. From then on, the microgrid operates in islanded mode.

There are two cases studied for the single-phase microgrid. In the first case, DG1 acts as a grid-forming generator, while DG2 and DG4 act as grid-supporting generators, and DG3 powered by solar power variation (PV) acts as grid-feeding generator which extracts maximum power from the solar resource. In the second case, DG1 still acts as grid-forming generator, DG3 still powered by solar panel acts as grid-feeding generator which extracts maximum power from solar panel, only DG2 acts as a grid-supporting generator, while DG4 is to simulate a planned battery charging/discharging according to predicted irradiance levels.

6. Results for the first case

The results for Case 1 are shown in **Figures 32–37**. **Figures 32–36** show the real and reactive power output from DG1 through DG4 and also from the reactive power compensator. After islanding occurs, each of the grid-forming and grid-supporting generators works collaboratively to ensure the voltage at each bus in the microgrid as close as possible to the rated value (240 Vrms). Stabilizing on a new equilibrium after switching-in or switching-off of loads takes different durations for different transients. Nevertheless, voltages at each bus can quickly settle down around the rated value after each switching. This can be seen from **Figure 37**, from which one can see that the voltage deviation from its rated value is within ±5% most of the time. It only deviates out of this range in a very short while when the load 3 with a significant amount of reactive power is switched in. The variation of solar energy does not influence the voltage at each bus as its change is relatively slow.

The reactive power compensator as installed at point *P*5 in **Figure 31** compensates reactive power demanded in the direction of current I_{cp} as shown in the figure. At the instant 4.5 s, when load 3 with 15 Kvar reactive power is switched in, the reactive power compensator quickly increases its reactive power output to meet this demand as shown in **Figure 36**. From 5 s, when load 3 is switched off, the reactive power output from the compensator reduces quickly as less reactive power is demanded in the compensation direction.

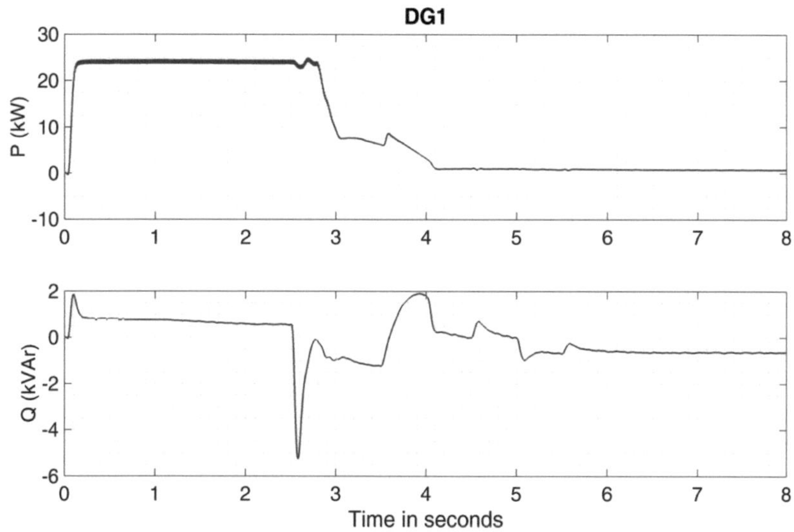

Figure 32. Real and reactive power output from DG1 in **Figure 31** for the first case.

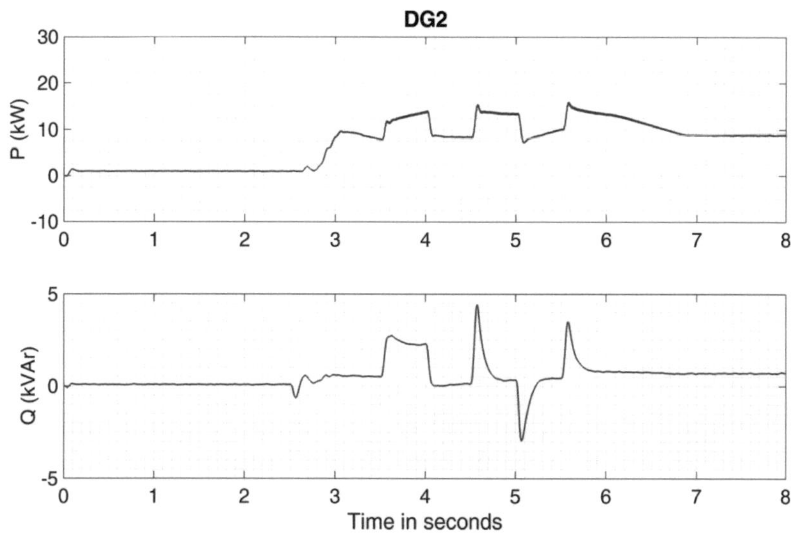

Figure 33. Real and reactive power output from DG2 in **Figure 31** for the first case.

As the dynamic load changes and the change of renewable energy generation happens at the coupling points *P*2, *P*3, and *P*4 in **Figure 31**, the real power output of the grid-forming generator does not vary much after reaching a stable point. Nevertheless, the two grid-supporting generators DG2 and DG4 change their output real power dynamically to stabilize the voltage at each bus around rated values when the loads and renewable energy generation change.

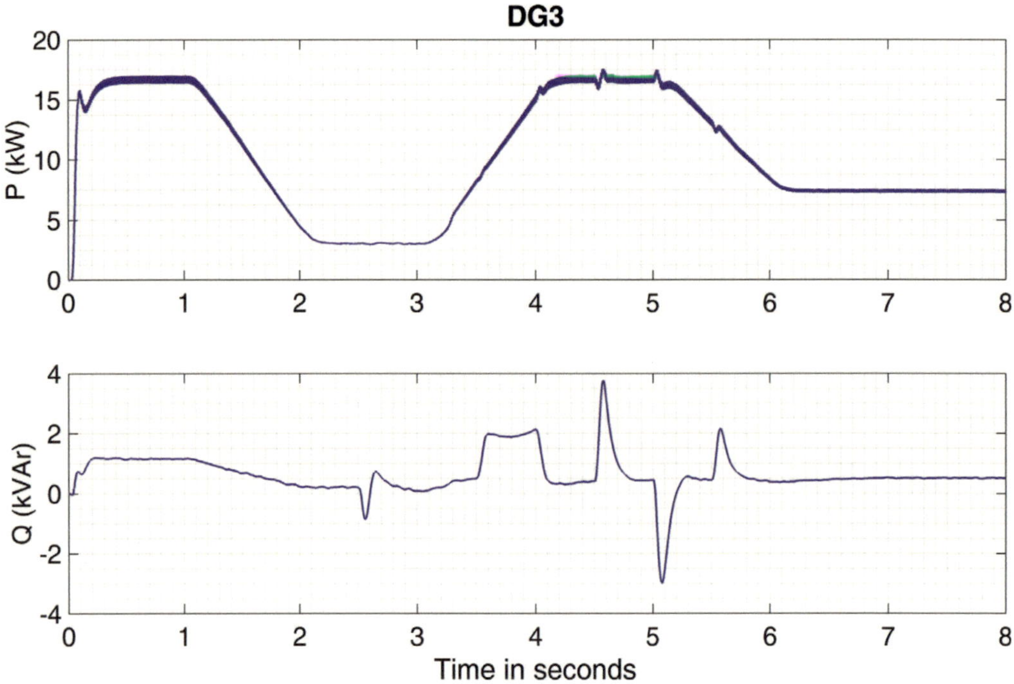

Figure 34. Real and reactive power output from DG3 in **Figure 31** for the first case.

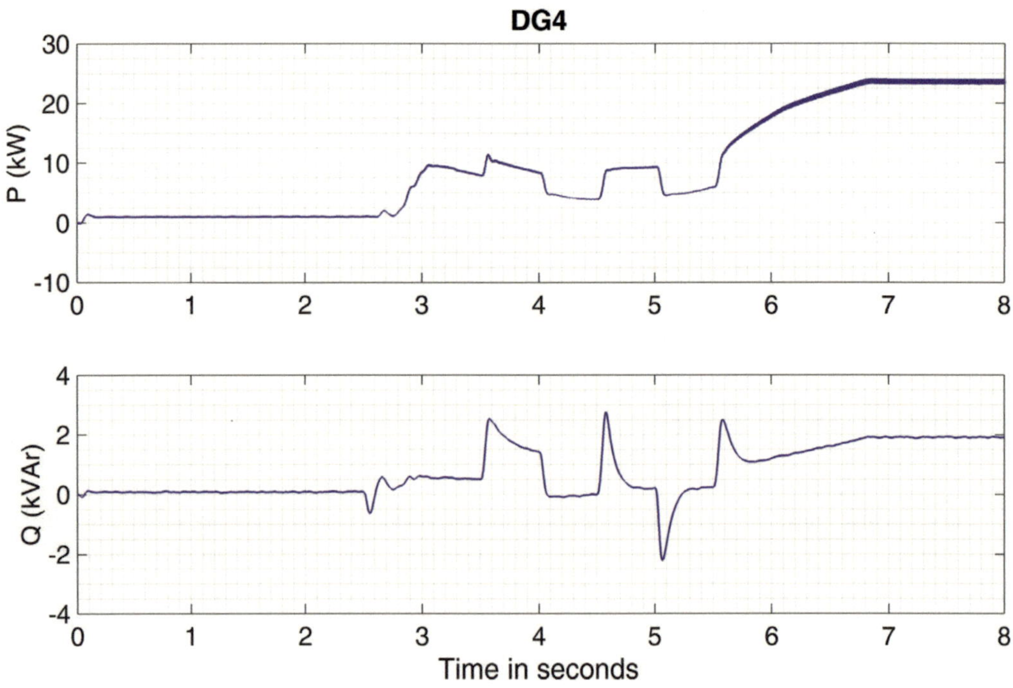

Figure 35. Real and reactive power output from DG4 in **Figure 31** for the first case.

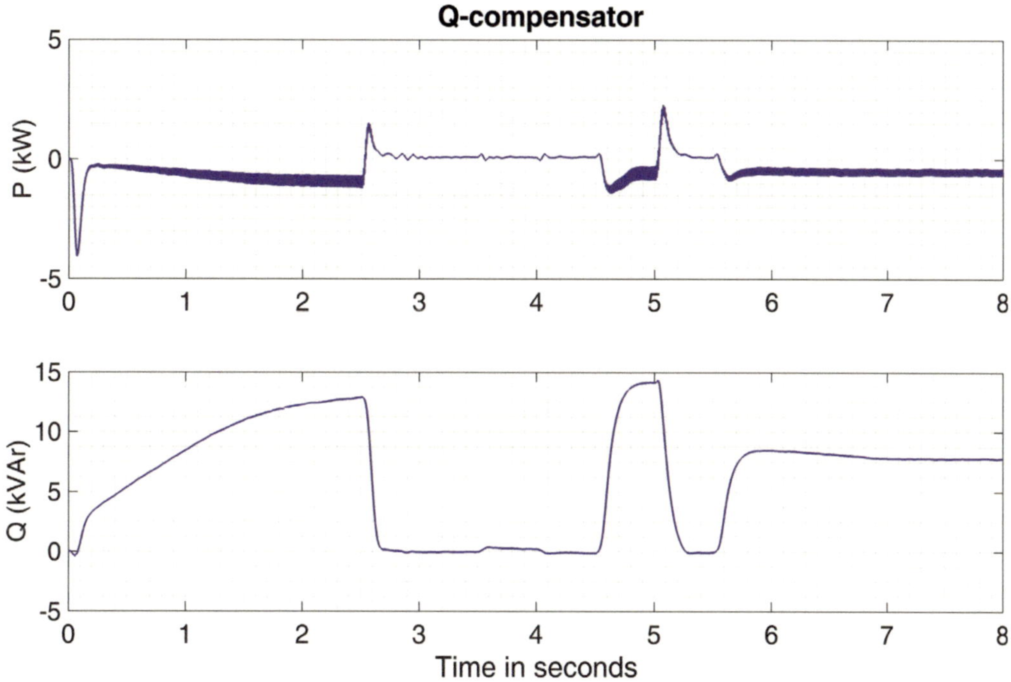

Figure 36. Real and reactive power output from the reactive power compensator in **Figure 31** for the first case.

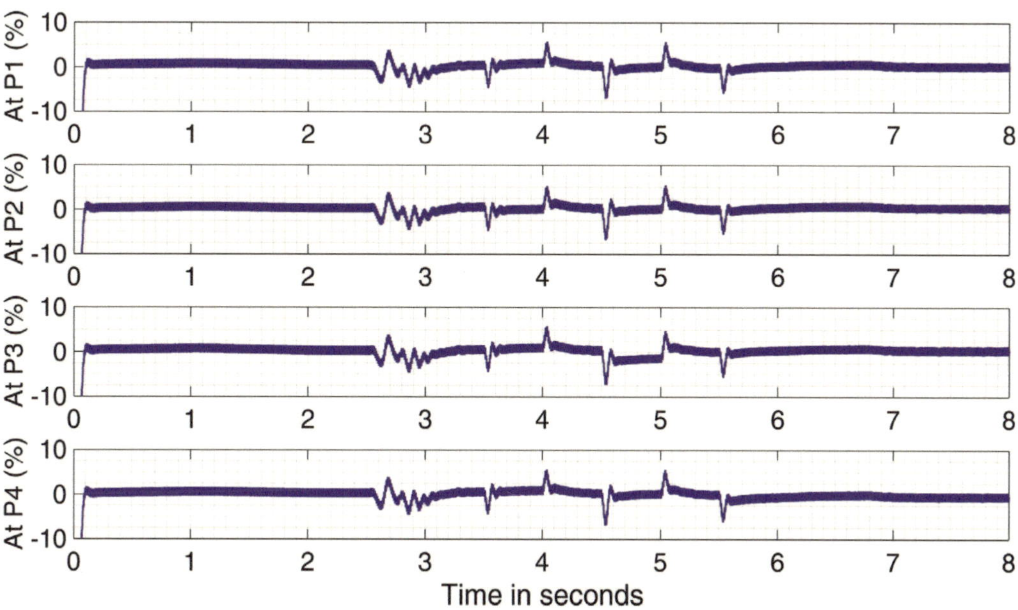

Figure 37. Deviation of voltage at *P*1–*P*4 from the rated value for the first case.

At 4.0 s, load 2 that is connected at *P*2, local to DG2, is turned off. After this moment, DG2 continues to output a certain amount of real power working with the second grid-supporting generator DG4 to stabilize the voltage at each bus close to the rated value. In the following transients, either due to the switch-on and switch-off of loads or due to the change of renewable energy generation, DG2 and DG4 work with DG1 collaboratively to keep the voltage profiles at each bus within acceptable limits ±5%, as shown in **Figure 37**.

7. Results for the second case

The results for Case 2 are shown in **Figures 38–44**. **Figures 39–44** show the real and reactive power output from DG1 through DG4 and also from the reactive power compensator. After islanding occurs at 2.5 s, the grid-forming and grid-supporting generators work collaboratively to ensure the voltage at each bus in the microgrid as close as possible to their rated value. For this case, only DG2 serves as the grid-supporting generator. DG4 acts as a planned battery charger and discharger. The charging and discharging power pattern as shown in the lower waveform in **Figure 38** is determined by the predicted irradiance pattern as shown in the top waveform in **Figure 38**. Compared with the first case, the grid-forming generator DG1 more actively participates in the real power generation as there is only one grid-supporting generator DG2, though it is located quite far away from points *P*3 and *P*4. Due to the collaborative

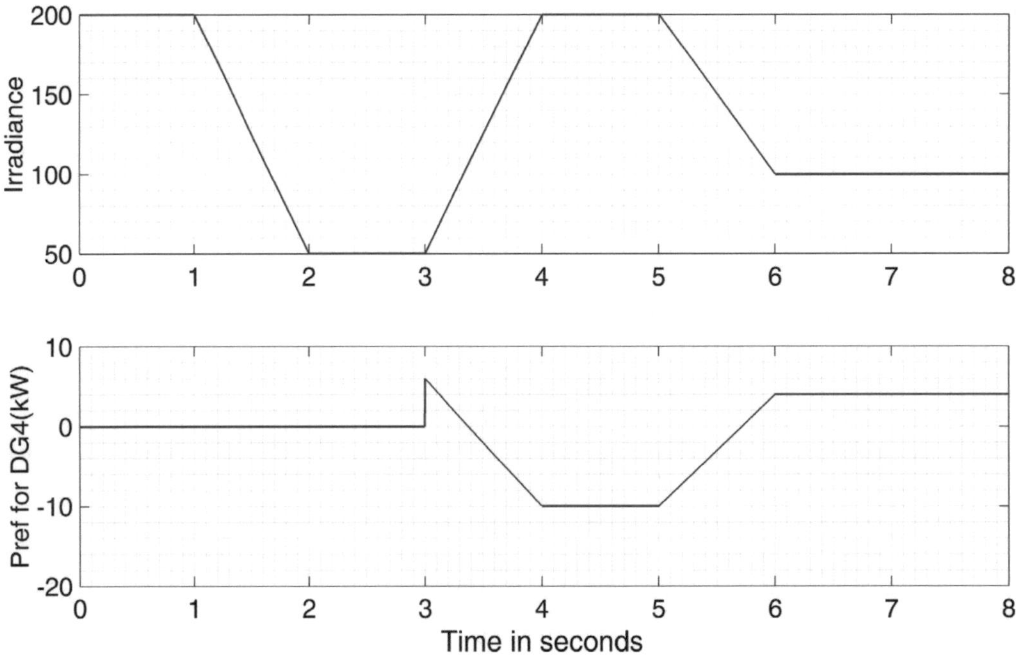

Figure 38. Variation of irradiance and planned charging and discharging for DG4: positive—discharging; negative—charging.

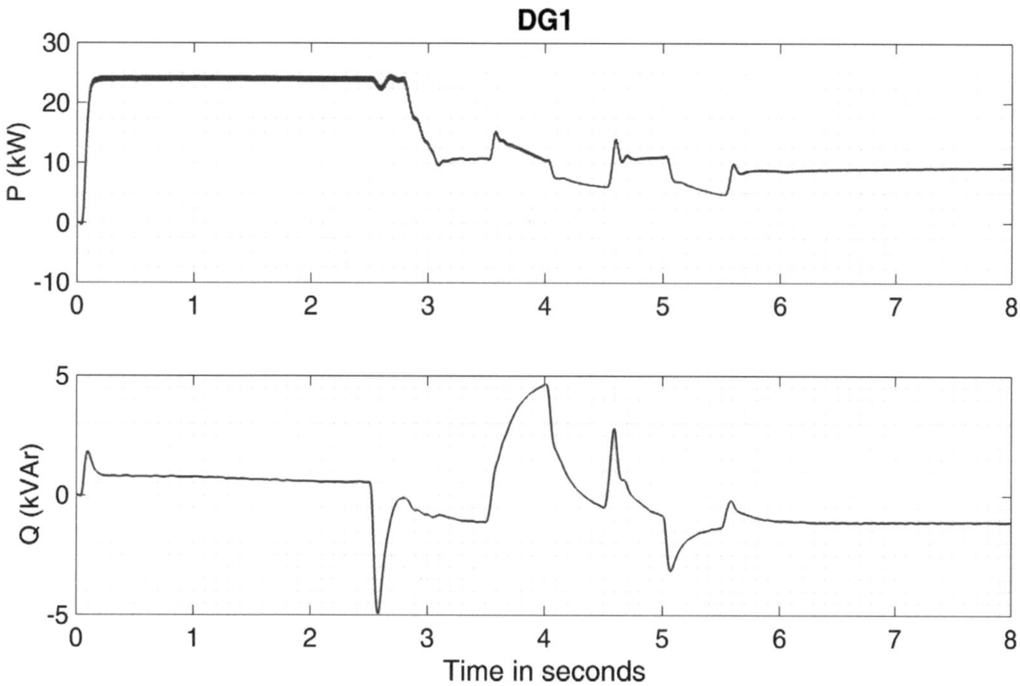

Figure 39. Real and reactive power output from DG1 in **Figure 31** for the second case.

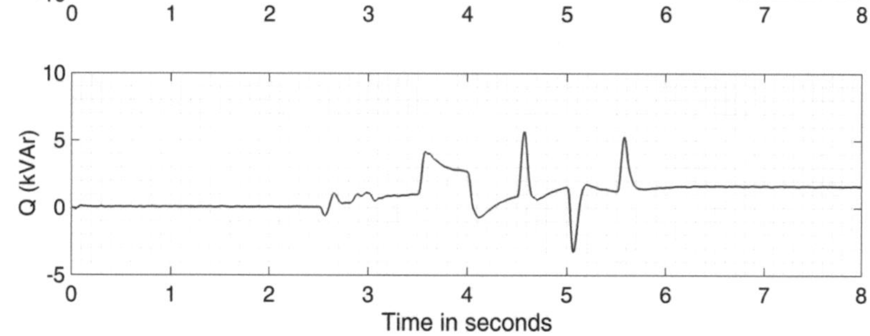

Figure 40. Real and reactive power output from DG2 in **Figure 31** for the second case.

effort of DG1 and DG2, the voltages at points of *P*1, *P*2, *P*3, and *P*4 are well kept around the rated voltage and their deviation from the rated value is within ±5% most time. It only deviates out of this range for a very short duration when load 3 with a significant amount of

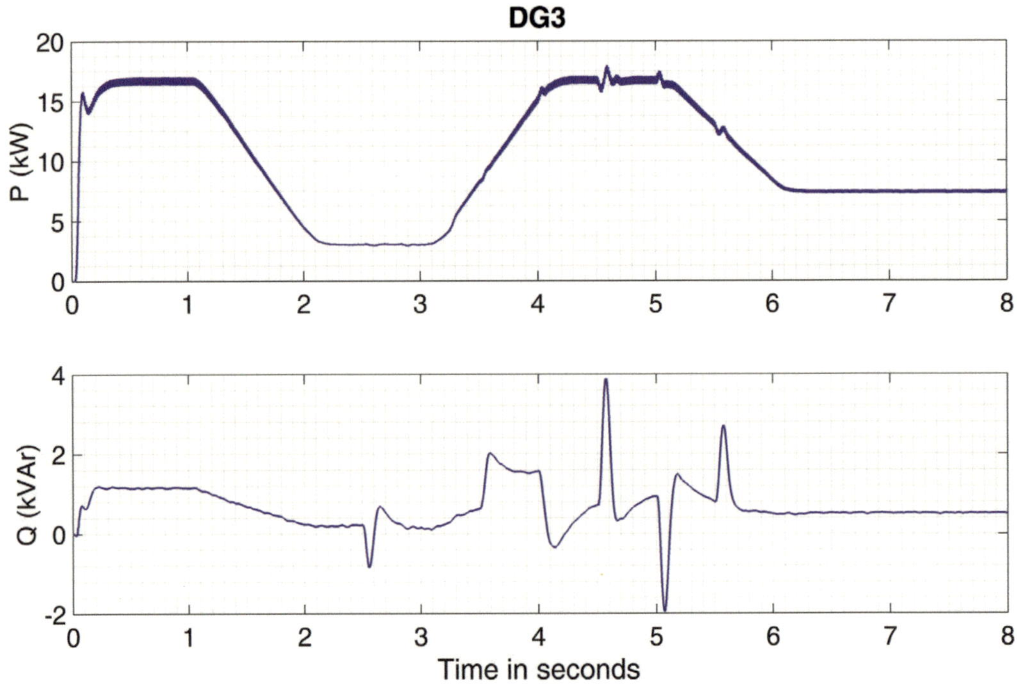

Figure 41. Real and reactive power output from DG3 in **Figure 31** for the second case.

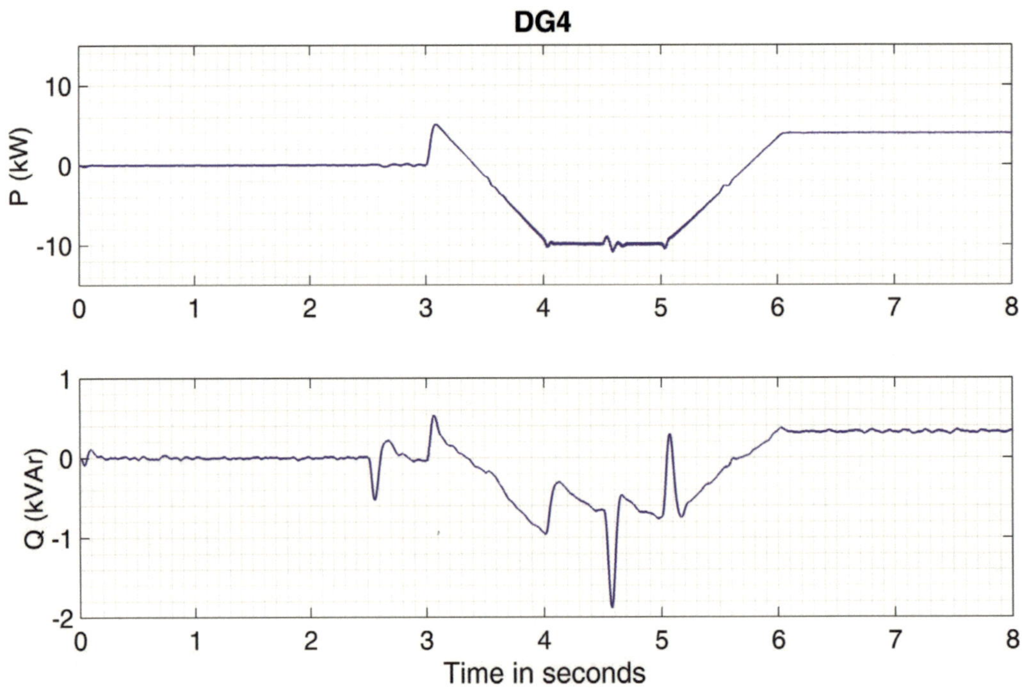

Figure 42. Real and reactive power output from DG4 in **Figure 31** for the second case.

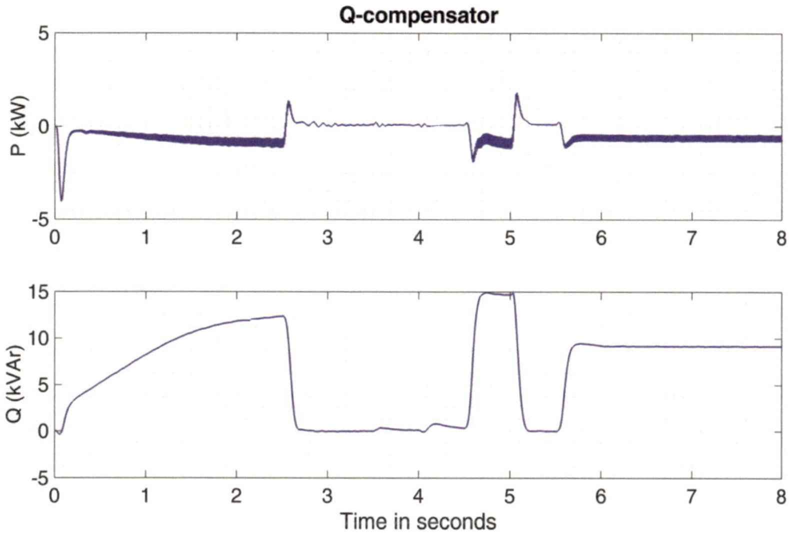

Figure 43. Real and reactive power output from the reactive power compensator in **Figure 31** for the second case.

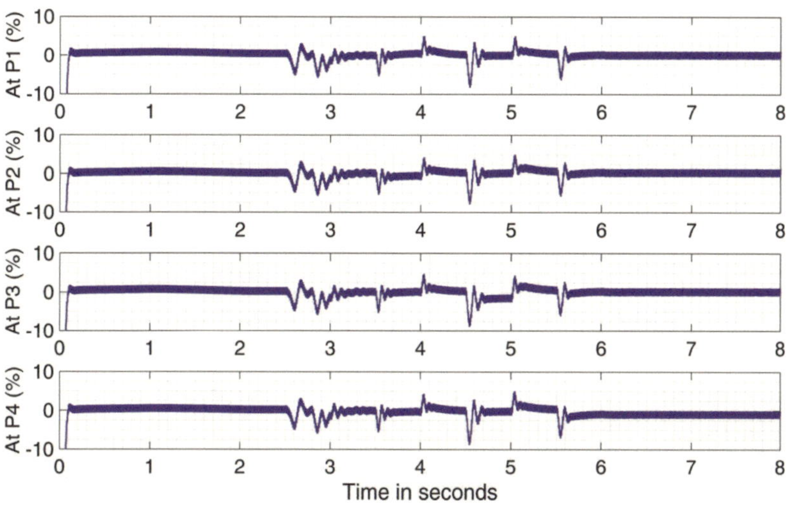

Figure 44. Deviation of voltage at *P*1–*P*4 from the rated value for the second case.

reactive power is switched in. The variation of solar energy does not influence the voltage at each bus.

The reactive power compensator as installed at point *P*5 in **Figure 31** again compensates reactive power demanded in the direction of current I_{cp} as shown in the figure. At 4.5 s, when load 3 with 15 Kvar reactive power is switched in, the reactive power compensator quickly increases its reactive power output to meet this demand as shown in **Figure 43**. From 5 s, when load 3 is switched off, the reactive power output from the compensator reduces quickly as less reactive power is demanded in the direction of compensation.

The planned battery discharging and charging starts at 3.0 s as shown in **Figure 38** after the islanding occurs at 2.5 s. During the charging period between 3.4 s and 5.7 s, both DG1 and DG2 output more real power to balance the system compared with the first case. This is naturally true as in the second case, DG4 acts as an equivalent load during the charging mode.

In summary, in the second case, the grid-forming generator DG1, and grid-supporting generator DG2 can work collaboratively to stabilize the voltages at each bus around the rated voltage.

8. Conclusion

A method has been presented in this chapter for overcoming the drawback of droop control based operation of islanded microgrids. The method generates a real power reference for the grid-forming generator and grid-supporting generators based on their respective terminal voltage variation. The grid-forming generator takes an extra role of being a reactive power sensor when the line impedance is small. Its reactive power is quickly transferred to and from its accompanying instantaneous var compensator. It is found that so long as the reactive power from such a grid-forming generator is small, its real power output can follow its reference accurately and easily and quickly adapts to changes in real power demand in the system to reach a new equilibrium. The effectiveness of the proposed method has been validated in a three-phase microgrid, which contains one grid-forming generator, one grid-supporting generator, and one renewable energy powered generator. It is found that grid-forming and grid-supporting generators are able to output real power to keep the microgrid in a balanced state when the load and renewable energy generations change. The dynamic balance of power demand and power generation is always achieved, and a voltage at each bus is kept at its rated voltage.

Furthermore, a multiplying factor is taken to reduce the response time of reference power generation. With such a factor, the system does not need to change the settings of real power and reactive power references, or alternatively, the system can operate with the temporary loss of a communication link. One more pronounced feature of the proposed control method is that the system operates at a constant frequency or 50 Hz.

This chapter has further developed a method for designing a practical microgrid system: using a single-phase microgrid to replace three-phase microgrid system. Such an approach can be adopted to achieve two main purposes, one being to identify the location where each necessary grid-supporting generator needs be installed, and the other being to size each of them to ensure the voltage profile at each bus is maintained at the rated value under all possible operating conditions, including sudden load change and sudden change in renewable energy generation.

With a sufficient number of grid-supporting generators distributed in the microgrid, it is foreseeable that the voltage at each bus in the microgrid system can operate close to its rated voltage.

http://dx.doi.org/10.5772/intechopen.69401

Author details

Daming Zhang* and John Fletcher

*Address all correspondence to: daming.zhang@unsw.edu.au

School of Electrical Engineering and Telecommunication, University of New South Wales, Sydney, Australia

References

[1] Claudio A, Canizares, Olivares DE, Hatziargyriou ND. Trends in microgrid control. IEEE Transactions on Smart Grid. 2014;**5**(4):1905-1919

[2] Keyhani A, Marwali M. Smart power grids. USA: Springer Press; 2012. p. 594

[3] Lidula NWA, Rajapakse AD. Microgrids research: A review of experimental microgrids and test systems. Renewable and Sustainable Energy Reviews. 2011;**15**:188

[4] Haichuan N, Meng J, Daming Z, John F. Autonomous micro-grid operation by employing weak droop control and PQ control. In: Australasian Universities Power Engineering Conference (AUPEC); Perth, Australia. 2014

[5] Zhang D, Dutta R. Application of partial direct-pole-placement and differential evolution algorithm to optimize controller and LCL filter design for grid-tied inverter. In: Australasian Universities Power Engineering Conference (AUPEC); Perth, Australia. 2014

[6] Lee CT, Chu CC, Cheng PT. A new droop control method for the autonomous operation of distributed energy resource interface converters. IEEE Transactions on Power Electronics. 2013;**28**(4):1980-1992

[7] Coelho EAA, Cortizo PC, Garcia PFD. Small-signal stability for parallel-connected inverters in stand-alone AC supply systems. IEEE Transactions on Industry Applications. 2002;**38**(2):533-542

[8] Zhang DM, Niu HC, Jiang M. Modeling of Islanding Detection by Sensing Jump Change of Harmonic Voltage at PCC by the Combination of a Narrow Band-pass Filter and Wavelet Analysis. Melbourne, Australia: ECCE Asia; 2013

[9] Zhang D. Operation of islanded microgrid at constant frequency with distributed grid supporting generators. IEEE 2nd Annual Southern Power Electronics Conference (SPEC); Auckland, New Zealand. 2016

[10] Teodorescu R, Liserre M, Rodriguez P. Grid Converters for Photovoltaic and Wind Power Systems. Hoboken, New Jersey: John Wiley and Sons, Ltd; 2011

[11] Zhang D, Tseng K. A Universal Controller for Grid-tied DC/AC Converters for Harnessing PV Panel Based Solar Energy and PMSG Based Wind Energy. Sydney, Australia: IEEE PEDS; 2015

[12] Blasko V, Kaura V. A novel control to actively damp resonance in input lc filter of a three-phase voltage source converter. IEEE Transactions on Industry Applications. 1997;**33**(2):542-50

[13] Malinowski M, Bernet S. A simple voltage sensorless active damping scheme for three-phase PWM converters with an LCL filter. IEEE Transactions on Industry Electronics. 2008;**55**(4):1876-1880

[14] Peña-Alzola R, Liserre M, BlaabjergF, Ordonez M, Kerekes T. A self-commissioning notch filter for active damping in a three-phase LCL-filter-based grid-tie converter. IEEE Transactions on Power Electronics. 2014;**29**(12):6754-6761

[15] Wang C, Hashem NM, Shaw SR. Dynamic models and model validation for PEM fuel cells using electrical circuits. IEEE Transactions on Energy Conversion. 2005;**20**(2):442-451

[16] Zhang D. Circuits of voltage source DC/AC converters with LCCL or LCC filter and other modified forms, and operation of microgrid with such circuits. Australia Patent Bureau December 2016. PCT International Patent, filed

Chapter 6

Detection and Operation of Unintentional Islands in the Presence of Distributed Generation Units

Narayanan Krishnan, Shahbaz A. Siddiqui and Manoj Fozdar

Additional information is available at the end of the chapter

http://dx.doi.org/10.5772/intechopen.68859

Abstract

The complexities and challenges for reliable operation of power system have increased due to various types of Distributed Generators (DG) in the Distribution Network (DN) to supply the increasing load demand. It necessitates a comprehensive approach in planning the system towards effective and reliable operation of the system. During the operation of the system, detection of unintentional islanding is critical as non-detection of islanding event could lead to cascaded failure of the system due to active or reactive power imbalance leading to frequency, angle or voltage instability. If undetected, the instability in the islanded part can cascade into the stable part of the system resulting in complete failure of the system. A robust Modified Islanding Detection Technique (MIDT) has been proposed for identifying the islanding event early and accurately in the distribution networks with DGs installed for multiple objectives and is compared with existing passive Islanding Detection Techniques (IDT). A rank-based load shedding scheme is proposed for stable and reliable operation of the identified island, which sheds only the most vulnerable loads in the island for regaining the frequency and voltage stabilities. The proposed MIDT and rank based load shedding schemes were tested on 11kV IEEE 118 Bus Test system.

Keywords: distributed generation, islanding, modified islanding detection technique (MIDT), rank-based load shedding, frequency stability, voltage stability, reliability

1. Introduction

Distributed generator (DG) is an electric power source connected directly to the distribution networks (DNs). The various definitions and technologies of DGs are described in Refs. [1, 2]. The importance of the DG units in the network is more profound with the emphasis on green energy technology and environmental concerns. DG plays a crucial role for the security, reliability, and efficiency of the modern power systems [3]. An exhaustive analysis of methods and models for optimal placement of DG units is given in Ref. [4]. The various techniques proposed for optimal placement of the DG units in the DNs, using various AI techniques for realizing the benefits of DGs like improvement of voltage profile, loss minimization, power transfer capability, uncertainties due to load and fuel prices, planning of dispatchable and non-dispatchable DG units, network security, reliability, etc. are described in Refs. [5–9].

The electrical isolation distribution system from the power system due to abnormal conditions, while being energized by the DG connected to it, is known as islanding [10]. During operation of the system, the detection of islanding event is critical in effective operation of the system. A comprehensive survey of islanding protection with renewable DG is reported in Ref. [11]. The islanding detection techniques are broadly classified into active and passive techniques [12]. Active islanding detection techniques introduce small perturbations in the system for detecting the islanding event. They have smaller nondetection zones (NDZ) but have large time of detection. As a result of the perturbations being introduced, the performance of the system degrades. Active techniques have been reported to work satisfactorily for single DG unit only and their response at multiple DG units is not guaranteed. The passive scheme utilizes local measurements of voltage and current signals. Many techniques ranging from usage of voltage variations and its derivatives, frequency variations and its derivatives, intelligent devices, etc. have been proposed for islanding detection in the presence of DGs in the system. The algorithms of passive scheme include under/over frequency and voltage, rate-of-change of frequency and power, vector surge, and harmonic distortion indices [13–18]. The passive methods have small time of detection and do not degrade the system power quality. However, the passive methods suffer from large NDZ.

The power mismatch may lead to collapse of the system, in the islanded part of the system, if proper corrective actions are not initiated timely. Load shedding is a commonly used emergency control action during mismatch of power in the island. Since the control action cannot be based on under-voltage, under-frequency load-shedding method is the commonly used method [19–22]. Most of the existing schemes need to be reinvestigated in the presence of DG units as load shedding depends on economic reasons along with technical reasons. In some cases higher amount of load shedding than required has been done since the load shedding has been performed on discrete basis in distribution systems to regain the stability.

The proposed modified islanding detection technique (MIDT) is based on utilizing the advantages of the existing passive methods of islanding detection with appropriate threshold values being identified for early and effective detection of the islanding event. In the proposed MIDT, in addition to the existing parameters used in the passive islanding detection techniques (IDTs), a new parameter is utilized for effective identification of the islanding event. The proposed parameter is based on the voltage sensitivity to the active power (voltage-active power sensitivity parameter (Δ_{VP})). A rank-based load-shedding scheme is proposed in the island to alleviate the power mismatch and to identify the island. The effectiveness of the proposed load-shedding scheme has been determined by a quantitative reliability analysis from the available failure data and reliability indices of the system.

http://dx.doi.org/10.5772/intechopen.68859

2. Proposed modified islanding detection technique (MIDT)

The complete blackout of the system may occur due to faults upstream or failure of the grid. The major requirement of installation of DG units in the distribution system is the ability to operate the system on islanded mode and also reducing the amount of load shedding needed under contingencies to achieve stable operation of the system. However, active or reactive power imbalance leading to frequency, angle, or voltage instability may take place in the unintentional islands or improper partition of the system. This further leads to instability in the interconnected network due to unwanted tripping of interconnected tie-lines. Hence, the unintentional islanding event has to be detected early and accurately to assist the system operator for ensuring appropriate control actions being initiated and avoid a blackout of the islanded region of the system.

The passive techniques have inherent disadvantage of large NDZ and requires accurate setting of threshold values of different parameters. Unwanted tripping occurs for low threshold values and higher threshold values may result in failure of detection of islanding event. The less cost of implementation of passive islanding detection techniques (IDTs) along with early detection of islanding makes it a most preferred method for IDT. Since the DG can supply only a small amount of load, the islanding has to be detected early and accurately. If undetected, the instability in the islanded part could cascade into the stable part of the system resulting in complete failure of the system. In the presence of DG units, the accurate identification of islanding becomes difficult due to complexity in monitoring the system parameters. Hence, the existing methods need to be reinvestigated for early and accurate detection of the islanding event. For accurate and early detection of islanding, the existing passive IDTs are modified by utilizing more robust parameters along with the existing parameters used in the passive IDTs. The existing parameters are used as alarm signals for the impending islanding event and the system enters into the alert state. The following parameters are utilized in the existing passive detection techniques:

The deviation in voltage at each bus is measured for every time instant as:

$$\text{Variation of voltage} = dV(\text{Volts}) \tag{1}$$

To avoid any errors in measurement, the voltage parameter is computed by averaging over five continuous cycles and is expressed in (V/s).

$$\text{Voltage parameter}(\delta_{vt}) = \left|\frac{dV}{dt}\right| < \sigma \text{ for 5 cycles} \tag{2}$$

σ is the predefined threshold value for the parameter and is taken as 160 V/s [17]. The frequency at each bus is measured and the deviation in frequency is calculated for every time instant as:

$$\text{Variation off requency} = df(\text{Hertz}) \tag{3}$$

The rate of change of frequency (ROCOF) is computed at every bus for each cycle and is expressed as frequency parameter in (hertz/seconds)

$$\text{Frequency parameter } (\delta_{ft}) = \left|\frac{df}{dt}\right| < \in \tag{4}$$

The ROCOF is used for fast islanding detection as the rate of change of frequency is a sensitive measurement. The ROCOF is calculated over few cycles and is set between 0.1 and 1.2 Hz/s for a 60 Hz system. The ROCOF relays may become ineffective if the power imbalance in the islanded system is <15% and is set as 2.18 Hz/s for 60 Hz system [14].

The variation of net active power is monitored at each bus for every cycle. The variation will be less in DG buses since the power available from DG units is fixed. The buses farther away from the DG bus will have more variations in active power for load variations

$$\text{Variation of active power} = dP(\text{MW}) \tag{5}$$

The ROCOP is calculated at each bus for every time instant in (megawatt/seconds)

$$\text{Rate of change of active power } (\delta_{Pt}) = \left|\frac{dP}{dt}\right| < \Upsilon \tag{6}$$

Υ is the predefined threshold limit and is fixed as 0.64 MW/s [14]. In this work, the parameters computed by Eqs. (3)–(5) are monitored and checked for threshold violation and the operator is alerted for an impending islanding event. Two additional proposed parameters are also monitored and computed for violations in the alert state. The voltage-active power sensitivity parameter (Δ_{VP}) is calculated by dividing Eq. (2) by Eq. (6) and the variation of voltage to real power parameter at a bus and is calculated in (volts/megawatt)

$$\text{Votlage active power sensitivity } (\Delta_{vp}) = \left|\frac{dV}{dP}\right| < \mu \tag{7}$$

μ is the threshold value of the proposed voltage-active power sensitivity parameter and set at 10%. During simulations, it was observed that for "μ" values less than 10%, false triggering of islanding event set in and for greater threshold values some islanding events were not detected. If either the voltage parameter (δ_{vt}) or frequency parameter (δ_{ft}) or ROCOP (δ_{Pt}) violates the predefined threshold limit, a case of islanding is suspected and the system operator is alerted for an impending islanding event. Mathematically it can be expressed as*****:

$$\text{Islanding suspicion} = \left|\frac{dV}{dt}\right| > \sigma(or)\left|\frac{df}{dt}\right| > \in(or)\left|\frac{dP}{dt}\right| > \Upsilon \tag{8}$$

Subsequently, if the voltage-active power sensitivity parameter (Δ_{vp}) also violates the threshold limit at any bus when the system is in alert state, it is classified as an islanding event and the bus at which these proposed parameters initially violate the limit is identified for islanding.

Mathematically it can be expressed as

$$\text{Islanding detection} = \text{Islanding suspicion and } \left|\frac{dV}{dP}\right| > \mu \tag{9}$$

A flowchart of the proposed MIDT is shown in **Figure 1**.

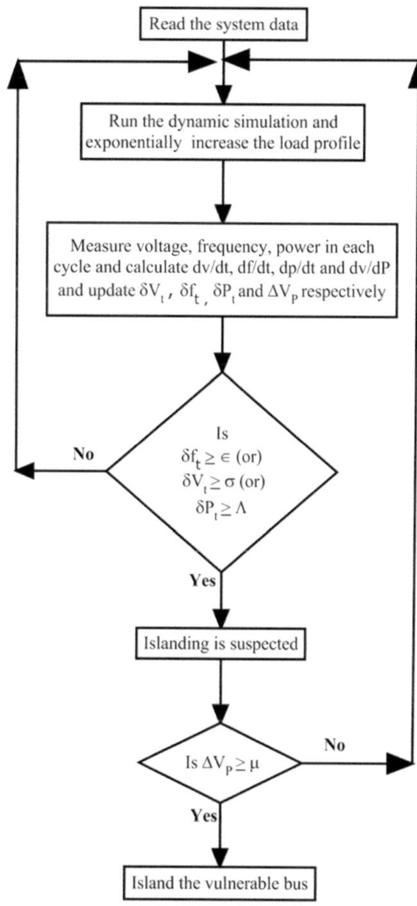

Figure 1. Flowchart for the proposed MIDT.

3. Rank-based load-shedding scheme with quantitative reliability analysis

The power mismatch in the islands makes the operation of islands a challenging task. The load shedding is an extensively used countermeasure for the stable operation of the island under such conditions. The quantitative reliability analysis gives a comprehensive idea to evaluate the merits of investing in various reinforcements for the system planner. Since it is based on the number of customers being affected, standard reliability indices help in measuring the effect of the load. A measure of number of customers affected by load shedding can also be obtained by the computed reliability indices. A rank-based load-shedding scheme is proposed for maintaining the power balance in the detected island when the power demand exceeds the power output of the DG sources in the island. The ROCOF and rate-of-change of voltage is utilized for ranking of the loads to ensure that the most vulnerable bus with overload for tripping is ranked higher. This ranking ensures that the DG bus does not participate in the load shedding as the frequency variation is not significant. The load shedding process continues till the frequency and voltages are brought back within threshold limits in the island. The total amount of load shed from each bus can be calculated as:

$$\text{Load shed} = DPC_{\mathrm{i}} \times P_{\mathrm{load,i}} \tag{10}$$

where $P_{\mathrm{load,i}}$ is the load at bus "i," DPC is the definitive participation coefficient for a particular bus and is calculated as:

$$\mathrm{DPC} = K_{\mathrm{i}} \times \zeta_{\mathrm{f}} \times \zeta_{\mathrm{v}} \tag{11}$$

where ζ_{f} and ζ_{v} are the coefficient of frequency and voltage components, respectively, of the buses and are calculated as follows:

$$\zeta_{\mathrm{f}} = \frac{f_{\mathrm{i,t}}}{f_{\mathrm{init,0}}} \tag{12}$$

$$\zeta_{\mathrm{v}} = \frac{V_{\mathrm{i,t}}}{V_{\mathrm{init,0}}} \tag{13}$$

where $f_{\mathrm{init,0}}$ is the frequency of the islanding bus when the islanding is detected $f_{\mathrm{i,t}}$ is the frequency at bus "i" when the proposed rank-based load-shedding process is initiated, $V_{\mathrm{i,t}}$ is the voltage at bus "i" when the proposed rank-based load-shedding process is initiated, and $V_{\mathrm{init,0}}$ is the voltage of the islanding bus when the islanding is detected. The value of K_{i} for the buses is given as:

$$K_{\mathrm{i}} = \begin{cases} 0, \text{ if DG is available} \\ 1, \text{ if DG is not available} \end{cases} \tag{14}$$

By the proposed method of load shedding, the amount of load being shed will be less than the total power demand in the island since loads of the most vulnerable buses are shed

$$\sum_{i=1}^{c} P_{\mathrm{Load\ shed,i}} < P_{\mathrm{L,island}} \tag{15}$$

considering "c" is the total number of buses in the island, $P_{\mathrm{Loadshed,i}}$ is the amount of load needed to be shed in the island to regain the stability, and $P_{\mathrm{L,island}}$ is the total demand in the island. As load shedding in the distribution system is performed in discrete values, the value of

K_i is considered as either 0 or 1. At each step of the load-shedding process, the frequency, voltage of the islanded bus, and the power flow limits of the identified island are monitored during each step of the load-shedding process and mathematically expressed as:

$$f_{min} \leq f_i \leq f_{max} \tag{16}$$

$$V_{min} \leq V_i \leq V_{max} \tag{17}$$

$$P_{min} \leq P_i \leq P_{max} \tag{18}$$

The values of f_{min} and f_{max} are taken to be 57.5 and 60.5 Hz. The values of V_{min} and V_{max} are considered as 0.95 and 1.05 pu. The quantitative reliability of the islanded system is computed before and after the load shedding to measure the effectiveness of the proposed load-shedding scheme. The reliability analysis is performed through standard reliability indices and from the failure rate and repair time of the lines in the system. From the customer failure statistics [23], the standard quantitative reliability indices system average interruption duration index (SAIDI), system average interruption frequency index (SAIFI), customer average interruption duration index (CAIDI), energy not supplied (ENS), and average ENS (AENS) are computed. A flowchart of the proposed rank-based load-shedding scheme is shown in **Figure 2**.

SAIDI is system average interruption duration index and expressed in terms of hours per customer

$$\text{SAIDI} = \frac{\sum \text{of Customer interuption durations}}{\text{Total number of customers}} = \frac{\sum U_i N_i}{\sum N_i} \tag{19}$$

where U_i is the annual outage time and N_i is the number of customers of load at point i.

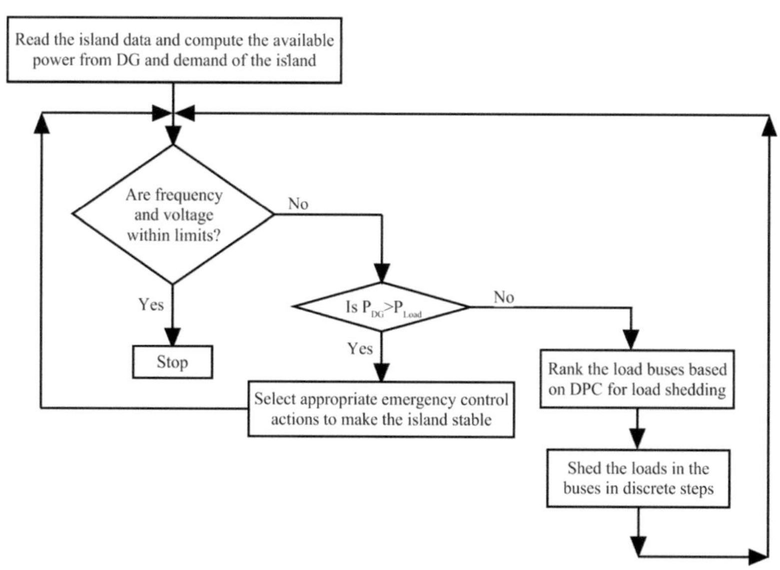

Figure 2. Flowchart of the proposed rank-based load-shedding scheme.

SAIFI is system average interruption frequency index and expressed as interruptions per customer

$$\text{SAIFI} = \frac{\sum \text{of Customer Interuption durations}}{\text{Total number of customers}} = \frac{\sum \lambda_i N_i}{\sum N_i} \tag{20}$$

where λ_i is the failure rate and N_i is the number of customers of load at point i.

CAIDI is customer average interruption duration index and expressed as hours per customer interruption

$$\text{AIDI} = \frac{\sum \text{of Customer interuption durations}}{\text{Total number of customer Interruptions}} = \frac{\sum U_i N_i}{\sum \lambda_i N_i} \tag{21}$$

ENS is energy not supplied index and is measured as kilowatt hour

$$\text{ENS} = \text{Energy not supplied} = \sum L_{a(i)} U_i \tag{22}$$

$L_{a(i)}$ is the average load connected to load at point i

AENS is the average ENS index and is expressed in kilowatt hour per customer

$$\text{AENS} = \frac{\text{Total energy not supplied}}{\text{Total number of customers served}} = \frac{L_{a(i)} U_i}{N_i} \tag{23}$$

4. Results and discussion

All simulations have been performed in MATLAB R2010a [24] and PSAT [25] Intel Core i5, 2.5 GHz, 4 GB random access memory machine. The proposed methods were tested for different loading conditions (base load and 140% of base load) on standard 11 kV IEEE 118 Bus Test system [5]. The DG units have been installed in the system by QOTLBO as mentioned in Ref. [5] and the loads are increased exponentially from the base load. The proposed MIDT is compared with existing passive IDT using the rate-of-change of voltage as given in Eq. (2) [3], ROCOF as given in Eq. (4) [15], and ROCOP as given in Eq. (6) [15]. In the proposed method, the rate-of-change of voltage with active power known as voltage-active power sensitivity is calculated along with voltage, frequency, and active power variations at each bus for every instant of time. The results of base case and extreme loading condition (i.e., 140% of base case) are considered for discussion in this work. The results obtained are shown in **Table 1**.

From this table, it can be seen that the proposed MIDT is effective in identifying the islanding event early and accurately. The frequency parameter initially triggers the MIDT. In the existing passive methods, since only one parameter is used for the identification of islanding events, the bus nearer to the grid is the vulnerable bus for islanding erroneously. In the passive method-II, at least three averaged values are needed for the identification of the islanding event as it uses averaging of voltage measurements for islanding detection, thus making a delayed identification

http://dx.doi.org/10.5772/intechopen.68859

Loading level	Islanding detection											
	Passive method–I			Passive method–II			Passive method–III			Proposed method		
	Time of detection	Islanded bus	No. of buses in island	Time of detection	Islanded bus	No. of buses in island	Time of detection	Islanded bus	No. of buses in island	Time of detection	Islanded bus	No. of buses in island
Base load	1.052	4	38	1.083	23	Wrong trigger	1.0187	29	37	1.0187	31	18
140% of Base load	1.052	4	38	1.083	24	Wrong trigger	1.0187	29	37	1.0187	31	18

Table 1. Islanding analysis of 118 bus system with DG units.

of the islanding event. The proposed MIDT works satisfactorily when the minimum base load in the system is increased to 140% of the base load. The islanding event is simulated by increasing loads from $t = 1$ s and the islanding event is identified at time $t = 1.05$ s. As the effect of DG penetration cannot be effective on buses away from the DG buses, the islanded bus moves away from the grid as the minimum load in the system is increased. The MIDT as the variations of voltage to active power is considered for each bus overcoming the problem of NDZ prominent in the frequency parameter. The islanded bus is identified accurately by the proposed MIDT and is not triggered due to switching of capacitors or switching events. This occurs as the variations of all the parameters are considered in every bus. Moreover, as the cross-coupling of parameters, namely, the voltage-active power sensitivity parameter does not vary much during nonislanding events. The identification of islanding event is effective as the cross-coupling of parameters exhibit large variations only under actual islanding conditions. A comparison of the formation of the island by different methods is shown in **Figure 3**.

In the identified islands the proposed rank-based load shedding is initiated. The load shedding is initiated at $t = 1.084$ s, i.e., after two cycles (in 60 Hz) system after the islanding is detected. The effect of the DPC parameter on the load shedding is measured when load shedding is performed without the DPC parameter. The load shedding is performed by shedding the loads from the buses farther away from the DG bus when the DPC parameter is not considered leading to more amount of load shedding to regain the voltage and frequency stability. The results of the proposed load-shedding scheme in the islands are shown in **Table 2**.

The effect of the load-shedding scheme is measured by quantitative reliability analysis, based on the failure data of the line and the repair time of the lines. The results of the reliability analysis are shown in **Table 3**. It can be seen from the indices that the number of customers affected by the proposed load-shedding scheme is less by considering the DPC parameter for load shedding.

The variations of frequency and voltage at the islanded bus, before and after the proposed load shedding, are shown in **Figures 4** and **5**.

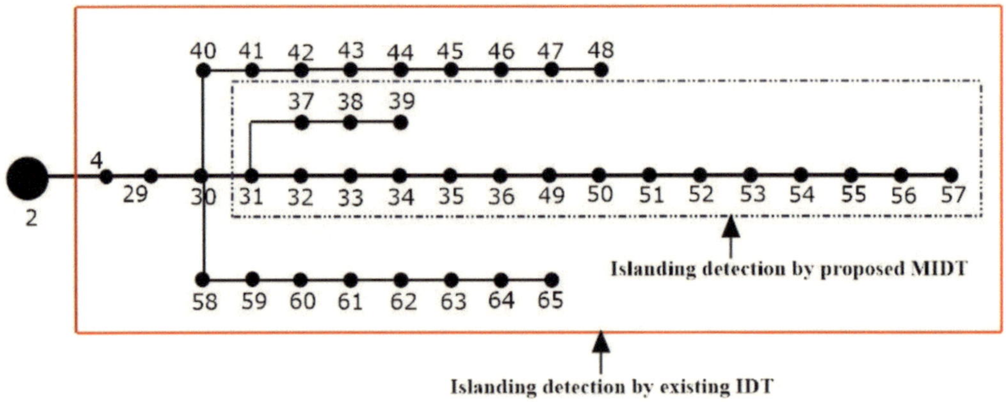

Figure 3. Identification of islands by proposed MIDT and existing IDT.

Load shedding technique	Islanded bus	Number of buses in the island	Power available, MW	Load, MW	Actual load shed, MW	Amount of load shed, %
With DPC	31	18	4.8302	7.612		
Without DPC	31	18	4.8302	7.612		

Table 2. Load shedding in island at base load.

Cases	Before load shedding					After load shedding				
	SAIDI	SAIFI	CAIDI	ENS	AENS	SAIDI	SAIFI	CAIDI	ENS	AENS
With DPC	0.3463	0.4090	0.8466	1.38	0.0511	0.3403	0.3993	0.7983	0.72	0.036
Without DPC	0.3463	0.4090	0.8466	1.38	0.0511	0.3419	0.4035	0.8428	0.9	0.05

Table 3. Reliability analysis of the island.

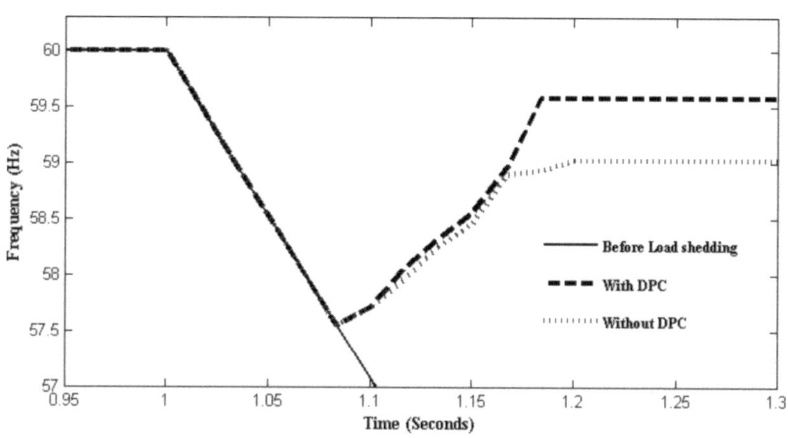

Figure 4. Comparison of frequency at islanded bus before and after load shedding.

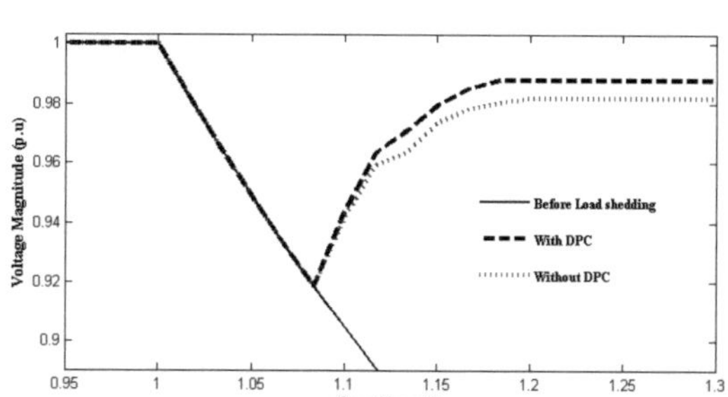

Figure 5. Comparison of voltage at islanded bus before and after load shedding.

As seen from the figures, the regain of frequency and voltage in the islanded bus is faster and effective by the proposed rank-based load-shedding method which considers the DPC parameter as the most vulnerable loads are shed from the islanded part of the system.

5. Conclusion

For identifying the islanding event in the presence of DG units, an additional voltage-active power sensitivity parameter which measures the variation of voltage with active power at a bus is proposed along with the existing parameters utilized in the passive IDTs. The additional voltage-active power parameter ensures against the false triggering of islanding event and also identifies the vulnerable bus accurately. Due to cross-coupling of voltage-real power and frequency-reactive power, the proposed parameter is more sensitive to sudden large load variations or disturbances and does not trigger the islanding event due to sudden switching of loads or capacitor switching events.

The proposed rank-based load shedding identifies the vulnerable buses for load shedding by using a dynamic DPC parameter. The amount of load shed by the proposed method is lesser compared to the conventional load shedding strategy for regaining the frequency and voltage stability in the island. The availability of DG unit, frequency, and voltage variations in a bus is taken into account in the proposed DPC parameter for identifying a bus for load shedding. The effectiveness of the proposed load shedding is analyzed by a quantitative reliability analysis in the islands before and after the load shedding. The reliability indices are improved as the number of customers affected by the load shedding is less. This is because only the vulnerable loads are shed when DPC parameter is considered. The proposed DPC parameter is effective in improving the reliability of the island as the reliability indices also depend on the number of customers being affected. Further investigations are needed for proper control actions when the DG power available is more than the demand in the island.

Author details

Narayanan Krishnan[1*], Shahbaz A. Siddiqui[2] and Manoj Fozdar[3]

*Address all correspondence to: narayanan.mnit@gmail.com

1 Department of Electrical & Electronics Engineering, SEEE, Sastra University, Thanjavur, Tamil Nadu, India

2 Department of Electrical Engineering, Manipal University, Jaipur, Rajasthan, India

3 Department of Electrical Engineering, MNIT, Jaipur, Rajasthan, India

References

[1] Zeineldin HH, Kirtley JL, Jr. A simple technique for islanding detection with negligible nondetection zone. IEEE Transactions on Power Delivery. 2009;**24**(1):779-786. DOI: 10.1109/PES.2009.5275939

[2] Laaksonen H. Advanced islanding detection functionality for future electricity distribution networks. IEEE Transactions on Power Delivery. 2013;**28**(4):2056-2064. DOI: 10.1109/TPWRD.2013.2271317

[3] Pourbabak H, Kazemi A. A new technique for islanding detection using voltage phase angle of inverter-based DGs. International Journal of Electrical Power and Energy Systems. 2014;**57**:198-205. DOI: http://dx.doi.org/10.1016/j.ijepes.2013.12.008

[4] Georgilakis PS, Hatziargyriou ND. Optimal distributed generation placement in power distribution networks: Models, methods, and future research. IEEE Transactions on Power Systems. 2013;**28**(3):3420-3428. DOI: 10.1109/TPWRS.2012.2237043

[5] Sultana S, Roy PK. Multi-objective quasi-oppositional teaching learning based optimization for optimal location of distributed generator in radial distribution systems. International Journal of Electrical Power & Energy Systems. 2014;**63**(2014):534-545. DOI: http://dx.doi.org/10.1016/j.ijepes.2014.06.031

[6] Kansal S, Kumar S, Tyagi B. Optimal placement of different type of DG sources in distribution networks. International Journal of Electrical Power and Energy Systems. 2013;**53**(2013):752-760. DOI: http://dx.doi.org/10.1016/j.ijepes.2013.05.040

[7] Rao RS, Ravindra K, Satish K, Narasimham SVL. Power loss minimization in distribution system using network reconfiguration in the presence of distributed generation. IEEE Transactions on Power Systems. 2013;**28**(1):317-325. DOI: 10.1109/TPWRS.2012.2197227

[8] Aman MM, Jasmon GB, Bakar AHA, Mokhlis H. A new approach for optimum DG placement and sizing based on voltage stability maximization and minimization of power losses. Energy Conversion and Management. 2013;**70**:202-210. DOI: http://dx.doi.org/10.1016/j.enconman.2013.02.015

[9] Othman MM, El-Khattam W, Hegazy YG, Abdelaziz AY. Optimal placement and sizing of distributed generators in unbalanced distribution systems using supervised Big Bang-Big Crunch Method. IEEE Transactions on Power Systems. 2015;**30**(2):911-919. DOI: 10.1109/TPWRS.2014.2331364

[10] IEEE Application Guide for IEEE Std. 1547. IEEE Standard for Interconnecting [Internet]. 2008. Available from: www.ieee.org

[11] Chowdhury SP, Chowdhury S, Crossley PA. Islanding protection of active distribution networks with renewable distributed generators: A comprehensive survey. Electric Power Systems Research. 2009;**79**(6):984-992. DOI: http://dx.doi.org/10.1016/j.epsr.2008.12.012

[12] Shayeghi H, Sobhani B. Zero NDZ assessment for anti-islanding protection using wavelet analysis and neuro-fuzzy system in inverter based distributed generation. Energy

Conversion and Management. 2014;**79**:616-625. DOI: http://dx.doi.org/10.1016/j.enconman.2013.12.062

[13] Narayanan K, Siddiqui SA, Fozdar M. Identification and reduction of impact of islanding using hybrid method with distributed generation. In: IEEE Proceedings of PES General Meeting; 27-31 July 2015; Denver, Colorado. Denver, CO: IEEE; 2015. pp. 1-5. DOI: 10.1109/PESGM.2015

[14] Samantaray SR, El-Arroudi K, Joos G, Kamwa I. A fuzzy rule-based approach for islanding detection in distributed generation. IEEE Transactions on Power Delivery. 2010;**25**(3):1427-1433. DOI: 10.1109/TPWRD.2010.2042625

[15] Samui A, Samantraray SR. New active islanding detection scheme for constant power and constant current controlled inverter-based distributed generation. IET Generation, Transmission & Distribution. 2013;**7**(7):779-789. DOI: 10.1049/iet-gtd.2012.0607

[16] Narayanan K, Siddiqui SA, Fozdar M. An improved hybrid method to reduce the effect of islanding in the presence of optimally located DGs. In: IEEE INDICON; 17-20 December 2015; New Delhi. New Delhi: IEEE; 2015. pp. 1-6. DOI: 10.1109/INDICON.2015.7443690

[17] Garmrudi M, Nafisi H, Fereidouni A, Hashemi H. A novel hybrid islanding detection technique using rate of voltage change and capacitor tap switching. Electric Power Components and Systems. 2012;**40**(10):1149-1159. DOI: 10.1080/15325008.2012.682249

[18] Samantaray SR, Babu BC, Dash PK. Probabilistic neural network based islanding detection in distributed generation. Electric Power Components and Systems. 2011;**39**(3):191-203. DOI: 10.1080/15325008.2010.526986

[19] Rudez U, Mihalic R. Monitoring the first frequency derivative to improve adaptive underfrequency load-shedding schemes. IEEE Transactions on Power Systems. 2011;**26**(2):839-846. DOI: 10.1109/TPWRS.2010.2059715

[20] Rudez U, Mihalic R. Predictive underfrequency load shedding scheme for islanded power systems with renewable generation. Electric Power Systems Research. 2015;**126**:21-28. DOI: http://dx.doi.org/10.1016/j.epsr.2015.04.017

[21] Laghari JA, Mokhlis H, Karimi M, Bakar AHA, Mohamad H. A new under-frequency load shedding technique based on combination of fixed and random priority of loads for smart grid applications. IEEE Transactions on Power Systems. 2015;**30**(5):2507-2515. DOI: 10.1109/TPWRS.2014.2360520

[22] Ketabi A, Fini MH. An underfrequency load shedding scheme for hybrid and multiarea power systems. IEEE Transactions on Smart Grid. 2014;**6**(1):82-91. DOI: 10.1109/TSG.2014.2349999

[23] Billinton R, Allan RN. Reliability Evaluation of Power Systems. 2nd ed. New York: Springer; 1994

[24] The Math Works, Inc., MATLAB programming; 2008

[25] Milano F. An open source power system analysis toolbox. IEEE Transactions on Power Systems. 2005;**20**(3):1199-1206. DOI: 10.1109/TPWRS.2005.851911

Section 3

Interconnection and Integration of Microgrid Systems

Design of an Energy Management System for Secure Integration of Renewable Energy Sources into Microgrids

Luis I. Minchala, Youmin Zhang and Oliver Probst

Additional information is available at the end of the chapter

http://dx.doi.org/10.5772/intechopen.69399

Abstract

This chapter presents the design and development of an energy management system (EMS), which guarantees a secure operation of an islanded microgrid under possible imbalances between generation capacity and loads demand. The EMS performs an optimal calculation of low priority loads to be shed, as well as charging and discharging cycles of batteries within the microgrid. A nonlinear model-predictive control (NMPC) algorithm is selected for implementing the EMS, which processes a data set composed of loads measurements, generation capacity, batteries state of charge (SOC), and a set of operation constraints. The EMS is designed under the assumption of having an advanced metering infrastructure (AMI) installed in the microgrid. The EMS is tested in a simulation platform that integrates models of the microgrid components, as well as their distributed controllers (DCs). Simulation results show the effectiveness of the proposed approach, since critical variables as the microgrid's frequency and voltage magnitude operate within a secured interval even under the presence of faults in one of the DCs.

Keywords: energy management, microgrid, fault-tolerant control, smart grids, power flow

1. Introduction

Many inventions have evolved over time from their initial conception, for example, the telephone. The telephone invention triggered a revolution in communications around the world that led to the powerful technology we have today. Alexander Graham Bell would be proud of his invention because of the impact and evolution of telephone, from landlines to satellite communications, the Internet, and so on. Another important invention is electricity. According to Ref. [1], electricity is the most important engineering invention of the last century. Electric networks' evolution led to a power system composed of three main systems: *generation, transmission,* and *distribution*. The main objective of the electric power system is to guarantee an efficient delivery of energy to end users. In the last 10 years, there has been a continuous search for emerging technologies, which foster green-energy integration into the main grid. In addition, stakeholders of the power grid have also changed their functionality in the system. Today, consumers are no longer just consumers, they can also be generators and their consumption and generation must be carefully regulated. This is the world of the smart grids (SGs) [2], where it is possible to have small isolated power systems operating as standalone islands purposefully. These islands are called *microgrids*.

A key objective of SGs is to provide reliable power supply through a combination of monitoring, control, and response. Electric utilities are expected to provide continuous and high-quality services to their customers at reasonable prices by making economical use of available infrastructure. The most powerful force shaping the future of power industry is economics, but environmental protection is receiving growing attention today. Global warming trend could be damped through energy conservation. Therefore, any new innovations/trends will probably be adopted only if they reduce costs and CO_2 emissions.

The development of an intelligent power distribution system requires new approaches. Reference [3] provides a French vision of SGs, which shows that distributed generation (DG) has strongly increased in the last 10 years. By 2020, the European Union targets 20% of the energy consumed to be generated by renewable energy sources (RESs). A technical report on the development of an intelligent distribution automation system in Korea [4] describes the following system features: remote operation, management of low voltage (LV) and medium voltage (MV) networks supported on a geographic information system (GIS), loss minimization, volt/var control for integrated DG, and power quality monitoring. Reference [5] details the features of an automatic power restoration system, which integrates DCs, distributed intelligence, and peer-to-peer communication to isolate faults and restore power to unfaulted sections. Further information about active microgrids projects around the world, including details on technical operational aspects and design criteria, can be found in the survey papers [6, 7].

2. Smart grids overview

The SG is defined as an electrical network that integrates in a smart way every action of the users plugged into it—generators, consumers, and those that can perform both actions, delivering electricity efficiently, sustainably, economically, and safely.

2.1. Distributed generation

A DG unit is a small-scale generation source with its output ranging from 1 KW to several MW and usually installed at the distribution level. A particular characteristic of DG is that the generating sources are near end users. As RESs penetrate the utility systems, the power industry

http://dx.doi.org/10.5772/intechopen.69399

undergoes a paradigm shift that will change the industry to the use of DG systems. **Figure 1** illustrates the DG concept by showing three different locations with installed DG units.

Small generators, which transform energy from RES, can be incorporated in the electric distribution network, for example, wind turbine generator (WTG) and photovoltaic (PV) arrays. These energy sources present many challenges to researchers and designers regarding power quality and economic issues. Today's power systems rely on spinning reserve and drooping frequency-load characteristics. Future systems, on the other hand, will rely on RES, which operate at peak power in order to displace as much fuel consumption as possible. This peak power constraint imposed by wind and sun complicates the frequency/load control of the entire system [8]. In order to compensate for this intermittent and changing power, short- and long-term storage devices should be deployed. Storage devices can be charged during periods of low-power demand and can supply power during high-power demand. This concept is shown in **Figure 2**.

2.2. Microgrids

Microgrids are small-scale LV/MV power systems with distributed energy resources (DERs), storage devices, and controllable loads, connected to the main power network or islanded, in

Figure 1. DG overview.

Figure 2. Example of an energy dispatch strategy for a microgrid.

a controlled and coordinated way [9]. Microgrids have different operating characteristics than bulk power systems (BPSs). A comparison, between microgrids and BPS, is shown in **Table 1**. Microgeneration units, typically located at users' sites, have emerged as a promising option to meet growing customer needs for electric power with an emphasis on reliability, power quality, and contribution to different economic, environmental, and technical benefits. However, the impact of microgeneration at LV levels on power balance and grid frequency is still a great challenge.

2.3. Advanced metering infrastructure

Smart meters (SMs) are important components of SGs. These devices have the following main features: full-duplex communication, ability to connect or disconnect consumer's loads, and recording capabilities for capturing waveforms for supervising voltage and current. SMs are gradually replacing traditional meters currently in operation and are also being installed in new microgrids. SMs transmit information to different information clients via SCADA systems and other networks.

Among the benefits that SMs offer to consumers, there is the possibility to know in real-time rates and pricing policies, allowing users to decide wisely how to use electric energy. Several research papers are devoted to household scheduling using AMI in order to reduce power consumption during peak consumption hours [10, 11]. **Figure 3** shows a possible architecture for an AMI. Due to the large number of SMs that will be available in distribution systems, the potential ability of SMs to provide additional information for outage management is also being investigated [12].

2.4. Fault-tolerant control

Critical-safety and operability issues with a defined performance in technological systems, such as electrical, industrial, aircraft control, nuclear generation, and so on, cause them to rely on complex control systems. Classic control schemes are suitable for guaranteeing a desired system performance in a specific operating range. However, these control strategies are unable to maintain the system performance under faulty scenarios. Therefore, there is a necessity of fault-tolerant control (FTC) strategies to improve the reliability and availability of critical-safe systems.

Quick detection of faults avoids serious damages to machines and humans, while allowing online reconfiguration of fault-tolerant controllers. Many books and research papers related to the field of FTC coincide on a two-step methodology for making a system fault-tolerant [13, 14]:

	RES penetration	Energy storage	Voltage levels	Dispatch objects	Operation mode
Microgrid	High	Yes	LV and MV	Controllable generators and loads	Grid-connected and autonomous
BPS	Low	Negligible	HV	Controllable generators	Independent operation

Table 1. Comparison between microgrids and BPS.

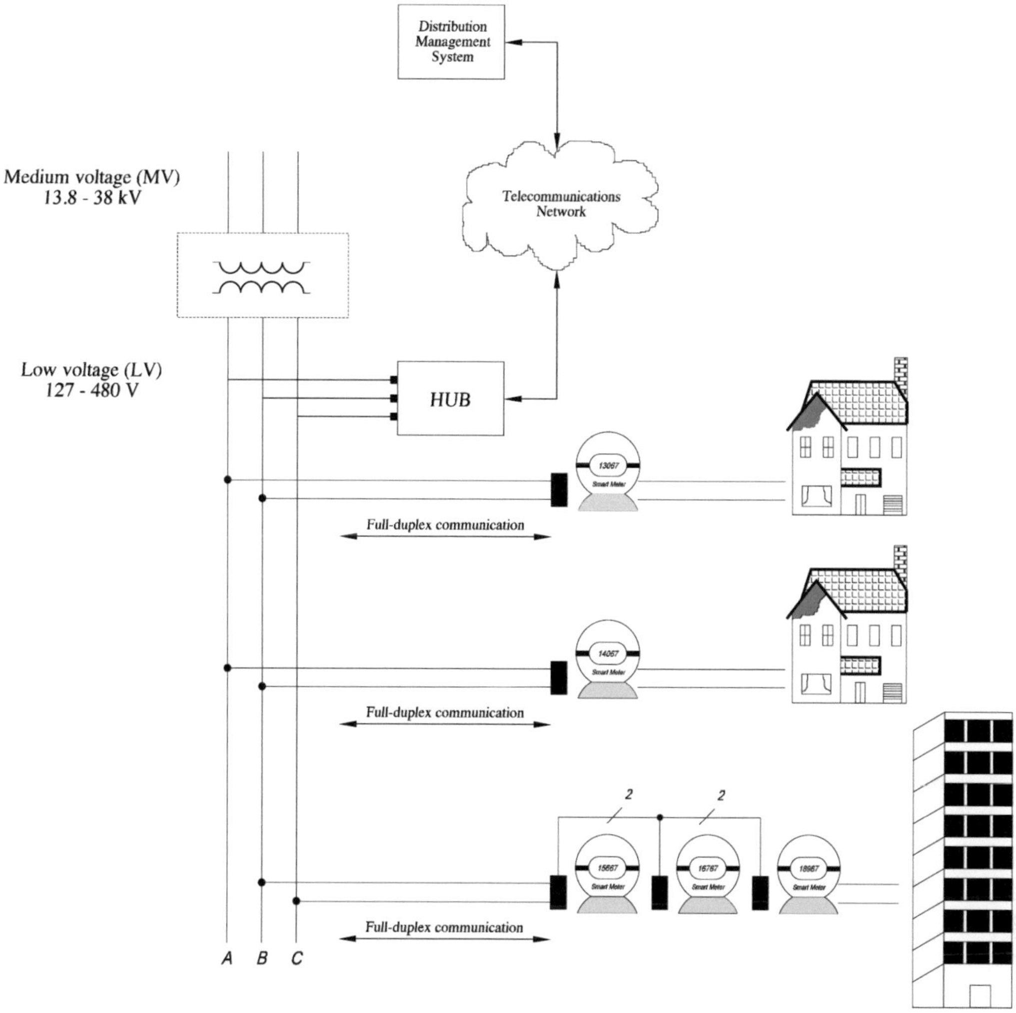

Figure 3. Remote management system for electricity measurement.

1. Fault diagnosis: Whenever a fault is present in the system, it has to be detected and identified.

2. Control reconfiguration: The controller has to be designed with the ability to accommodate faults on components automatically.

This methodology is an active research field, mainly due to the variety of possibilities for executing the abovementioned steps. Fault diagnosis is performed by a fault detection and diagnosis (FDD) module, while control reconfiguration could be done by many different control approaches, such as model-based, intelligent, gain scheduling, and so on. Such an FTC system, which relies on the fault information obtained from the FDD module, is called an active fault-tolerant control system (AFTCS). **Figure 4** shows an architecture of an AFTCS [14].

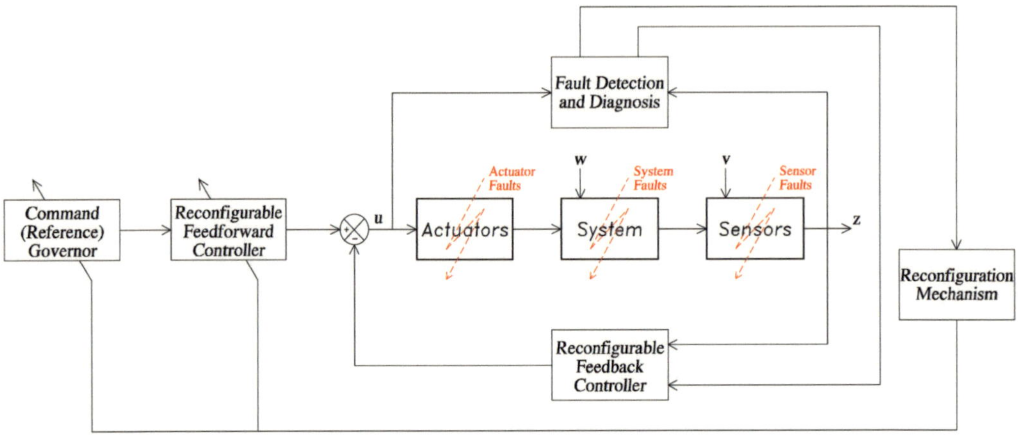

Figure 4. A general structure of an AFTCS [14].

3. Microgrid modeling

Control engineering most of the times is model dependent. Understanding the process, system, or plant to be controlled is fundamental for proposing proper control strategies. This section presents modeling procedures of the microgrid components used throughout this research: diesel engine generator (DEG), WTG, PV array, battery storage system (BSS), and power converters. At the end of this section, a microgrid benchmark model is presented, which integrates the microgrid components in one single simulation environment. **Figure 5** shows the aforementioned elements in a microgrid configuration.

The rationale for the SGs lies in the integrative analysis of DERs, many of which will be intermittently operating, with the deployment of short-term and long-term storage systems. Current strategies on load sharing will not work to integrate RES due to its peak-power and intermittent operation. Therefore, new control strategies for voltage/reactive-power and load-sharing/frequency need to be developed; microgrid modeling is the first step prior to advanced controllers design.

3.1. DG units modeling

A DG unit is conformed mainly of three components:

1. Microgeneration unit. Typical choices are batteries, PV, WTG, flywheels, fuel cells, and so on.

2. Power conditioning system (PCS). PCS is related with power conversion, AC/DC or DC/AC, and its control techniques.

3. Coupling circuit. Interface elements, most of the times a filter, for coupling the DG unit with the network.

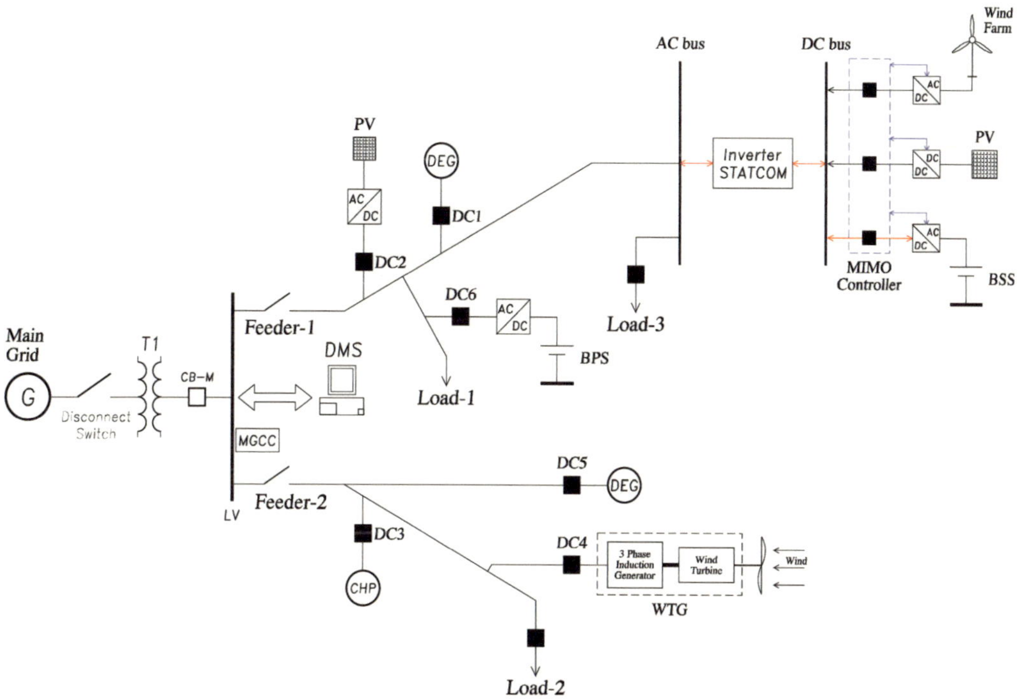

Figure 5. Microgrid control and management system architecture.

3.1.1. Power electronic converters

Proper integration technology has to be developed for using DG units in a microgrid configuration. The majority of DG units is integrated to the grid through the use of power electronics-based interfaces, which convert the power, firstly, to DC and then puts the energy into the grid by using an inverter. Adequate control strategies of the PCS allow maximum extraction of the energy from RES.

Power electronic converters performing conditioning have to be highly efficient, flexible, and reliable. It is well known that improving the performance of power converters increases system's efficiency. According to **Figure 5**, there are mainly three power electronic circuits that need to be implemented to control voltage, power, and frequency of a DG unit: AC/DC converter, DC/DC converter, and voltage source inverter (VSI).

3.1.2. Diesel engine generator

Diesel generating sets are typically used in power systems without connection to the power grid, as emergency power supply if the grid fails, as well as for more complex applications such as peak-shaving, grid support, and energy export to the power grid. This section presents models for the diesel engine (DE) components: synchronous generator and diesel engine.

3.1.2.1. Synchronous machine model

The synchronous generator has two circuits magnetically coupled: the first one is static and has the shape of a hollow cylinder with longitudinal slots and an armature winding; the second component is the rotor whose winding is supplied with DC current. The DC current is supplied to the field winding by an exciter, which may be a generator mounted on the same shaft or a separate DC source connected to the field winding through brushes bearing on slip rings [15].

As a prime mover drives the machine shaft, the magnetic field generated by the field winding links the stator coils to induce voltage in the armature windings.

As presented in Ref. [16], a state-space model, which uses the *dq* dynamic equations of the electrical circuit of a synchronous generator with a pure resistive load (RL) connected to its terminals, can be represented as follows:

$$\mathbf{L}\frac{d\mathbf{x}}{dt} = \mathbf{A}\mathbf{x} + \mathbf{B}v_F$$

$$\mathbf{x} = [i_d \quad i_q \quad i_F]^T \tag{1}$$

$$\mathbf{A} = \begin{bmatrix} -(R_s + R_L) & \omega L_s & 0 \\ -\omega L_s & -(R_s + R_L) & -\omega L_m \\ 0 & 0 & -R_F \end{bmatrix} \qquad \mathbf{L} = \begin{bmatrix} L_s & 0 & L_m \\ 0 & L_s & 0 \\ L_m & 0 & L_F \end{bmatrix} \qquad \mathbf{B} = \begin{bmatrix} 0 \\ 0 \\ 1 \end{bmatrix}$$

where $[i_d \; i_q \; i_F]^T$ are the *dq* stator and field currents, respectively; R_s and R_F are the stator and field resistances; L_s, L_m, and L_F are the stator, magnetizing, and field inductances; ω is the electrical speed; v_d and v_q are the *dq* stator voltages; and v_F is the field voltage which will be used as a control input.

3.1.2.2. Diesel engine model

The DE contains the combustion system and is responsible for the movement of the pistons, consequently the movement of the crankshaft that generates the output torque *T*(s). **Figure 6** shows a block diagram of a DE. A first-order system models the actuator dynamics of the DE. The time delay $e^{-\tau s}$ and a torque constant K_b model the combustion system. The flywheel block

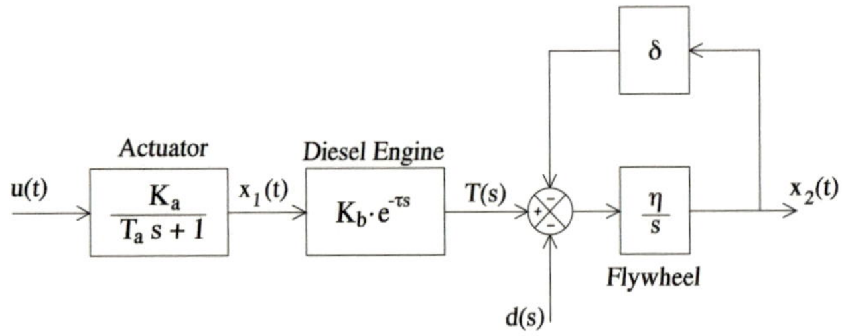

Figure 6. DE block diagram [6].

http://dx.doi.org/10.5772/intechopen.69399

System parameter	Value range	Nominal range
Actuator gain constant K_a (pu)	1.0	1.0
Actuator time constant T_a (s)	0.05–0.2	0.125
Engine torque constant K_b (pu)	0.8–1.5	1.15
Engine dead time t (s)	0–1	0.5
Plant and flywheel acceleration δ (s^{-1})	0.1–0.5	0.3
Friction coefficient η (pu)	0.1	0.1

Table 2. System parameters of a typical DE.

models the inertia generated inside the machine, η represents the flywheel acceleration constant, and the coefficient δ represents friction. The fuel injected to the DE is represented by $x_1(t)$, whereas $x_2(t)$ represents the angular velocity of the engine's shaft. d(s) models load changes in the rotor's shaft.

$$\begin{aligned} \dot{x}_1(t) &= -\frac{1}{T_a}x_1(t) + \frac{K_a}{T_a}u(t) \\ \dot{x}_2(t) &= \eta K_b x_1(t\text{-}\tau) - \eta x_2(t) \end{aligned} \tag{2}$$

Table 2 shows the characteristic values of the DE constants of model (2).

3.1.3. Wind-driven generation system

This section presents details on the modeling of a wind-driven electricity generation system (WEGS). A horizontal axis wind turbine (WT) is chosen as prime mover, while an induction generator performs the energy conversion. The wind turbine induction generator is an attractive DG unit in a deregulated electric energy market since wind energy is a non-polluting source.

Wind energy is air in motion whose energy is derived from sun, because about 2% of the solar flux falling on earth's surface is transformed into wind due to uneven heating of the atmosphere. Wind energy has some limiting characteristics such as non-schedulability, uncontrollable, and so on.

3.1.3.1. Wind turbine model

The WT model used throughout this chapter is a lumped mass model. The WT is pitch controlled through the blade pitch angle, β. The power coefficient, C_p as shown in Eq. (3), characterizes the WT and depends on the tip speed ratio, $\lambda = \Omega R/V_w$, and β, where R is the WT rotor radius, Ω is the mechanical angular velocity of the WT rotor, and V_w is the wind velocity. The pitch angle β is only varied to limit the over-speed of the generator

$$C_p(\lambda,\beta) = 0.5716(116\kappa\text{-}0.4\beta - 5)e^{-21\kappa} + 0.0068\lambda$$
$$\kappa = \left(\frac{1}{\lambda + 0.08\beta} - \frac{0.035}{\beta^3 + 1}\right) \tag{3}$$

The dynamic output mechanical torque of the WT, T_m, is expressed as

$$T_m = \frac{\rho ARC_p V_w^3}{2\lambda} \tag{4}$$

where ρ is the air density and A represents the swept area of the blades.

3.1.3.2. Induction generator model

The electrical equations of the induction generator model in the dq reference frame can be expressed in pu as

$$\begin{aligned}
v_{qs} &= r_s i_{qs} + \frac{\omega}{\omega_b}\psi_{ds} + \frac{p}{\omega_b}\psi_{qs} \\
v_{ds} &= r_s i_{ds} - \frac{\omega}{\omega_b}\psi_{qs} + \frac{p}{\omega_b}\psi_{ds} \\
v'_{qr} &= r'_r i'_{qr} + \left(\frac{\omega\text{-}\omega_r}{\omega_b}\right)\psi'_{dr} + \frac{p}{\omega_b}\psi'_{qr} \\
v'_{dr} &= r'_r i'_{dr} + \left(\frac{\omega\text{-}\omega_r}{\omega_b}\right)\psi'_{qr} + \frac{p}{\omega_b}\psi'_{dr} \\
\frac{p}{\omega_b}\omega_r &= \frac{1}{2H}(T_e\text{-}T_0) \\
T_e &= \psi'_{qr} i'_{dr} \text{-} \psi'_{dr} i'_{qr}
\end{aligned} \tag{5}$$

where v_s, v_r, i_s, i_r, Ψ_s, and Ψ_r represent the voltage, current, and flux (subscript s stands for stator and subscript r for rotor); r_s and r_r are stator and rotor resistance, respectively; ω_r is the rotor angular speed; ω_b is the base electrical angular velocity; H represents inertia moment, T_0 is load torque, and p denotes a time derivative operation.

3.1.4. Photovoltaic generation system

A PV cell is represented as a single-diode mathematic model, which is composed of a current source I_{ph}, a nonlinear diode, and internal resistances, R_s and R_{sh}. **Figure 7** shows the PV cell model.

A PV array is composed of the combination of N_p parallel and N_s serial PV cells. The total current produced by a PV array is expressed as follows:

$$i(t) = N_p i_{ph}\text{-}N_p I_s\left(e^{\frac{q}{AkT}\left(\frac{V}{N_s}+\frac{iR_s}{N_p}\right)} - 1\right) - \frac{N_p}{R_{sh}}\left(\frac{V}{N_s}+\frac{iR_s}{N_p}\right) \tag{6}$$

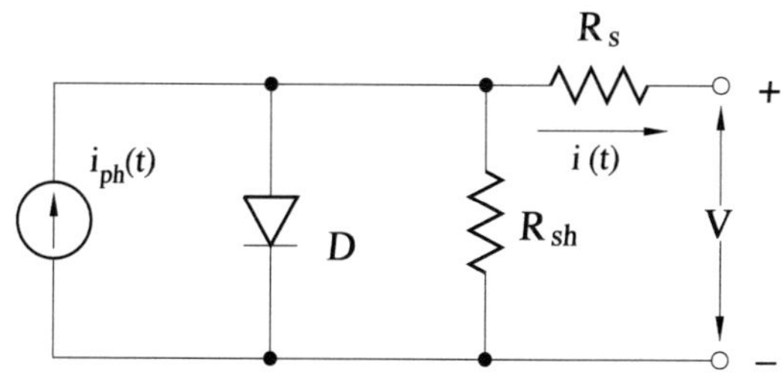

Figure 7. Single diode PV model.

3.1.5. Battery model

An electrical battery is one or more electrochemical cells that convert stored chemical energy into electrical energy. The lead-acid battery is the oldest type of rechargeable battery with a low energy-to-weight ratio. Lead-acid batteries are fully charged if it is possible to measure an open-circuit voltage of fully discharged battery cell(s). The term discharged means that all free charges are zero and the only voltage source is the cell(s) voltage, V_0 [17].

References [18, 19] present a simple nonlinear Thevenin model for the lead-acid battery. This model considers the dynamic response of the battery, which is influenced by the capacitive effects of the battery plates and also by the charge-transfer resistance. **Figure 8** shows the equivalent circuit of a lead-acid battery

$$\begin{gathered} V_0 = R_d C\frac{dV_p(t)}{dt} + \frac{1}{C}i_B(t) + \frac{1}{R_d C}V_0 \\ \frac{dV_p(t)}{dt} = -\frac{1}{R_d C}V_p(t) - \frac{1}{C}i_B(t) + \frac{1}{R_d C}V_0 \\ V_B(t) = V_p(t) - R_B i_B(t) \end{gathered} \tag{7}$$

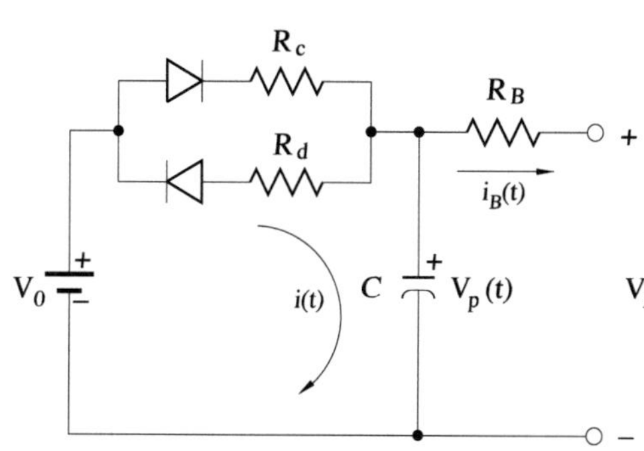

Figure 8. Nonlinear lead-acid battery equivalent circuit.

3.2. Microgrid model

Operating a microgrid within the limits of the established operation standards requires the development of novel control strategies. According to the standard ANSI C84.1, utilities are required to maintain voltage at the customer's service panel between 114 and 126 V (±5%)

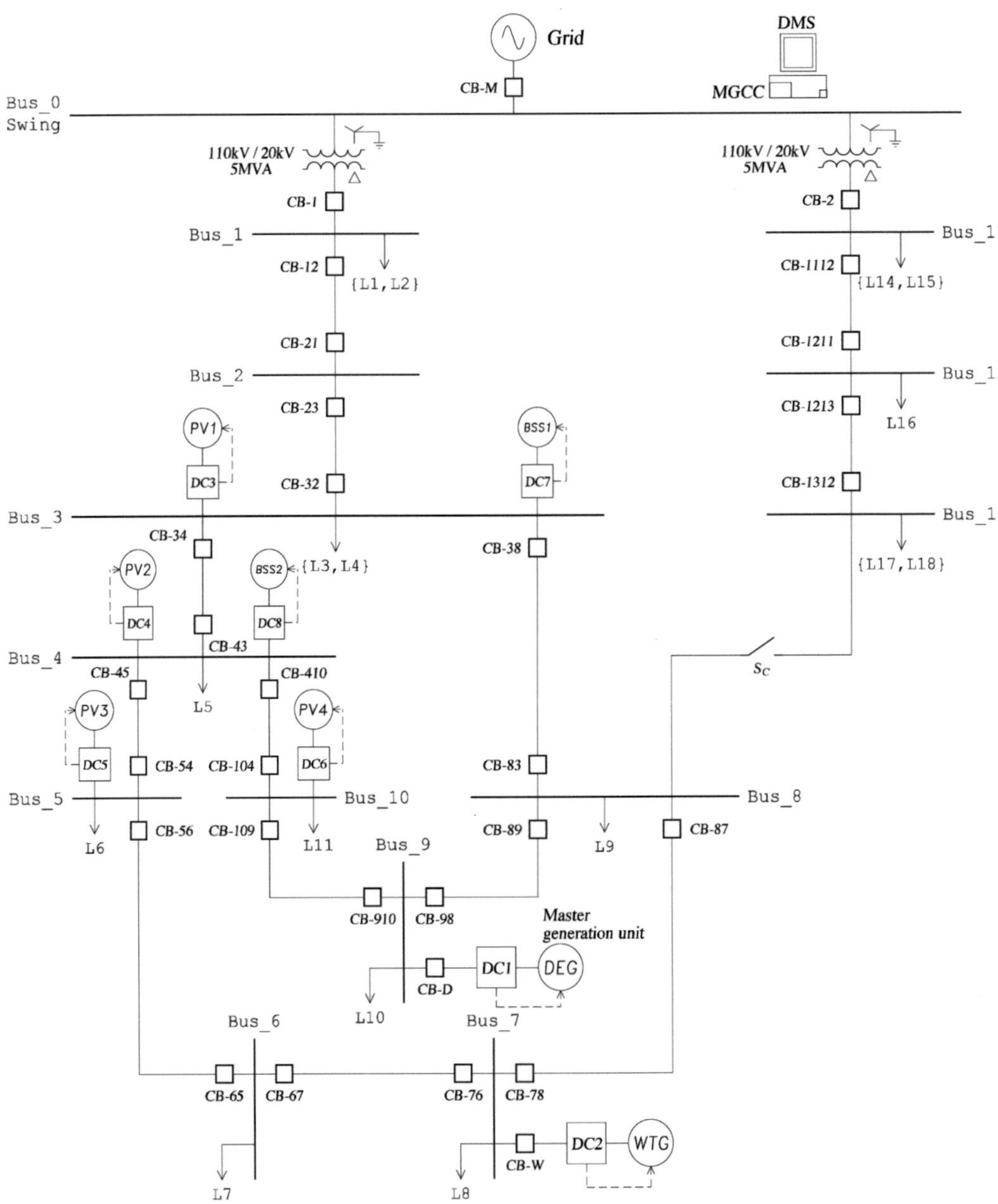

Figure 9. Microgrid MV benchmark model CIGRE TF C6.04.02.

based on a 120-V nominal secondary voltage. Standard IEEE 1547-2-2011 [20] recommends that for interconnecting DERs with electric power systems, the total time should be less than 0.15 s when the magnitude of the frequency variation exceeds 0.5 Hz and the magnitude of the voltage variation exceeds 5%.

Figure 9 shows the microgrid benchmark model configuration, consisting of two feeders supplied by a distribution substation. A grid of DG units is connected to the left-side feeder, including four PV array units, one WTG, two BSS, and one DEG. Every DG unit has a DC. The rated voltage level of the network is 20 KV, which is supplied from a 110 KV transformer station. The parameters of the network, the load and the DG units (in pu), were taken from Ref. [21] and are summarized in **Tables 3** and **4**.

Although maximum values for active and reactive power loads are considered in the network parameters shown in **Table 3**, variable load profiles have been generated for loads L_1, L_2, L_6, L_7, and L_9.

The features of DG units are as follows

- DEG: As previously described, a DE is used as the prime mover of a synchronous generator. The system frequency is determined by the DE mechanical speed ω_m, and the synchronous

Load no.	Load type	P_{max} (pu)	Q_{max} (pu)
1	Residential	0.15000	0.03100
2	Industrial	0.05000	0.01000
3	Residential	0.00276	0.00069
4	Industrial	0.00224	0.00139
5	Residential	0.00432	0.00108
6	Residential	0.00725	0.00182
7	Residential	0.00550	0.00138
8	Industrial	0.00077	0.00048
9	Residential	0.00588	0.00147
10	Industrial	0.00574	0.00356
11	Industrial	0.00068	0.00042
12	Residential	0.00477	0.00120
13	Residential	0.00331	0.00083
14	Residential	0.15000	0.03000
15	Industrial	0.05000	0.01700
16	Industrial	0.00032	0.00020
17	Industrial	0.00330	0.00020
18	Residential	0.00207	0.00052

Table 3. Load parameters of the microgrid model.

From	To	R	X	C	L
Node	Node	(Ω/km)	(Ω/km)	(nF/km)	(km)
0	1				
1	2	0.579	0.367	158.88	2.82
2	3	0.164	0.113	6608	4.42
3	4	0.262	0.121	6480	0.61
4	5	0.354	0.129	4560	0.56
5	6	0.336	0.126	5488	1.54
6	7	0.256	0.130	3760	0.24
7	8	0.294	0.123	5600	1.67
8	9	0.339	0.130	4368	0.32
9	10	0.399	0.133	4832	0.77
10	11	0.367	0.133	4560	0.33
11	4	0.423	0.134	4960	0.49
3	8	0.172	0.115	6576	1.3
0	12				
12	13	0.337	0.358	162.88	4.89
13	14	0.202	0.122	4784	2.99

Table 4. Transmission line parameters.

generator field current sets the voltage magnitude. The rated output power of the DEG is 0.3125 pu over a P_{base} of 5 MW.

- WTG: The characteristics of WTG are assumed in Section 3.1.3. The maximum output power of the WTG is 0.2 pu.
- PV array: The features of the PV modules 330 SunPower (SPR-305) are used. PV_1 array consists of 66 strings of five series-connected modules connected in parallel, providing 0.02 pu. PV_2, PV_3, and PV_4 have 0.02, 4×10^{-3}, and 5×10^{-3} pu of power generation capacity, respectively. The PCS of every PV array is composed of a boost converter and a VSC.
- BSS: The lead-acid batteries are combined with bidirectional DC/AC converters with a maximum output power of 0.02 pu for BSS1 and 0.015 pu for BSS2. The charging power for every BSS is 0.01 pu.

DERs do not provide system frequency regulation. Therefore, for power flow calculations, the DG units (except for the DEG) are considered as load nodes with negative power consumption. During grid-connected operation, the main grid controls voltage and frequency. Islanded operation demands that local microgrid generation controls voltage and frequency.

http://dx.doi.org/10.5772/intechopen.69399

4. Energy management system

The system states for the islanded section (left-side feeder) of the distribution system in **Figure 9** are defined as

$$\mathbf{x} = \begin{bmatrix} \mathbf{x_1} \\ \mathbf{x_2} \end{bmatrix} \quad \begin{matrix} \mathbf{x_1} \in \mathbb{R}^{10} & \mathbf{x_1} = [V_i]^T & i = 1, 2, \ldots, 10 \\ \mathbf{x_2} \in \mathbb{R}^{10} & \mathbf{x_2} = [\delta_i]^T & i = 1, 2, \ldots, 10 \end{matrix} \tag{8}$$

where V_i and δ_i are node's voltage and angle of the bus i.

Additionally, more variables and vectors are needed for the controller formulation, such as power at the nodes $S_i = P_i + jQ_i$, admittance matrix Y, and power generated by the DG units, P_{DGi}:

$$\begin{matrix} Y = [Y_{i,j}] \\ S = \begin{bmatrix} \mathbf{P_{load}} \\ \mathbf{Q_{load}} \end{bmatrix} \quad \begin{matrix} \mathbf{P_{load}} \in \mathbb{R}^{10} & \mathbf{P_{load}} = [P_{Li}]^T & i = 1, 2, \ldots, 10 \\ \mathbf{Q_{load}} \in \mathbb{R}^{10} & \mathbf{Q_{load}} = [Q_{Li}]^T & i = 1, 2, \ldots, 10 \end{matrix} \end{matrix} \tag{9}$$

$$\mathbf{P}_{DG} = \left[P_{DG_i}\right]^T \qquad i = 1, 2, \ldots, 8 \begin{cases} i = 1, \text{ Diesel engine power } (P_{DE}) \\ i = 2, \text{ Wind turbine power } (P_{WT}) \\ i = 3, PV\ power\ array\ 1\ (P_{PV_1}) \\ i = 4, PV\ power\ array\ 2\ (P_{PV_2}) \\ i = 5, PV\ power\ array\ 3\ (P_{P3}) \\ i = 6, PV\ power\ array\ 4\ (P_{PV_4}) \\ i = 7, Power\ in\ BBSS\ 1\ (P_{BSS_1}) \\ i = 8, Power\ in\ BBSS\ 2\ (P_{BSS_2}) \end{cases} \tag{10}$$

$$S_i = V_i \sum_{m=1}^{N} (Y_{im} Vi)^* \tag{11}$$

Equation (11) is solved iteratively through the Newton-Raphson (NR) power flow algorithm, with prior knowledge of P_{DGi}, and current load consumption of every power system node. P_{DE} is estimated in a prediction horizon of length N. An important modification of Eq. (11) is the inclusion of the reactive power consumed by the WTG at bus 7, which is calculated as follows [22]:

$$\begin{matrix} Q_{WT} = -\dfrac{V_7^2}{z_p} + \dfrac{-V_7^2 + \sqrt{V_7^4 - 4P_7 z^2}}{2z} \\ z = z1 + z2 \qquad z_p = \dfrac{z_c z_m}{z_c\text{-}z_m} \end{matrix} \tag{12}$$

where the negative sign of Eq. (12) represents reactive power consumption of the WTG induction generator from the network; z_m, z_c, z_1, and z_2 represent the excitation reactance, reactance of a capacitor bank installed at the terminal of the induction generator, and the stator and rotor reactance, respectively.

The control objectives of the proposed strategy are as follows:

- To manage the connection and disconnection events of batteries;
- To shed low priority loads every time the load demand is greater than the generation capacity. This control action anticipates any harmful operation of the system through the predicting model, which predicts potential load imbalances;
- To keep the voltage magnitude with a maximum variation of ±5%.

A microgrid-centralized controller (MGCC) implements an NMPC algorithm. Loads management as well as batteries connecting and disconnecting events are performed by a control vector, **u**, defined by Eq. (13). The control vector is calculated in real time by the MGCC, and transmitted to the DCs in the microgrid. **Table 5** presents the relationship between each bit of **u** and its corresponding load controller for switching purposes, that is, for $u_i = 1 \rightarrow L_i$ load is connected, and for $u_i = 0 \rightarrow L_i$ load is disconnected

$$\mathbf{u} = [u_i] \qquad i = 1, 2, \ldots, 13 \qquad u_i \text{ is a binary signal} \tag{13}$$

An important requirement in the design of NMPC is the availability of a model for predicting the output variables. The NMPC algorithm requires predicted values of the power generated by the DEG to optimally decide which load has to be shed. Trip commands are sent from the MGCC to proper loads. **Figure 10** shows the EMS architecture. This control architecture is deeply analyzed in Ref. [23].

Control signal	Load	Observations
$u1$	L1* = {L1 U L2}	Variable loads
$u2$	L3 U L4	Constant loads
$u3$	L5	Constant load
$u4$	L6	Variable load
$u5$	L7	Variable load
$u6$	L8	Constant load
$u7$	L9	Constant load
$u8$	L10	Constant load
$u9$	L11	Constant load
$u10$	BSS1	Charge mode
$u11$	BSS1	Discharge mode
$u12$	BSS2	Charge mode
$u13$	BSS2	Discharge mode

Table 5. Control vector correspondence with loads and BSS.

http://dx.doi.org/10.5772/intechopen.69399

The NR power flow algorithm is used to predict the microgrid's system states, and consequently the P_{DE} in a specific prediction horizon N. To find the optimal steady-state operation of the microgrid, the connection and disconnection commands of the control vector **u** are accounted within the NR algorithm. An initial data set Z_k composed of the batteries' SOC, load demand, and the active power generated by every DER is required prior to the execution of the NR calculation. The data set Z_k does not consider load variations within the prediction horizon. This fact is considered, and two approaches were tested for the initial iterative load values of the NR algorithm in order to predict P_{DE}:

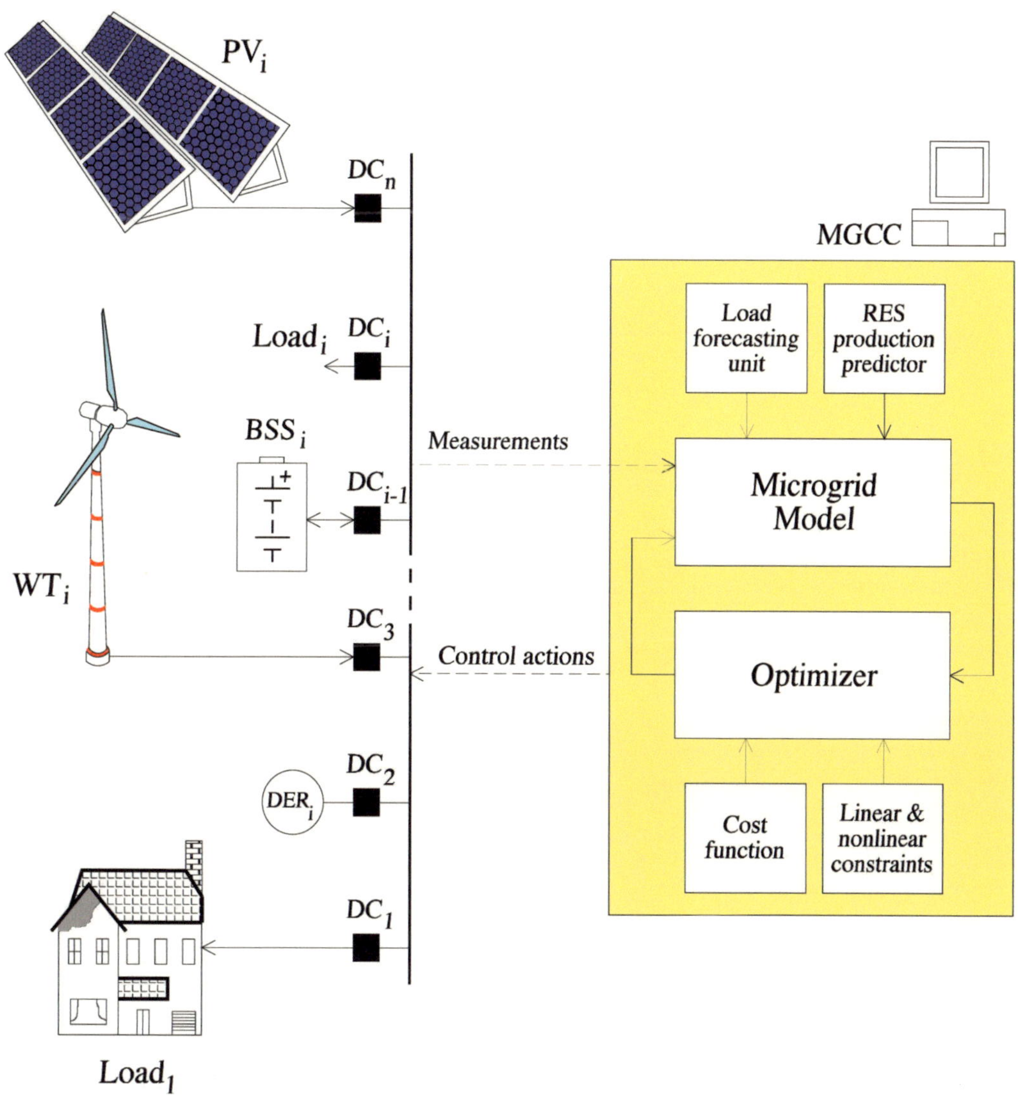

Figure 10. NMPC architecture for a centralized load shedding strategy.

1. Consider the load measurements as constants during the entire prediction horizon N;

2. Use a load predictor based on an artificial neural network (ANN). For this purpose, 20 load profiles from different days of the week for every variable loads in the microgrid were used as training set for configuring a three-layer ANN.

Another approach for predicting the P_{DE} was developed with an autoregressive model with external input (ARX) through a data-based modeling using an adaptive neuro-fuzzy inference system (ANFIS). As in the case of the ANN training algorithm, 20 different generation profiles of the DEG for different days of the week were used as training set for the ANFIS. The ARX configuration developed is the one detailed in Ref. [24]. This modeling procedure does not imply an NR calculation, therefore reducing the computing time of the control algorithm.

Since the voltage magnitude of the microgrid is to be kept within a ±5% range of variation, a static voltage stability index, presented in Ref. [22], is used for defining a secure range of operation of the

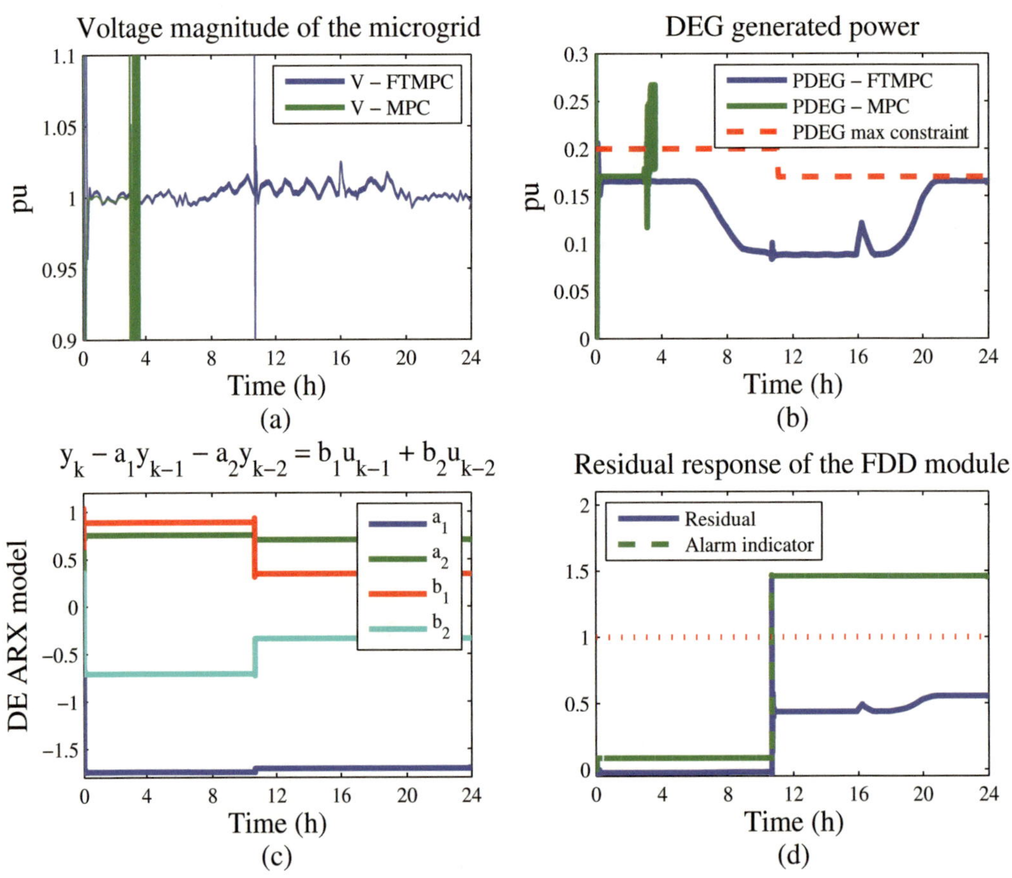

Figure 11. Simulation results of the EMS combining an NMPC algorithm running at the MGCC, and an FTMPC installed at the DEG. (a) Voltage magnitude response comparison between the FTMPC and the MPC systems. (b) Power output of the DEG comparison. (c) ARX model parameters adaptation of the DEG. (d) Residue calculated by the FDD module.

DEG. Consider average generation values of DERs: $P_{WT} = 0.15$, $\Sigma P_{P\ V1,2,3,4} = 0.049$ and $P_{BSS1,2} = 0.035$ in discharge mode and $P_{BSS1,2} =$ -0.02 in charge mode, and that the voltage magnitude limits are $0.95 \leq |V| \leq 1.05$, an optimization procedure in which a sweep of the values of P_L, Q_L, and $|V|$ within allowable ranges is performed in order to obtain a secure margin of operation of the microgrid. The secure range of operation for the DE, estimated by the optimization process, is

$$P_{DE} \leq P_{DE}^{+} = 0.2 \tag{14}$$

The control vector **u** calculated by the NMPC algorithm is restricted to be binary. Optimization problems of this type are called mixed-integer nonlinear programming (MINLP) problems. The MINLP package of TOMLAB for MATLAB was used for solving this optimal control problem.

Figure 11 shows the transient response of the voltage magnitude of node-1 of the microgrid in islanded operation. Near 04h00 AM, the DE presents a loss of effectiveness in the servo-mechanism used for fuel injection. Based on the detection of the fault and estimation of the post-fault model of the DE by using a combination of the parity space and Kalman filter methods, a fault-tolerant model-predictive control (FTMPC) has been implemented (see Ref.

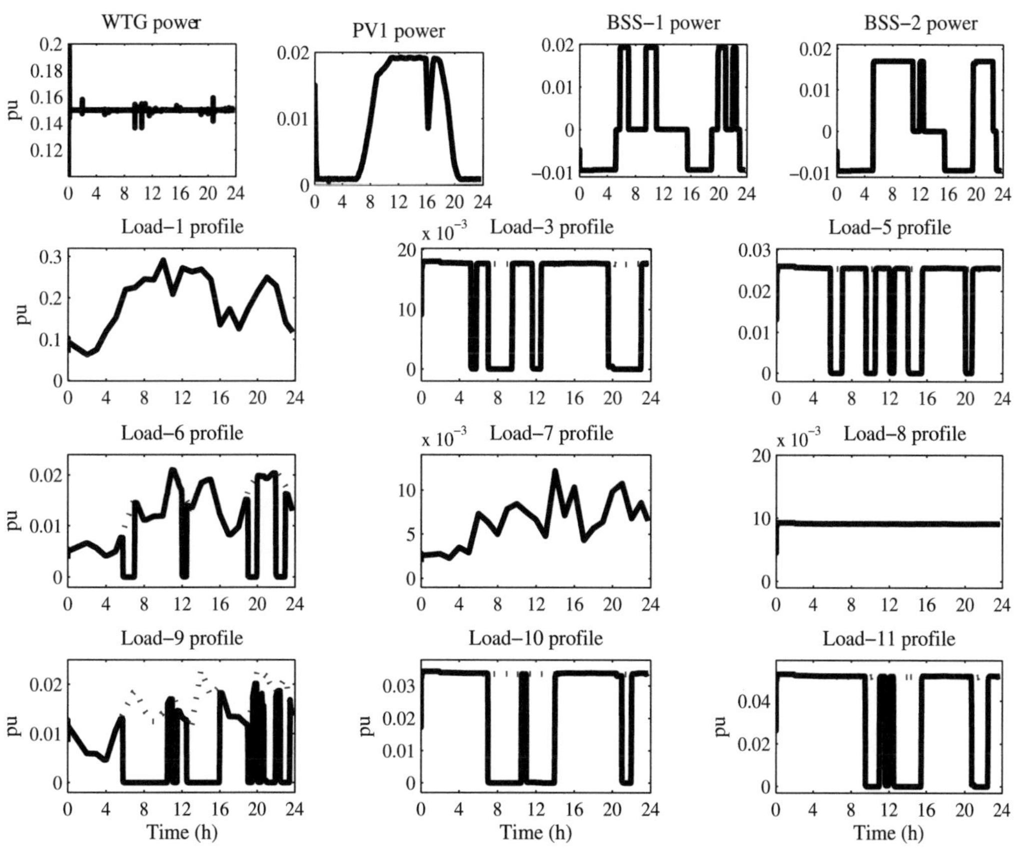

Figure 12. Profiles of RES generated power and load power consumption.

[25] for further details). A simple MPC has also been tested, which causes microgrid instability due to its lack of adaptability to faulty situations.

Figure 12 provides additional information on the NMPC performance by showing the voltage and frequency of the microgrid, and the loads and batteries switching due to the NMPC calculation. Voltage magnitude maintains within the ±5% band (in the weaker bus/node 1) when the microgrid operates in islanded mode. In this operation mode, when no control action is performed, the voltage magnitude constraint is violated. On the other hand, constraints included in the NMPC algorithm are not violated. From **Figure 12**, it is seen that high priority loads L_1, L_7, and L_8 were not disconnected, and at least one load of the low priority loads group kept connected, as it was programmed in the NMPC algorithm. Batteries charge at off-peak times, when there is availability of power from the generation units ($P_{gen} > P_{load}$). Batteries deliver power to the grid when there is a power deficit due to peak consumption. The inclusion of load shedding and battery management in the NMPC algorithm improves the microgrid's overall performance by guaranteeing a reliable and secure operation.

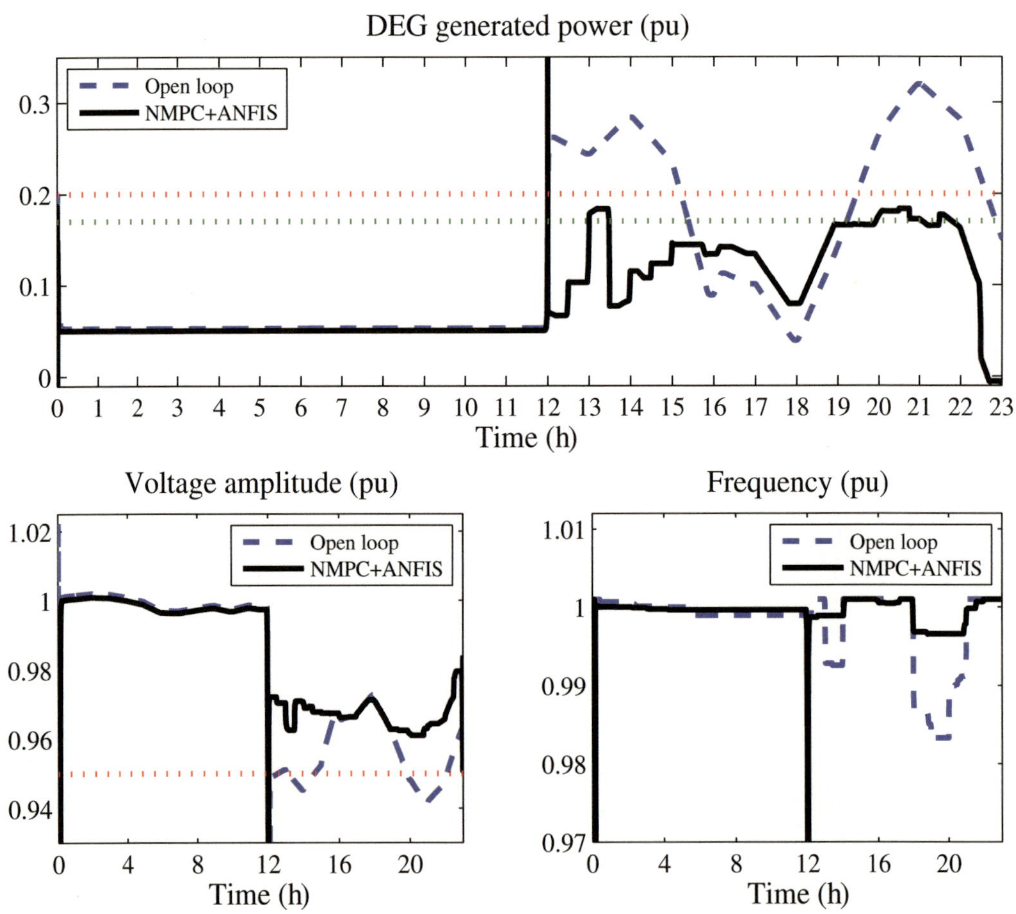

Figure 13. (Upper) P_{DE} when the load shedding NMPC strategy is used (islanding event at 12h00). (Lower left) Voltage magnitude variation; (lower *right) frequency variation.*

In addition to the previous results, **Figure 13** shows the operation of the microgrid under similar conditions of **Figures 11** and **12**. The islanding event occurs at 12h00, and as expected the voltage magnitude variation does not overpass the ±5% variation when the EMS takes decisions on the batteries energy management, and low priority loads to be shed. A similar situation is observed with the frequency deviation from 1 pu, which remains in an acceptable range of variation of maximum 0.05%. By contrast, the simulation of the microgrid operating without the EMS shows an unsecure behavior of the system, violating the $|\Delta V| < \pm 5\%$ operation constraint, and also presenting a much larger frequency variation, as also shown in **Figure 13**.

5. Conclusions

A predictive control scheme that prevents from unbalances between the load demand and the capacity of generation installed in an islanded microgrid is analyzed in this chapter. The NMPC calculates, within an optimization framework, load shedding when necessary as well as the energy management from the batteries in the microgrid. Therefore, an optimal control problem is established, where all the operating conditions of the microgrid are integrated, that is, load priorities for disconnection and batteries charging and discharging cycles. A comparison of some simulation results of the microgrid working with and without the MGCC shows improvements in the reliability of the microgrid when it operates in islanded mode, since simulation results showed the capability of the control strategy of maintaining within safe limits voltage and frequency of the microgrid, as well as a correct balance of generated power and load demand.

Author details

Luis I. Minchala[1,2]*, Youmin Zhang[3] and Oliver Probst[1]

*Address all correspondence to: ismael.minchala@ucuenca.edu.ec

1 Instituto Tecnológico y de Estudios Superiores de Monterrey, Monterrey, NL, México

2 Universidad de Cuenca, Cuenca, AZ, Ecuador

3 Concordia University, Montreal, QC, Canada

References

[1] Constable G, Somerville B. A Century of Innovation—Twenty Engineering Achievements that Transformed Our Lives. Washington, DC: Joseph Henry Press; 2003

[2] Mahmoud MS, Azher Hussain S, Abido MA. Modeling and control of microgrid: An overview. Journal of The Franklin Institute. 2014;**351**(5):2822–2859

[3] Mamo X, Mallet S, Coste T, Grenard S. Distribution automation: The cornerstone for smart grid development strategy. 2009 IEEE Power & Energy Society General Meeting, Calgary, AB, 2009, pp. 1–6

[4] Ha B, Lee S, Shin C, Kwon S, Park S, Park M. Development of intelligent distribution automation system, 2009 Transmission & Distribution Conference & Exposition: Asia and Pacific, Seoul, 2009, pp. 1–4

[5] Craig D, Befus C. Implementation of a Distributed Control System for Electric Distribution Circuit Reconfiguration, 2005/2006 IEEE/PES Transmission and Distribution Conference and Exhibition, Dallas, TX, 2006, pp. 342–347

[6] Hossain E, Kabalci E, Bayindir R, Perez R. Microgrid testbeds around the world: State of art. Energy Conversion and Management. 2014;**86**:132–153

[7] Haritza C, Octavian C, Aitor E, Alvaro L, Amélie HP. Research experimental platforms to study microgrids issues. International Journal on Interactive Design and Manufacturing. 2016;**10**(1):59–71

[8] Masoum M, Dehbonei H, Fuchs E. Theoretical and experimental analyses of photovoltaic systems with voltage and current-based maximum power-point tracking. IEEE Transactions on Energy Conversion. 2002;**17**(12):514–522

[9] Paire D, Miraoui A. Power management strategies for microgrid—A short review, 2013 IEEE Industry Applications Society Annual Meeting, Lake Buena Vista, FL, 2013, pp. 1–9

[10] Minchala-Avila LI, Armijos J, Pesántez D, Zhang Y. Design and implementation of a smart meter with demand response capabilities. Energy Procedia. 2016;**103:**195–200

[11] Chen X, Wei T, Hu S. Uncertainty-aware household appliance scheduling considering dynamic electricity pricing in smart home. IEEE Transactions on Smart Grid. 2013;**4**:932–941

[12] He Y, Jenkins N, Wu J. Smart metering for outage management of electric power distribution networks. Energy Procedia. 2016;**103**:159–164

[13] Blanke M. Diagnosis and Fault-Tolerant Control. Berlin, Heidelberg: Springer-Verlag GmbH; 2003

[14] Zhang Y, Jiang J. Bibliographical review on reconfigurable fault-tolerant control systems. Annual Reviews in Control. 2008;**32**(2):229–252

[15] Grainger JJ, Stevenson WD. Power System Analysis. New York, NY: McGraw-Hill; 1994

[16] Munoz-Aguilar RS, Doria-Cerezo A, Fossas E, Cardoner R. Sliding Mode Control of a Stand-Alone Wound Rotor Synchronous Generator, in IEEE Transactions on Industrial Electronics, 2011;**58**(10):4888–4897

[17] Fuchs EF, Masoum MA. Power Conversion of Renewable Energy Systems. Boulder: Springer; 2011.

[18] Chiasson J, Vairamohan B. Estimating the state of charge of a battery. IEEE Transactions on Control Systems Technology. 2005;**13**(3):465–470

[19] Farrell J, Barth M. Battery state-of-charge estimation. Proceedings of the American Control Conference. 2001;**2**:1644–1649

[20] IEEE Guide for Design, Operation, and Integration of Distributed Resource Island Systems with Electric Power Systems, in IEEE Std 1547.4-2011, vol., no., pp. 1–54, July 20 2011

[21] Rudion K, Orths A, Styczynski ZA, Strunz K. Design of benchmark of medium voltage distribution network for investigation of DG integration, 2006 IEEE Power Engineering Society General Meeting, Montreal, Que, 2006, p. 6

[22] Chen H, Chen J, Shi D, Duan X. Power flow study and voltage stability analysis for distribution systems with distributed generation. In: 2006 IEEE Power Engineering Society General Meeting; 2006. p. 8

[23] Minchala-Avila LI, Garza-Castanon L, Zhang Y, Ferrer HJA. Optimal energy management for stable operation of an islanded microgrid. IEEE Transactions on Industrial Informatics. 2016;**12**(4):1361–1370

[24] Nourzadeh H, Fatehi A, Labibi B, Araabi BN. An experimental nonlinear system identification based on local linear neuro-fuzzy models, 2006 IEEE International Conference on Industrial Technology, Mumbai, 2006, pp. 2274–2279

[25] Minchala-Avila LI, Vargas-Martínez A, Zhang Y, Garza-Castañón LE. A model predictive control approach for integrating a master generation unit in a microgrid, 2013 Conference on Control and Fault-Tolerant Systems (SysTol), Nice, 2013, pp. 674–679

Chapter 8

FutureGrid: Use of Microgrids in Underserved Communities

Andrew Hubble and Taha Selim Ustun

Additional information is available at the end of the chapter

http://dx.doi.org/10.5772/intechopen.68622

Abstract

Rural electrification in developing countries—especially Sub-Saharan Africa—has trailed urban development drastically. The extreme costs associated with expanding traditional grid networks, and the relatively few people they serve, have proved to be a serious economic barrier. Being able to generate and distribute electricity at an affordable rate is crucial in order to effectively power homes, schools, health clinics, and private business. Through this continued cycle and lack of access to electricity, poverty only continues. If given access, quality of life increases through more educated, longer, and healthier lives as well as through developed entrepreneurship and business growth. Unfortunately, because of the remoteness of many communities they are often dismissed as unreachable. Furthermore, microgrids help address another global need: increased renewable energy penetration. Small-scale energy production lends itself to solar installations, but depending on the location and available resources, wind and hydropower can also play an important role.

Keywords: microgrids, distributed generation, energy storage, grid extension, rural communities

1. Introduction

Often when we look at how technology has changed since its inception, it is difficult to imagine how the creators would react to the relentless progress and improvements on their original idea. Over a hundred years after Alexander Graham Bell invented the telephone, today's smartphones are infinitely more complex, contain thousands of features, and possess processing power, Bell could never have imagined. Once luxury, cell phones are now found in every corner of the globe including the most remote villages in the developing world. If Thomas Edison and Nikola Tesla could see the state of electrification today, it is safe to assume they would be sorely disappointed. While electrification has certainly improved, it has severely lagged behind the growth of other technology. The generation and transmission of electricity looks much the same as it did over 130 years ago when the Vulcan Street Hydroelectric Plant in Appleton, Wisconsin began producing 12.5 kW of DC power. Over the next few years, more plants were constructed in both AC and DC, mostly powered by water or coal. While access to this electricity increased—as did the quality and economic viability—electricity never experienced the gigantic expansion in both availability and technology that other sectors did. Since its creation and original spurt of distribution, electricity has been slow to advance to a significant portion of the global population.

Unlike cell phones, electricity cannot be manufactured and shipped in discrete units. Because it is not a physical device, the infrastructure required to produce and distribute it is entirely unique. Due to the immense capital costs associated with electrification, individual business (and thus competition and natural advancement) have not developed in the same manner. As a result, electric utilities are slow to develop or expand, leaving no need for innovation.

Nowadays, power production finds its way into the public view as we battle the negative effects of climate change. Instead of the natural pressure in the industry to improve and outcompete other companies, utilities are now being pressured externally via the government and general public. While a shift to renewable energy is undoubtedly important, it does not represent the only problem in this area. Access to any form of electricity in developing and rural areas is severely limited. At face value, it may seem that this is not an immediate problem, but there are innumerable secondary effects all stemming from a chronic lack of access.

Imagine a rural town in a developing country. There are 300 homes, a few grocery stores, a pharmacy, a general store, a school, a carpentry workshop, and a coffee milling station. What they do not have is electricity. While the government and utility are aware that the town exists, plans to offer electrification have never gone further than a Master Plan written years ago and shelved. Extending transmission lines are expensive, and if the utility thinks that there is not sufficient demand, they will not invest the money. The utility may also be unaware of the current size of the town, and thus the potential customers. Even if they did extend the grid, there are production shortages. Blackouts lasting hours or even days. This is not an unreasonable scenario, and is an accurate descriptor of a large portion of the unelectrified developing world. This lack of electricity means that at night families burn candles or kerosene lanterns which have harmful effects on the respiratory system when used in enclosed spaces; refrigeration is impossible, and food cannot be saved for long; water is pumped manually from boreholes, or carried from the nearest stream. All these activities which have to be done manually take an immense amount of time. Frequently, children are required to help their families in these tasks, and their studies suffer. By expending all this energy on the day-to-day tasks, it is difficult to develop and remove oneself from this cycle. Lacking access to electricity keeps people impoverished and uneducated.

2. Rural electrification and national grid distribution

Often microgrids are seen as solutions creating a more stable and reliable interconnected grid in urban settings [1–3], however, they need not be limited to these uses. Electrification in developing countries has trailed industrialized nations drastically and even more so in rural settings [4]. The vast majority of those without an electrical connection live in rural developing settings, where their access to resources in general is scarce. The eradication of poverty is on nearly every government agenda around the globe, and while on the surface, the undertaking is targeted and defined, in reality it is not. Access to reliable electricity is one large step in the correct direction and can no longer be considered a luxury. Lack of electrification contributes to the continued cycle of poverty, child mortality, chronic but otherwise treatable health issues, as well as suppressing education. Electricity is necessary for providing lighting without risking unnecessary smoke inhalation, for pumping water on anything other than an individual scale, and for refrigeration—which allows families to reduce food waste. Providing reliable and affordable electricity should be the top priority in tackling poverty eradication.

Unfortunately, the extreme costs associated with electrifying rural developing areas, as well as the relatively few people served, have caused come countries to exclude entire regions from their electrification schemes. With transmission lines costing up to $20,000 per km [5] in rough and rural terrain, shortening the transmission distances through the use of distributed generation and microgrids brings down the cost of rural electrification significantly. To understand the scope of the problem, transmission line coverage can best be depicted through utility maps. In **Figure 1**, large portions of Brazil, Sudan, and South Sudan are not currently serviced by existing transmission lines. In order to electrify these regions, millions of dollars would

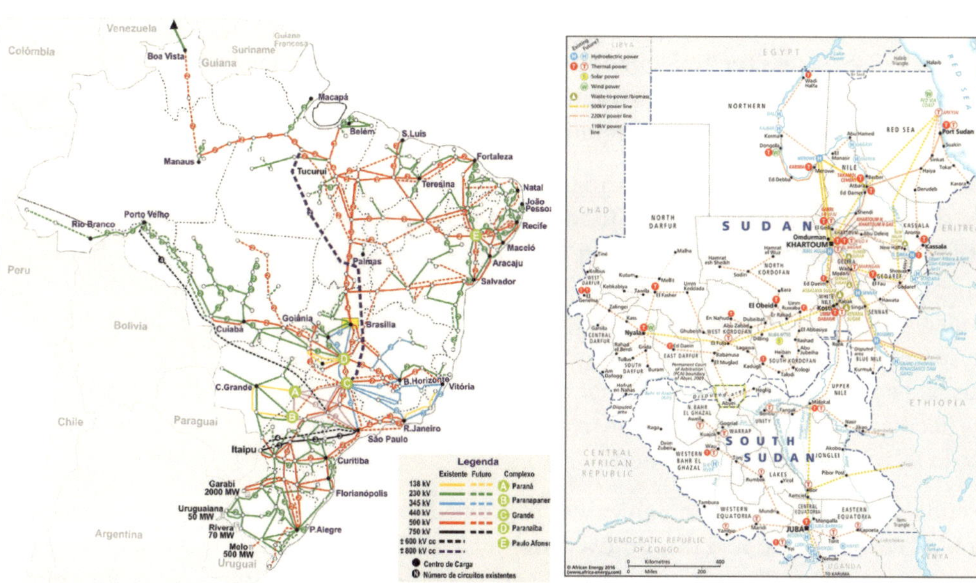

Figure 1. Brazil's, Sudan's, and South Sudan's existing electrical networks.

be required in infrastructure development. Fortunately, distributed generation microgrids can be utilized instead.

Immediately, it becomes apparent that a more economical solution skips the long transmission lines and produces power closer to the users. The lower capital investment and varying sizes of communities present a wide array of customizable solutions, and as a result, there exists no uniform microgrid design which is applicable to all or even most potential microgrid sites. Despite this microgrids still hold a place in the global electrification scheme. This chapter aims to demonstrate that not only are microgrids closer to wide scale deployment in rural developing areas than may be commonly believed, but that there exist methods and technology making microgrids uniquely suitable for rural electrification.

3. Economic feasibility of microgrid breakeven distances

Traditionally, there has been a singular approach to electrification: extend the national grid. When the utility considers this option for remote areas, little or no math goes into the determination. They are simply too far from the grid, and their demand is too low to justify the immense cost. With extension costs as high as $15,000 per km [6, 7] in rough terrain, the cost per kilowatt-hour to achieve any kind of payback would have to be prohibitively expensive. As an alternative, residents of these rural areas can sometimes afford a few solar panels and batteries—especially if the cost is shared. This simple setup can sustain a little lighting and offer a place to charge cell phones, but stops well short of an acceptable solution. Instead, there exists a medium between these two solutions. Something, larger and more robust than a few panels linked together, but less expensive than grid extension: the microgrid. The microgrid can be sized and built according to demand and expanded with usage. The costs can be kept in check because the electricity is produced and consumed in the same area—no need for the expensive transmission lines. As a solution to rural electrification, the microgrid is new on the scene, and being largely untested requires some analysis to determine its feasibility.

The first step in determining if a microgrid is suited to a particular rural site is to compare the cost of the stand-alone microgrid to the cost of extending the existing grid structure. The cost of a stand-alone system is dependent on the load generated from the community, but is also dependent on the available resources. Wind and solar are obvious options because of their availability in remote areas, but diesel for generators should also be considered due to its widespread availability and capacity for consistent energy generation. These three options, plus battery storage, are at the heart of the microgrid solutions examined here. While options like geothermal and hydroelectric are completely viable, they have been intentionally omitted due to their geographical restrictions.

A study of 200 fictitious communities was performed in 50 unique locations with varying load profiles. For simplicity, and to ensure a variety of locations, one location was chosen over each of the 50 countries. The countries are shown in **Figure 2**. The countries were furthermore subdivided into five categories based on their economic standing. The countries presented in **Table 1** rank the selected countries based on estimates of annual income generated by the

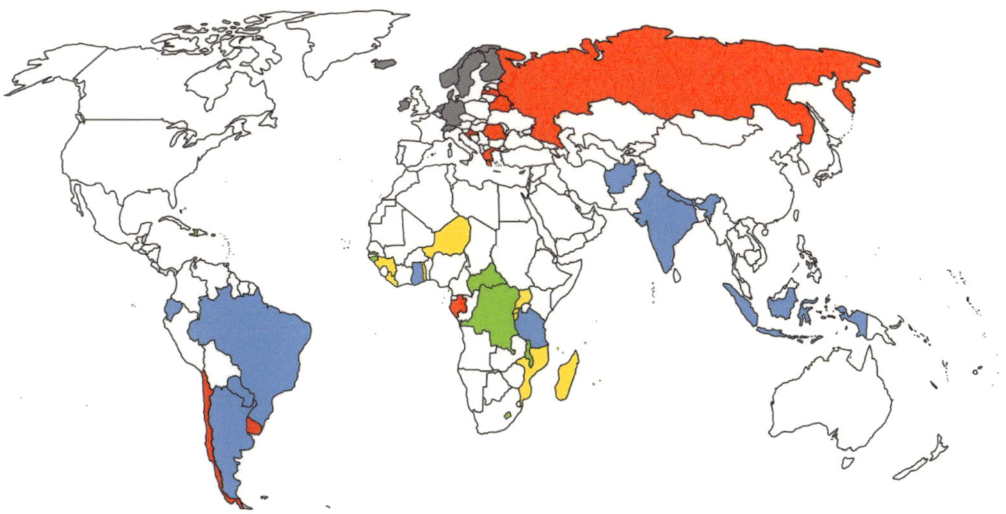

Figure 2. Selected countries.

poorest 10% of the population. This is calculated by taking the GDP produced by the poorest 10% [8] and dividing it by a tenth of the population.Class 1 countries generate less than $100 of income per year, whereas class 5 earns more than $10,000 per year. **Figure 2** highlights the countries chosen.

This distribution of communities better captures the purchasing power of lower income households. The World Bank currently draws the international poverty line at $1.90 per day [9]. This boundary encompasses all of Classes 1 and 2, as well as half of Class 3 in **Table 1**—in all, 46% of the communities listed can be considered to fall under the poverty classification.

In each of the simulations, four load profiles were constructed from two Rwandan villages. Rwamiko and Nyakabanda [10, 11] are used as representative load profiles and are shown in **Figure 3**. From these two profiles, two more were generated to increase the robustness of the model. The first was a combination of both original profiles into a fictitious village called Nyakamiko. This profile is simply the sum of the hourly originals. The final profile, titled Small Rwamiko, is a scaled down version of the Rwamiko profile. The four profiles weigh in at 164, 248, 433, and 74 kWh per day, respectively.

These load profiles are atypical of traditionally profiles where the peak occurs in the evening. The first distinction to recognize is that these are not home-level systems where occupants are away during much of the daylight hours. These systems account for the entire village including 200 households, 4 small grocers, 2 restaurants, 2 small shops, 1 dispensary, 1 office building, and 2 water pumps in Rwamiko; and 200 households, 1 coffee milling station and 1 restaurant in Nyakabanda. While daytime operations tower above the early morning and evening, it can be seen that the evening still carry a relative peak.

These profiles can then be loaded into HOMER—a microgrid optimization tool—to determine the best course of action for electrification. The microgrid design considers photovoltaic, wind,

Country	Poorest 10%	Class	Country	Poorest 10%	Class
Central African Republic	43.02	Class 1	Brazil*	1138.44	Class 3
Haiti	49.45		Indonesia*	1187.26	
Malawi	56.11		Argentina*	2001.52	
The Gambia	84.37		Chile	2469.82	Class 4
Guinea-Bissau	90.85		Gabon	2477.57	
Dem. Rep. Congo*	92.89		Russia	2975.52	
Lesotho*	93.08		Mauritius	3004.99	
Liberia	109.89	Class 2	Uruguay*	3193.29	
Mozambique	111.27		Belarus	3296.42	
Burundi	114.40		Croatia	3368.81	
Madagascar	116.84		Latvia	3458.23	
Togo	120.66		Greece	3654.73	
Comoros	121.51		Romania	3698.77	
Rwanda*	146.09		Seychelles	3730.37	
Fed. Sts. Micronesia	152.85		Singapore*	12,945.45	Class 5
Guinea	161.88		Germany	16,259.45	
Niger	162.40		Ireland	16,856.05	
Uganda	171.50		Denmark	16,998.03	
Afghanistan*	240.76	Class 3	The Netherlands	17,738.54	
Nepal*	245.59		Sweden	18,860.41	
Ghana*	273.91		Iceland	19,241.66	
Tanzania*	287.47		Finland	19,431.24	
India*	553.53		Switzerland	28,246.13	
Paraguay*	706.92		Luxembourg	31,499.45	
Ecuador*	1015.33		Norway	35,030.67	

*Countries were part of a previous study and not selected based on the same criteria discussed in the text.

Table 1. Rural GDP distribution.

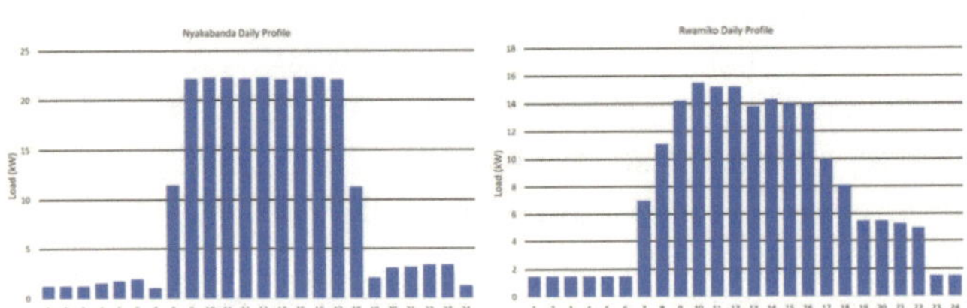

Figure 3. Nyakabanda (left) and Rwamiko (right) load profiles.

and diesel generation with battery storage. **Figure 4** illustrates this setup. It is important to note that the DC line carries two BWC turbine options. This allows for economies of scale to evolve by providing large-scale turbines (7.5 kW) for areas with high wind or smaller versions (1 kW) for low wind.

Besides the varying load profiles, the difference from site-to-site hinges on resource availability. The three resource factors are wind, diesel prices, and solar irradiation, which are based on longitude and latitude and can be obtained through HOMER [12, 13]. The alternative—grid extension—was priced at $15,000 per km and an exceedingly reasonable electricity price of $0.10 per kWh (it is not uncommon to see electricity prices more than double this figure).

The resultant optimization by HOMER reveals a wide spread of solutions. In **Table 2**, the solutions for three sites are shown, Ireland with its high renewable availability and diesel prices, Russia for its low diesel prices and poor renewable availability, and Guinea-Bissau for being a balance of the two. It is important to stress again that the communities named after their countries do not represent the renewable availability of the entire country, just at the coordinates selected for the rural site. A different site could have been selected for Russia which favored renewables more; this is the reason for choosing 50 countries to gain a diverse spread.

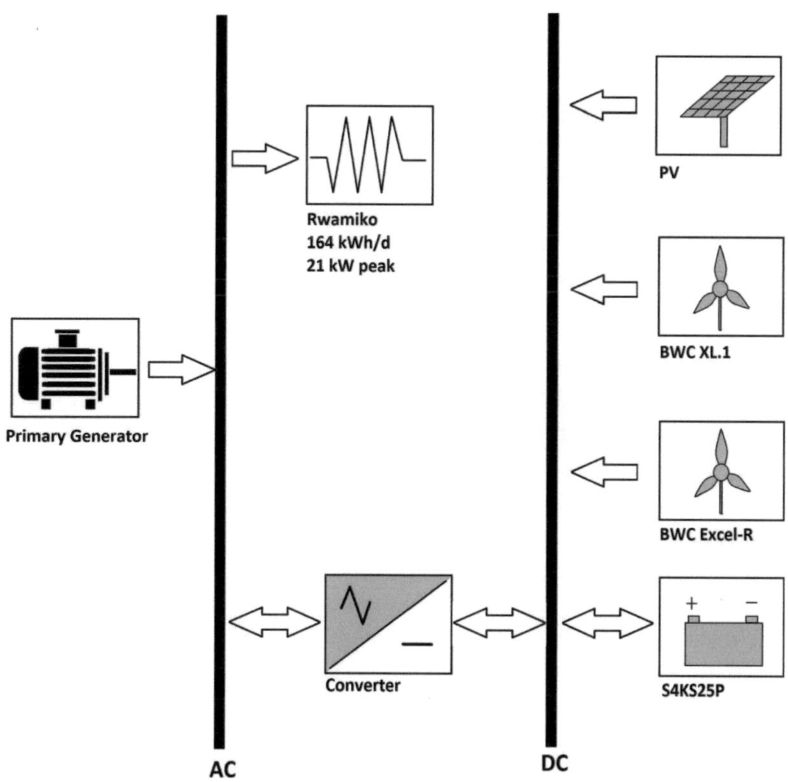

Figure 4. HOMER microgrid diagram.

Country	PV Array (kW)	Wind Turbines (kW)	Batteries* (Strings)	Diesel Generator (kW)
Ireland (Small Rwamiko)	5	7.5	3	0
Guinea-Bissau (Small Rwamiko)	30	1	4	0
Russian Federation (Small Rwamiko)	15	0	2	5
Ireland (Rwamiko)	30	15	2	0
Guinea-Bissau (Rwamiko)	60	0	6	0
Russian Federation (Rwamiko)	30	0	2	7
Ireland (Nyakabanda)	40	22.5	3	0
Guinea-Bissau (Nyakabanda)	90	2	3	1
Russian Federation (Nyakabanda)	40	0	3	10
Ireland (Nyakamiko)	90	30	4	0
Guinea-Bissau (Nyakamiko)	180	0	5	0
Russian Federation (Nyakamiko)	100	15	4	15

*Strings contain 3, 6, 9, and 12 batteries, respectively. Each battery with 1900 Ah or 7.6 kWh capacity.

Table 2. HOMER optimization solutions.

Some obvious and unsurprising patters begin to emerge. With its high wind speeds and decent solar activity, the Ireland site favors turbines and panels in place of diesel generation. However, it becomes apparent that Guinea-Bissau and Russia sometimes have more installed capacity than Ireland. This is indicative of Ireland having a greater renewable efficiency. Guinea-Bissau and Russia require spending more money to achieve the same level of electrification, and as a result they are less well adapted to utilizing their renewable resources.

Now that a microgrid solution has been obtained for each of the sites, these optimized microgrids need to be compared to the alternative: the national grid. The easiest method in this case is to compare the capital costs and maintenance plus the electricity which would be purchased—in this case—over a 20-year period. The location of the community does not matter to the microgrid because the microgrid is built in the community. Its proximity to other communities or the national grid is irrelevant. However, a large cost associated with electrifying thorough the national grid is the capital required to bring that electricity to the community. By comparing these two options, the breakeven distance is born. If the community is very close to a national grid tie-in point, then the cost to electrify through the national grid is low, and is thus the better option. At some distance, this flips. Eventually, the cost to extend the national grid outweighs the cost of the microgrid, and the microgrid becomes the better option. These breakeven distances are shown in **Figure 5**.

Each country on the x-axis corresponds to the four load profiles. As the load profile increases in size, so too does the breakeven distance. This is unsurprising since a larger microgrid would

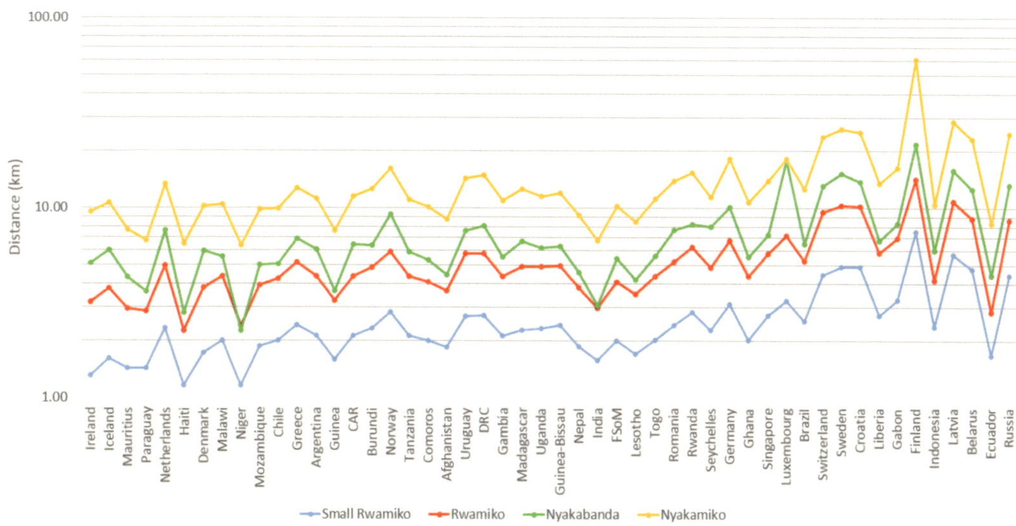

Figure 5. Breakeven distances for different load profiles.

be required to meet the demand, driving up costs. Many sites with the largest demand do not even exceed 20 km, a surprisingly low value. Additionally, the smaller loads can have breakeven distances in the 1–3-km range, rendering grid extension all but completely unreasonable. Those sites that do have unusually high breakeven distances, such as Finland, can be explained by a combination of high diesel prices and poor renewables, making it costly to build the microgrid no matter the combination.

To comprehend what proportion of a country this area may constitute, **Figure 6** highlights in yellow the areas of Brazil and Ghana within 20 km of the national grid. Communities similar to the Nyakamiko load profile in the white space of these figures are more likely to be better off with a stand-alone microgrid than with grid extension. Furthermore, the coverage map is realistically less than that depicted here. Distance to the grid and distance to a grid tie-in point are two separate ideas. The highlighted areas represent the corridor around the grid, but the grid cannot be tapped into anywhere. The voltage must be stepped down through a transformer, often requiring a complete substation to be erected.

When planning potential microgrid sites, this breakeven distance information can be translated into something more useful. The average breakeven distance from the above results can be estimated using Eq. (1). Since this is an average estimation, the equation and its constants were simply derived from a line of best fit through the case data in this work.

$$\text{Breakeven distance (km)} = 0.0316 \cdot \left(\frac{\text{kWh}}{\text{day}}\right) + 0.0565 \tag{1}$$

The average load in kWh per day of the proposed site can be plugged into the equation and an estimate of the breakeven distance can be obtained. If the community is closer to a grid tie-in point than the value returned by this equation, then grid extension is more likely viable. To

Figure 6. A 40-km wide corridor around existing transmission lines in Brazil (left) and Ghana (right).

increase the certainty, but decrease the rage of this equation, another version incorporating the third standard deviation can be generated. If the breakeven distance returned by Eq. (2) is less than the actual distance from grid to community, then it is extremely likely that a microgrid is the more viable option. Eq. (2) is derived through similar methods as Eq. (1), only including the third standard deviation as well. As a result, the larger the demand, the greater the uncertainty, hence the estimate becomes quadratic.

$$\text{Breakeven distance (km)} = 8.0 \times 10^{-5} \cdot \left(\frac{\text{kWh}}{\text{day}}\right)^2 + 0.0215 \cdot \left(\frac{\text{kWh}}{\text{day}}\right) + 1.6629 \tag{2}$$

The two equations are presented graphically in **Figure 7** become clear. Sites that fall below the orange line should be studied further to determine if they should be stand-alone microgrid systems. Sites above the orange line are highly likely to be better off as stand-alone microgrid systems.

These relatively short breakeven distances highlight the microgrid viability when compared directly to the national grid. Additionally, the parameters chosen in this setup are fairly conservative. This breakeven distance only decreases further when a more typical electricity price is considered, or when new generation to meet the demand is factored in. At a high level, the microgrid is clearly a viable option for areas that are not already being serviced by the national grid. Furthermore, this microgrid versus national grid decisions can be streamlined through the use of the breakeven distance prediction equations. Determining what type of

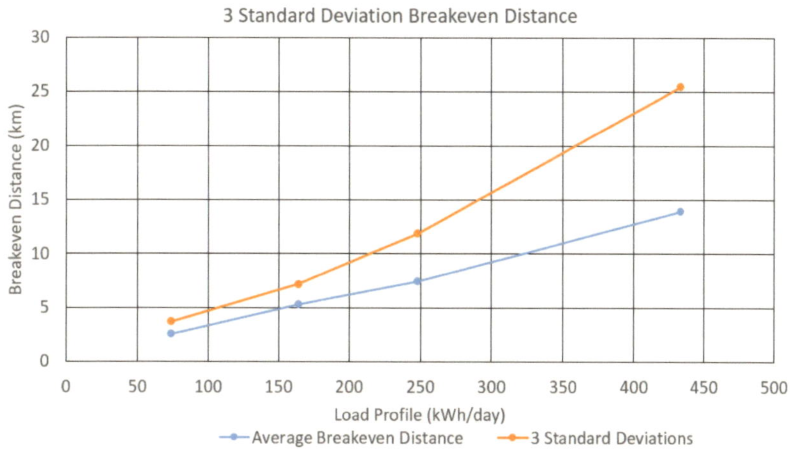

Figure 7. Average and three standard deviations breakeven distance.

electricity and where it comes from is the first step in rural electrification. The remainder of this chapter takes a closer look at several real-world examples, with a focus on how to identify potential sites, as well as what technology and process improvements can be made to better the reliability and decrease the cost of a microgrid.

4. Analysis of identification schemes for electrification in rural communities

Determining whether or not a community should be electrified with a microgrid or the national grid connection is only part of the problem. One of the larger issues is finding communities, measuring their load potential, and categorizing their current level of electrification, if any. The largest factor in determining the payback period for an investment is a substantial understanding of how much true demand exists. It is easy to ask someone if they want electricity, and they will most likely say yes. Since the individual does not front the capital costs, there is little risk to them. But the entity funding the project needs to calculate an electricity price based on supply and demand. If the price is too high, no one will use the electricity, and if it is too low, then the microgrid or extension will not pay for itself and thus continued electrification is dissuaded.

Since the data on rural electrification is sporadic, the United Nations developed the global tracking framework "United Nations Sustainable Energy for All" (UNSE4ALL). This allows for different levels of electrification to be compared directly independent of location. Traditionally, electrification has been measured solely by the number of people connected to a national grid. While this data is relatively easy to access and measure, it is not entirely accurate [14, 15]. Numerous other metrics already exist, such as measurement of minimum energy consumption thresholds [16], tracking of income-variant energy demand [17], the

"Multidimensional Energy Poverty Index" (MEPI) [18], and the "Total Energy Access" (TEA) [19]. These types of metrics treat electricity as an output of the system, instead of an outcome, and fail to truly capture the whole picture.

Before the use of the multidimensional approach, electrification was categorized into levels. Communities fell into one of eight steps based on the scope and source of electricity (if any) [20]. These steps are shown in **Table 3**.

While this method provides multiple levels of classification, it does not account well for emerging or combined technologies. If a system does not fit neatly into one of these categories, it must be either misrepresented or reported as an exception. In contrast, UNSE4ALL classifies electrification by usage instead of categorizing it by production. By defining electrification by attributes, similar technologies, which provide the same or similar effects, can be distinguished by quality of energy produced.

UNSE4ALL classifies electrification by the service provided (lighting, refrigeration, etc.) as well as peak capacity, duration, evening supply, affordability, formality, and quality [21]. These tiers are shown in **Table 4**.

What remains are five different tiers of electrification. Tier 0 not corresponding to any particular type, while tiers 1–5 evaluate on the criteria above. Tiers 4 and 5 constitute either a grid connection or a reliable but independent mini-grid with national grid backup. Power here can be used in the home for the majority of the day, and the capacity exists to run most appliances found in the home. Tiers 1–3 represent varying electricity access through power available, and duration of availability as detailed in **Table 4**.

The nonbinary nature of metrics like the UNSE4ALL electrification ranking system offers a substantial basis on which to define energy access. An alternative to the UNSE4ALL method is the Energy Sector Management Assistance Program (ESMAP) that also provides a multitiered approach to defining electrification. Unlike UNSE4ALL, ESMAP measures electrification independently of technology and seeks to measure the quality of access. ESMAP measures seven criteria: capacity, duration, reliability, (technical) quality, affordability, legality, and health/safety. **Table 5** outlines the tiers and criteria [22].

With this method, the performance for each criterion is evaluated and is then assigned the electrification classification tier based on the lowest performing attribute. The tier rating for each household is calculated by applying the lowest tier rating across all the criteria [21]. The effect of this classification is presented in Eq. (3).

$$\text{Index of access to energy} = 20 \times \sum_{k=0}^{5} P_i \times k \tag{3}$$

where P_i is the rate of households in the kth tier. This method allows for a level of customization whereby the organization employing this tool can easily set targets to be tracked. This multitiered framework is a significant shift from past binary systems and more accurately assesses the level of electrification in rural communities. There are limits to this approach as well; the information gathered does not directly segue into a solution for the best course of

http://dx.doi.org/10.5772/intechopen.68622

action. While further research is required to optimize these methods and formulate a transition into electrification solutions, UNSE4ALL, ESMAP, and other multitiered frameworks are the best course of action in identification.

Steps	Energy Source	Uses
Step 0	Candles, Kerosene	Lighting
Step 1	Battery powered torches	Lighting, mobile phone charging and radio
Step 2	Car and motorcycle batteries	Step 1 + small TVs and low wattage appliances
Step 3	PV lanterns/torches	Same as Step 2
Step 4	Solar home systems	Step 3 + small refrigerators
Step 5	Isolated Minigrids	Step 4 + fans, air conditioning, full size refrigerators, motors/electric pumps
Step 6	Grid-connected Minigrids	Same as Step 5
Step 7	Grid-based power	Same as Step 5

Table 3. Classic electrification rankings.

Energy Access According to UNSE4ALL	**No**	**Basic**	**Advanced**			
Attributes	**Tier 0**	**Tier 1**	**Tier 2**	**Tier 3**	**Tier 4**	**Tier 5**
Services		Task light and phone charging	General lighting, television, and fans	Tier 2 and any low power appliances	Tier 3 and any medium power appliances	Tier 4 and any higher power appliances
Peak Available Capacity (Watts)	–	>1	>20/50	>200/500	>2000	>2000
Duration (hours)	–	>4	>4	>8	>16	>22
Evening Supply (hours)	–	>2	>2	>2	>4	>4
Affordability	–		✓	✓	✓	✓
Formality (Legality)	–			✓	✓	✓
Quality (Voltage)	–			✓	✓	✓
Indicated Minimum Technology		**Nano-grids/ Micro-grids,** Pico-PV/Solar lantern	**Micro-grids/Mini-grids,** Rechargeable batteries, Solar home systems	**Micro-grids, Mini-grids,** Home systems	**Mini-grids,** And grid	**Mini-grids,** And grid

Table 4. UNSE4ALL global tracking framework tiers.

		Tier 0	Tier 1	Tier 2	Tier 3	Tier 4	Tier 5
1. Capacity	**Power***		Very low power ≥3 W	low power ≥ 50 W	Medium power ≥ 200 W	High power ≥ 800 W	Very high power ≥ 2 kW
	AND Daily Capacity		≥12 Wh	≥200 Wh	≥1 kWh	≥3.4 kWh	≥8.2 kWh
	OR Services		Lighting of 1000 lm-hrs per day and phone charging	Electrical lighting, air circulation, television, and phone charging are possible			
2. Duration	**Hours per Day**		≥4 hrs	≥4 hrs	≥8 hrs	≥16 hrs	≥23 hrs
	Hours per Evening		≥1 hrs	≥2 hrs	≥2 hrs	≥4 hrs	≥4 hrs
3. Reliability						≤14 disruptions per week	≤3 disruptions per week. Total ≤ 2 hrs
4. Quality						Voltage problems do not affect the use of desired appliances.	
5. Affordability					Cost of a standard consumption package of 365 kWh per annum is less than 5% of household income.		
6. Legality						Bill is paid to the utility, prepaid card seller, or authorized representative.	
7. Health and Safety						Absence of past accidents and perception of high risk in the future.	

*The minimum power capacity ratings in watts are indicative, particularly for Tier 1 and Tier 2, as the efficiency of end-user appliances is critical to the real level of capacity, and thus the type of electricity services that can be performed.

Table 5. ESMAP electrification tracking framework tiers.

Once communities lacking sufficient access to electricity have been identified, work can then progress to rectify the problem. Knowing which areas are in need of electricity and devising an electrification method are two incredibly independent functions. The next section delves deeper into two case studies—one in Brazil and the other in Rwanda—in order to ascertain what a customized solution for these might look like.

5. Case studies on rural loading in Brazil and Rwanda

With the understanding that microgrids are a viable option, and armed with the tools to identify communities and classify them based on their access to electricity, we turn our attention to two specific case studies. First, Brazil's Amazonian region contains an abundance of natural resources, but due to its expanse and remoteness, it remains largely underdeveloped. Providing access to electricity is one critical step in the direction of eradicating poverty in the area.

A population of more than 5.6 million people in the Brazil's Amazonian region are living outside of city centers, and approximately 155,000 rural households are unelectrified [23]. Access to many of these communities is not possible by road; the rainforest is too thick. Instead, inhabitants move via waterways or, when possible, air. Grid infrastructure improvements are severely limited to areas with road access, making expansion to these sites extremely difficult. A common solution in rural areas desperate for electrification is to use diesel generation. Typically, 10–100 kVA generators are sourced for this kind of application, but the high price of diesel and an unreliable supply make for a bad combination. Apart from the cost and access, diesel is a pollutant fossil fuel, whose affects are compounded by poor generator efficiencies from ill-maintained equipment. However, if properly tapped, there are an abundance of renewable energy sources (RESs) including solar, wind, hydroelectric, and biomass. These RESs can offset or completely replace the diesel fossil fuels currently utilized.

Brazil is in the midst of a push to offer electrification for these rural areas, which makes it an excellent case study. The program Brazil has adopted is called "Luz Para Todos" or "Lights for All" (LFA). Currently, LFA has provided electrification for 2.9 million households or 14.4 million people [24]. Unfortunately, while this scheme has been successful, it lacks the ability to penetrate into the more remote and rural areas. The initial phase of the plan was a national grid extension network through relatively populated areas. This equated to a large number of people benefited through a relatively short distance of grid extension. As the distances increase and population density decreases, this tactic becomes increasingly less effective.

Previously, **Figure 1** illustrates Brazil's transmission network—both existing and planned. In the figure, the solid lines represent existing power connections, whereas the dashed lines are proposed extensions [25]. This interconnected system is capable of generating and distributing 138 GW of installed power. Despite this there are still a large number of people lacking access in the dispersed settlements. In contrast to the integrated grid network, there are numerous independent isolated systems of both diesel generation and thermoelectric in the North-West region of Brazil, where the interconnected systems do not penetrate [26].

Brazil has a unique and widely varying electrification rate. The rural electrification rate of Brazil is approximately 97%, whereas specifically in the Amazonian region, rural electrification drops to 61.5% [4]. This drop illustrates the extent to which this region has been isolated and ignored. **Figure 8** presents this unequal distribution into focus by shading areas with less access a darker color. Immediately it becomes obvious the Amazonian region is seriously underelectrified [27].

Eirunepe city, located 1160 km from the region's capital Manaus, is an excellent example of an Amazonian electrification problem. While Eirunepe city has electricity through diesel generation, the surrounding communities do not. This is because the diesel comes from Manaus by boat. The journey is approximately 2400 km along the river. Because of these massive distances involved, the cost of electricity is increased, and thus expansion becomes difficult. Torre de

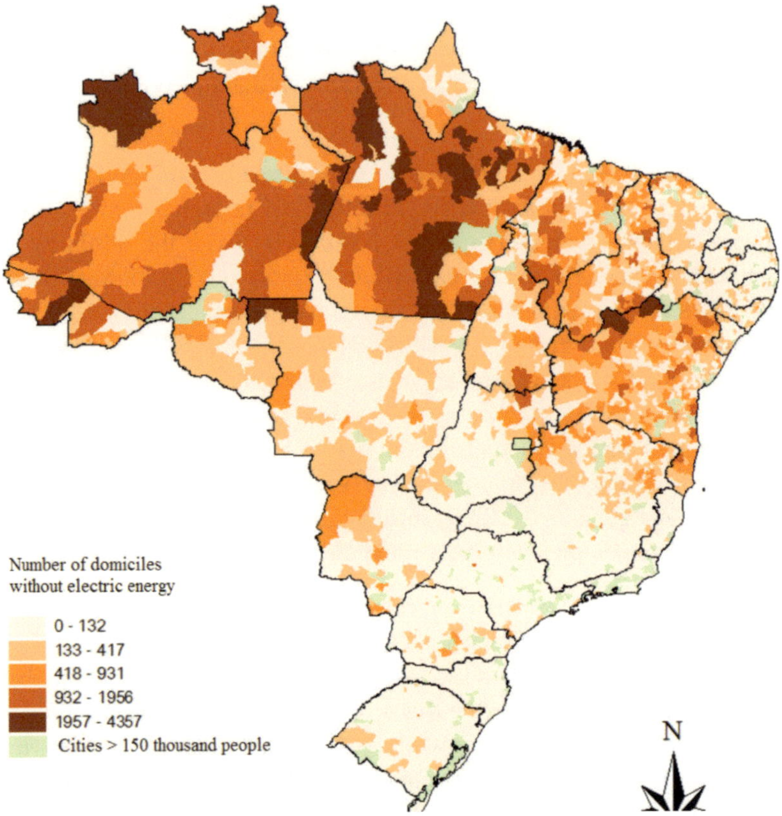

Figure 8. Areas lacking household access to electricity.

Lua, a community near Eirunepe city, consists of approximately 25 households near the river. The purpose of the remainder of this section is to determine the specific viability of a renewable energy-based microgrid system including solar and hydropower. Because the community is unelectrified, there exists no real-world data to base usage on. Instead, with the size and population of the community, and expected load profile can be estimated. **Table 6** shows the breaks down the power requirements for a community like Torre de Lua, and **Figure 9** shows the aggregated demand for the entire community.

Loads	Power consumption (W)	Number of items	Demand (W)	Hours per Day
LED Lights	9	5	45	4
Refrigerator	100	1	100	24
Radio	50	1	50	4
Television	100	1	100	4
Fan	120	1	120	4

Table 6. Single household load in Torre de Lua.

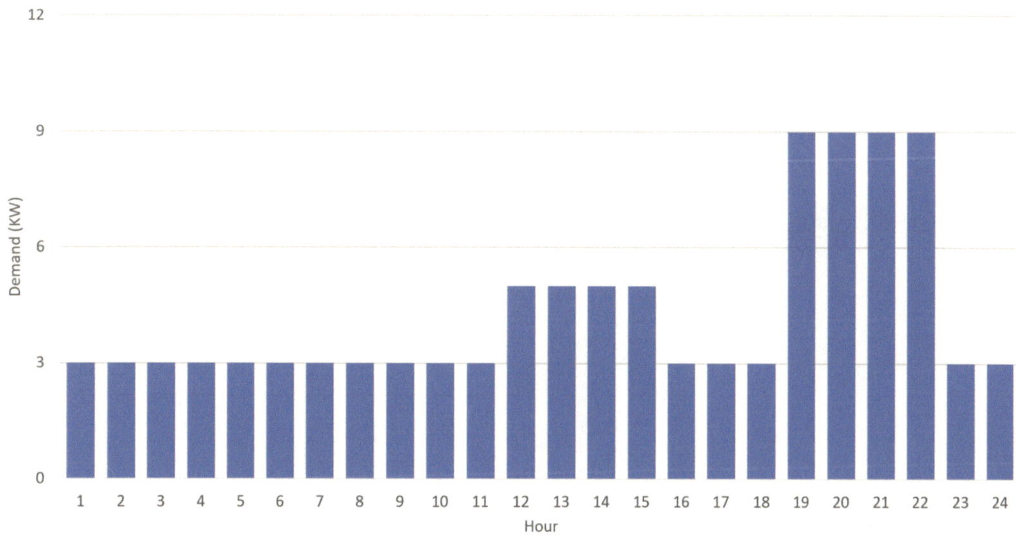

Figure 9. Torre de Lua community load profile.

Once a load profile has been established for the community, an optimization can be run using HOMER to determine the appropriate makeup of generation and storage. **Figure 10** shows the proposed generation and storage in connection with the demand. A converter is used to move power from the AC and DC buses as the batteries or solar generation power the load, or as the hydropower recharges the batteries.

The calculations for the available solar power are based on the NASA radiance database for Eirunepe city, which come from longitude and latitude. **Figure 11** illustrates the radiation expected on the monthly bases, which is directly translated into kWh per m^2 per day, which when coupled with the efficiency of solar panels, an exact power can be calculated.

The second resource—hydropower—can similarly be predicted based on location. The Water National Agency (ANA) in Brazil tracks flow rates of various rivers. The Taruaca River, on which Torre de Lua sites, has a station known as Envira (Station Code 12680000) which is the closest to Torre de Lua. **Figure 12** shows the average stream flows in the area, peaking in June [28]. This is complimentary to the solar resource, as July through October are the highest solar times, whereas there is a drastic dip in flow rates for these months.

The remaining inputs are cost and sizing options. Inputting a price list for HOMER and giving a range of options for sizing to be compared (e.g., solar sizing options: 0, 5, 10, and 15 kW), HOMER then optimizes the generation and storage solution. **Table 7** shows the cash flow expected for a solar-hydro system to be able to supply the energy required.

HOMER's optimal solution returns a PV-Hydro-Battery mix which proves to be revenue neutral by charging $0.091 per kWh. Compared to the price of electricity from the Amazonas Electrobras utility of $0.090 per kWh, this is a highly competitive solution.

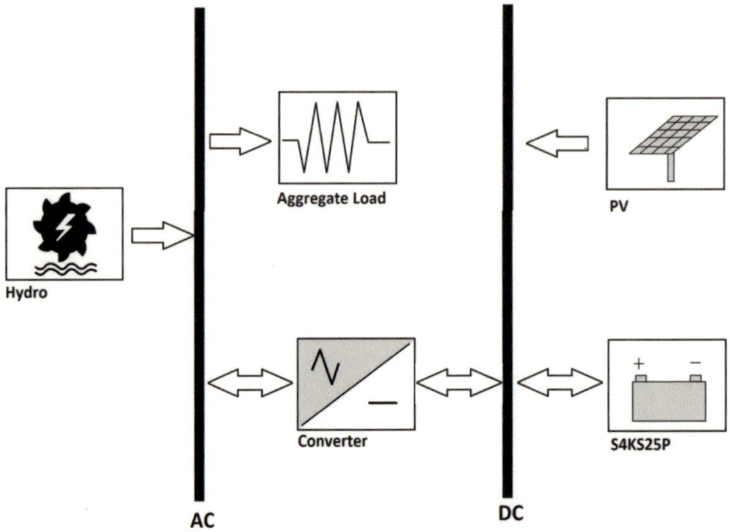

Figure 10. Torre de Lua HOMER microgrid schematic.

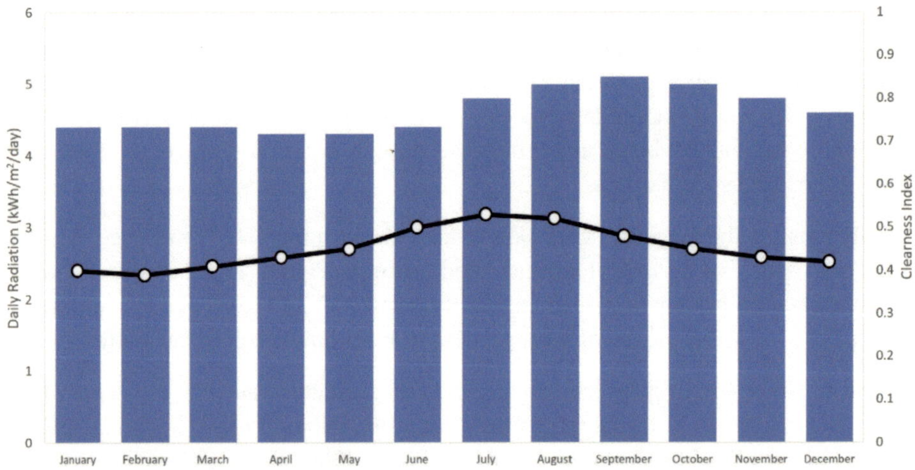

Figure 11. Torre de Lua solar radiance.

The first of the two case studies proves to be feasible, but this is one isolated view. For the second community, we travel thousands of miles and a continent away to Rwanda. Rwanda provides an excellent case study. It is a highly populated land-locked resource-poor country where 75% of the population lives in rural areas [29]. Similar to Brazil, Rwanda has also rolled out a version of LFA called the Electricity Access Roll-out Program (EARP); and again, the aim is to provide a higher rate of electrification for the rural population; however, the total installed capacity for Rwanda's 12 million people was 115 MW as of 2014 [30]. This shortage in supply means that simple grid extension will be grossly insufficient, as there is not enough power

Component	Capital	Replacements	O&M	Fuel	Salvage	Total
Generic Flat Plate PV	$18,000.00	$0.00	$0.00	$0.00	$0.00	$18,000.00
Hydro	$15,500.00	$0.00	$646.38	$0.00	$0.00	$16,146.00
Discover 12VRE-3000TF-L	$2200.00	$629.05	$2895.80	$0.00	($257.66)	$5467.20
Converter	$2750.00	$986.44	$0.00	$0.00	($185.66)	$3550.80
System	$38,450.00	$1615.50	$3542.10	$0.00	($443.31)	$43,164.00

Table 7. Cash flow for RES system.

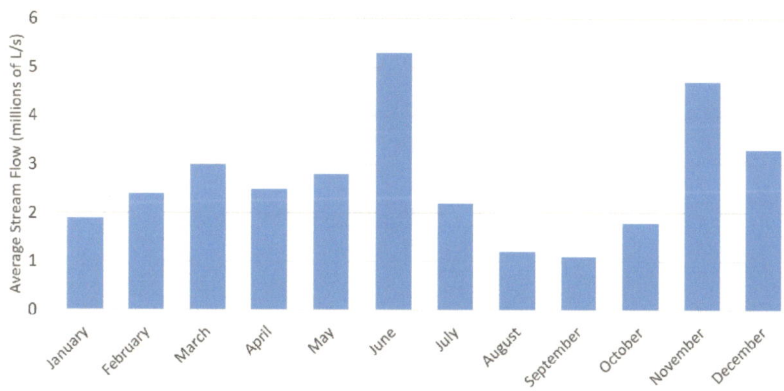

Figure 12. Torre de Lua river flow rates.

currently generated. Coupled with the fact that Rwanda has no ocean access for importing petroleum or other fossil fuels cheaply (Rwanda does not have natural reserves, with the exception of some methane production on Lake Kivu), a greater emphasis must be placed on the role of renewable energy.

A community called Nyakabanda is located 40 km west of the capital. Nyakabanda is an unelectrified community a little over 11 km away from the national grid, as shown in **Figure 13** [31]. Similar to the Brazilian case, the first step is to determine a load profile.

Because the region in which Nyakabanda is located is known for growing coffee, two milling stations are included in the load profile, along with household lighting shown in **Table 8**.

This may appear to be a very conservative, bare-bones electrical demand, however, because of its unavailability, current electricity usage is limited to running lights off of car batteries (if any electricity is used at all). This proposed system would be to run industrial equipment brought in by the government or private business (in the event electricity was available) and lighting. There is no economic room to account for larger loads, such as refrigerators or washing machines because the community would not immediately buy these appliances due to the high capital cost. Instead, the microgrid must be allowed to grow naturally.

Similar to the Brazilian case, a HOMER grid is generated. However, in this case, wind is considered along with solar and hydro, as shown in **Figure 14**. When the simulation is run,

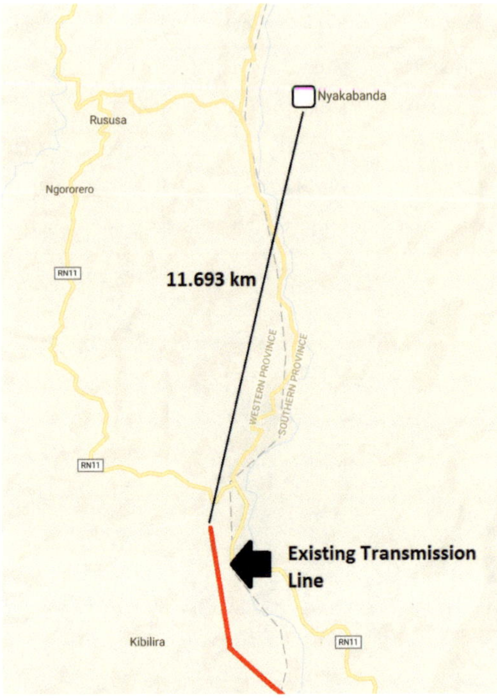

Figure 13. Relationship between Nyakabanda community and closest grid access point.

	Load	Power
Household	LED Lights (2)	5-Watts Each
	Phone Charger (2)	3-Watts Each
200 households		
Coffee milling station	Milling Machines (1)	11-kW Each
	Outdoor LEDs (2)	5-Watts Each
2 Milling Stations		
Restaurant	Indoor LEDs (3)	5
	Outdoor LEDs (1)	5
1 Restaurant		

Table 8. Estimated Nyakabanda loads.

the optimal solution contains 5 kW of solar, 11 kW of microhydro, and 16 kW of battery storage. The wind in this location was neither strong nor constant enough to financially justify a wind turbine at this scale. Compared to the grid price of approximately \$0.26 per kWh, the levelized cost of energy for this system is approximately \$0.24 per kWh. The breakeven distance for this system was approximately 1 km. Since the community is 11 km

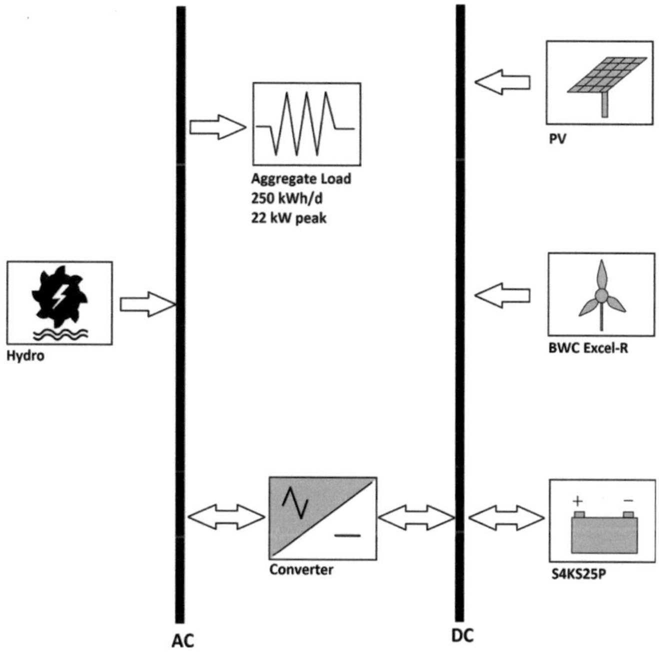

Figure 14. Nyakabanda community HOMER microgrid setup.

from the existing grid infrastructure, it stands to reason that the microgrid option is more economical.

Both the Rwandan and Brazilian case studies have shown that isolated microgrids are economically feasible. These are rough snapshots of what is required for electrification, but they highlight the major aspects. Thus far we have seen the viability of microgrids compared to grid extension, and the actual "per kWh" costs of implementing the microgrid. But these solutions represent an out-of-the-box electrification scheme not tailored to rural life. Because the way electricity is consumed in the West and in rural developing locations differs drastically, new technology, process improvements, and innovative uses can drastically reduce wasted energy and thus further bring down the costs. The next section in this chapter delves into these changes and analyzes how much of an impact they can have on a microgrid.

6. Innovative enhancements for microgrid optimization

So far the microgrid has been treated as a response to a static load profile. Generally, this is acceptable because the electrical loads receive no feedback from the supply as to the amount of over/under production. If a new load is added to an already saturated system, voltage drops or blackouts may occur. Since these microgrids are designed and built from scratch, extra measures can be taken to reduce peak loads and distribute power consumption—easing the burden on the microgrid.

Batteries are often used in microgrids as a means of load shifting. When renewable energy generation is producing power in excess, the batteries can be charged and when the load outweighs the supply, the power stored in the batteries is used. Unfortunately, traditional batteries can be expensive and have limited cycle use before they need to be replaced. Alternative power storage methods are available—such as pumped hydro or flywheels—but a more grassroots solution reshapes the problem. Rather than store energy to match supply to meet demand change the nonessential demand to match supply.

6.1. Smart energy management for health Centers

Examples of how loads can be rearranged are readily available. One system, known as "smart energy management for health centers" (SEMHCs) clearly analyzes how this process works [32]. Rural health centers in particular are more prone to suffer from chronic power outages because of the nature of their loading. Clinics often have high-power equipment that is only run for a short period of time. If many of these devices are powered simultaneously, it can overload the individual PV systems the run off. Since sufficient battery banks can be cost prohibitive at this scale, the clinics are at the mercy of whatever their arrays can produce. Alternatively, when the system is not being overloaded, energy that is not being used is lost. Instead, low-tech scheduling systems can organize and shift loads to use the electricity more wisely.

Figure 15 illustrates this point further. The dotted line represents the power curve generated by a PV array throughout the course of a day, and the gray shaded blocks represent the loads. From t_0 to t_1 and between t_4 and t_5 the demand exceeds the generation, while the white space between them represents energy generated by the array that is lost.

Scheduling patients' services without the knowledge of power available ultimately hinder the services provided. Traditionally, clinics operate on a first-come, first-served basis, and if overloaded, the power will cut out with waiting patients. With the already low density of health centers in rural developing areas, patients sometimes have to travel large distances in

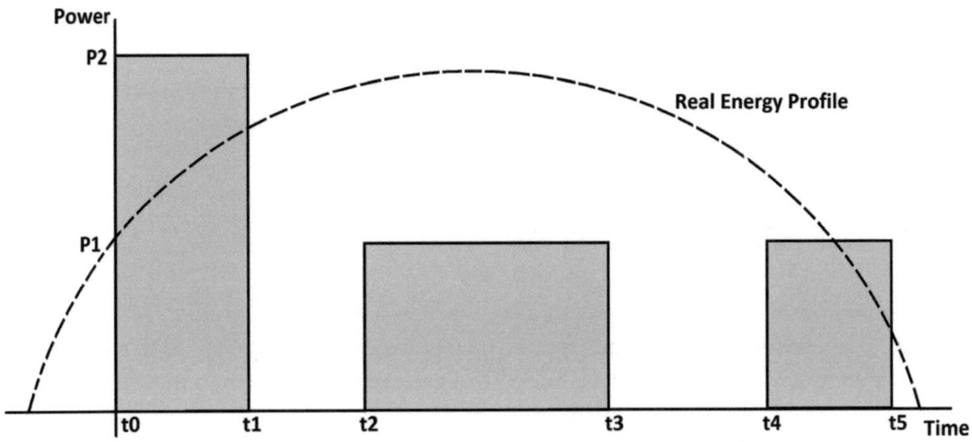

Figure 15. Unused energy in storage-less systems.

order to arrive at the clinic. When the clinic then does not operate due to power issues, access to health services is essentially nonexistent.

SEMHC addresses this issue by scheduling services based on available energy and the priority of the service. The program starts by assessing the solar power production available, defined in Eq. (4) where s is the surface of the PV array, k is a constant for in-line energy loss due to increased temperature, incident angle of radiation on the array, shading, and panel degradation, and $R(t)$ is the average solar radiation on the PV panels over a specific period of time t. The values for $R(t)$ can be pulled from satellite data from organizations such as NASA.

$$E(\Delta t) = s \cdot k \cdot R(T) \cdot \Delta t \tag{4}$$

Next, the algorithm assigns a value C_i to the service required, where C represents the power rating and i is the device ranking or priority. The device is expected to operate for a closed time interval d_i. The process is optimized mathematically with Eq. (5).

$$\max_{n\epsilon\mathbb{N}}\left(\int_{t_d}^{t_d+\min(d_i)} s \cdot k \cdot R(t) \cdot \Delta t - \sum_{i-1}^{n} C_i d_i - A\right) \tag{5}$$

The constant A represents the base load of the clinic—devices that are always on and consuming power. t_d represents the lowest time for which the generation power exceeds the nominal power of the devices. The final assurance is to guarantee that the demand does not exceed the power produced P_n by the panels, eliminating the possibility of overloading the system. Condition Eq. (6) then must be true at all times.

$$\sum_{i=1}^{n} C_i d_i \leq P_n \tag{6}$$

Since this is a priority-driven system, any excess power not used by critical loads can be used low low-priority low-power demands. The system is shown in **Figure 16** graphically. At the onset of the program, t_{min}, there are five loads already scheduled and running. Since the power available exceeds the demand, a new load NL3 is moved from standby to operating. NL3 is used because the higher priority loads would render condition (6) false.

After the completion of any load, the system reevaluates which waiting loads can be fit into the unused capacity. With this low-cost, low-tech solution to power management, unnecessary outages can be avoided and wasted energy can be minimized. This is only one specific case of a computer-based microgrid improvement. With the advent of smart meters, control at a household level can be more than binary (on/off) as seen with the health clinic example. With built-in wireless communication hardware, individual loads inside a household can be monitored and controlled.

6.2. The load attenuating stochastic simulator

The load attenuating stochastic simulator (LASS) utilizes the abilities of networked smart meters to control loads at a fine granularity, delivering or cutting power at the appliance level

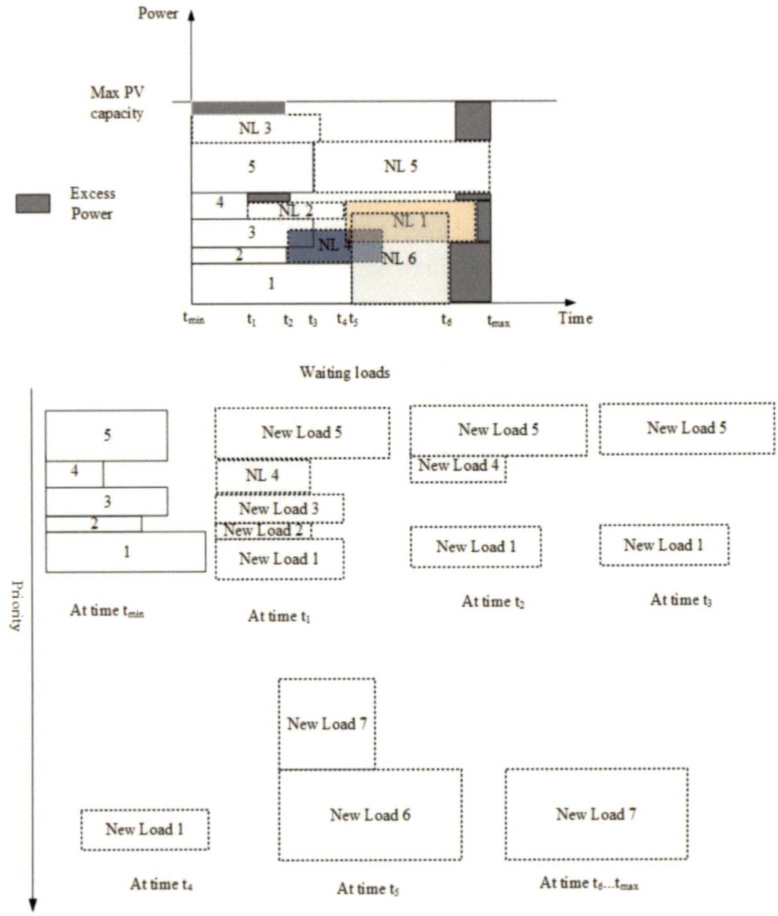

Figure 16. Rearrangement of loads to match SEMHC scheduling.

depending on the customer's desires or requirements. This creates a tiered electrification scheme in which customers can pay a reduced rate under the condition that their power may be throttled first in the event, the demand exceeds generation [33]. While the results in this field suggest that the most economical solution is to provide sufficient capacity to meet all demand, once there is a limitation in supply, customers are best served by having their loads clipped versus having the entire system overload and experience blackouts. This works in much the same way as SEMHC, except it is applied over an entire microgrid, not an individual consumer.

To prove that load clipping increases overall electricity distribution, a simulation is used whereby LASS uses a Probability Mass Function (PMF) as an input over each time step for the possible loads and generation over a microgrid system for a traditional weekday and Saturday [33]. The simulator generates demand and loads for each time step based on a fictional microgrid setup. Inside this simulation, there exist two tiers of customers: customers whose power can be clipped and customers whose power cannot be clipped. The goal is to reduce the probability of power outages by determining what percentage of customers must be in the clipped category.

The base case (where no customers are clipped) and four other scenarios where 70, 75, 80, and 100% of homes can be clipped are examined. In the control case with no clipping, there are instances where power will have to be cut for multiple hours per day for both weekday and weekend profiles. The severity of this can be seen in **Figure 17**. The reverse is obviously true as well. If you cut power to 100% of homes, then the probability that there will be an outage is essentially nonexistent, since instead of overloading the microgrid, it has just been shut down completely.

Whether or not power has been clipped, a power outage that has been avoided is only part of the problem. For any form of electricity distribution system to be viable, the maximum amount of electricity needs to be sold without shutting down the entire system. Protecting the grid by turning power off to everyone means that the utility (or distribution owner) loses income. Instead, there exists a balance between providing electricity to the maximum amount of people without increasing the risk of an outage. **Figure 18** highlights the amount of power consumed (or sold) respective to each clipping scheme. No clipping is still the worst performing, since without clipping an outage is nearly guaranteed, and therefore no power is consumed by anyone. Again, the inverse is not ideal. If you cut power to everyone, then again, no power is sold or consumed. When 75% of customers' power is clipped, the probability of an outage is less than 1%—nearly the same as any of the higher schemes. However, clipping only 75% of homes compared to 80% generates more revenue through more distributed power.

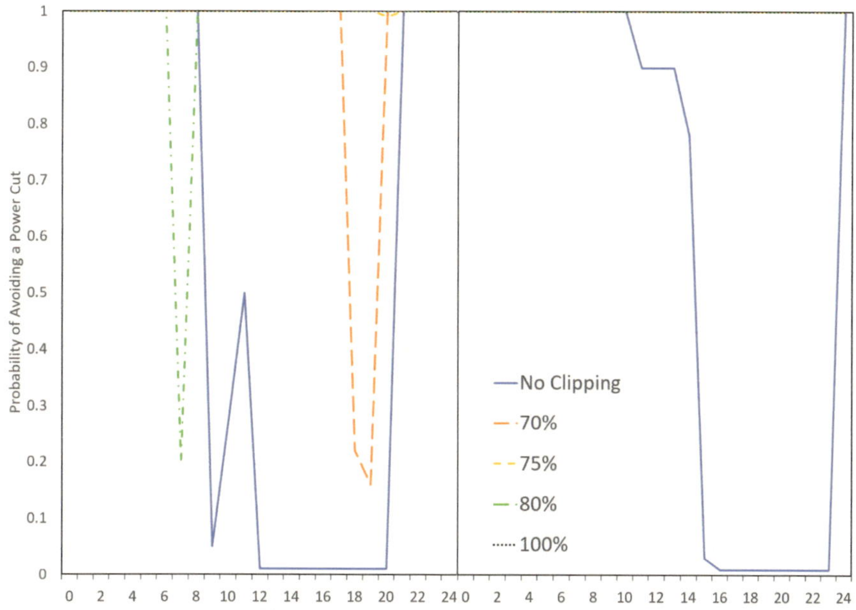

Figure 17. Probability of sufficient power for multiple clipping tiers.

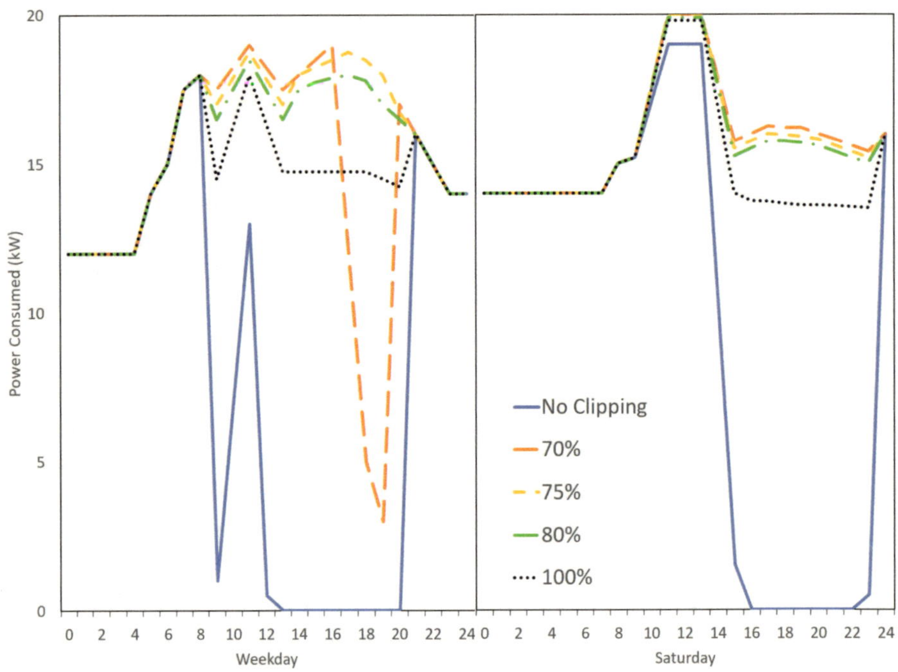

Figure 18. Power consumed based on an amount of home clipped.

The figures provide excellent visuals for hour-by-hour outage probability, but the root of the issue focuses on the total time, or overall percent chance of an outage. **Table 9** shows the solutions for the weekday and weekend times.

	Percentage of homes clipped	Expected total power cut duration (hours/day)	Expected energy sold (kWh/day)
Weekday	0 (No Clipping)	11.2303	197.17
	70%	1.9474	360.59
	75%	0.0137	392.31
	80%	0.0005	385.33
	100%	0.0005	356.29
Saturday	0 (No Clipping)	9.4835	240.20
	70%	0.0005	396.87
	75%	0.0005	392.89
	80%	0.0005	388.92
	100%	0.0005	373.13

Table 9. Optimum clipping rates.

Clipping to 75% on weekdays and 70% on Saturdays results in the highest amount of energy sold. While the power outages required for weekdays is higher than if more homes were clipped and the overzealous clipping actually leads to less power sold. This does not increase the robustness of the system, but instead takes an overly cautious approach and shuts down power unnecessarily.

Choosing the correct percentage of homes to clip is critical in the cost effectiveness and reliability of microgrids. While this study has provided an excellent starting point or generalization, each microgrid will have to be fine-tuned to its particular power generation and loading curves. The goal of this section was to demonstrate that demand-side changes can have a meaningful impact on both the reliability and affordability of microgrid setups.

7. Agricultural and biomass generation for baseloading and critical demands

The final section of this chapter explores a rural-specific energy source not often considered in microgrids: agriculture. About 78% of the world's poor lives in rural areas and relies on agriculture for both food security as well as household income [34]. Biomass is a large and often untapped rural resource that can provide a significant portion of on-demand power.

Burning agricultural waste in small steam furnaces allows for localized generation utilizing an abundant and proximal resource. Generation units 10 and 50 kW are already in production for exactly this kind of use [35]. Village industrial power (VIP) operates in East Africa and offers an off-the-shelf 10 kW unit, which can be transported in the back of a small pickup truck. Agricultural waste is burned directly and the self-contained unit can generate electricity. Used in tandem with solar or wind generation, a biomass unit can act in the same way that a diesel generator traditionally assists microgrids. Biomass can be stored and the unit brought online when demand is high.

The viability of biomass as a generation source primarily hinges on its availability, which is what makes this option suitable for rural and not urban use. Case studies have been carried out which are generally geographically limited. Specifically, the Punjab region of India was analyzed and it was determined to house vast untapped biomass resources. If the Punjab example is followed, the first step is determining what types of biomass are available. The six major biomass options from crops grown in the region are outlined in **Table 10**, and are categorized into four sections [36].

The A1–A4 category ratings simply separate the styles of biomass, where A1 represents generic unused dry biomass and A2 represents woody biomass. It is important to note that the energy reserves of biomass do not come directly from the sum of biomass itself, rather, the

Category	Type of biomass	Name of crop
A1	Straw	Wheat
		Barley
		Paddy
		Seasum
		Pulses
A2	Stalk	Cotton
		Maize
		Arhar
		Rapeseed and Mustard
A3	Bagasse*	Sugar Cane
	Tops and Leaves	Sugar Cane
A4	Cobs*	Maize
	Husks*	Paddy
	Shells*	Groundnuts

*Indicates processing residue.

Table 10. Identification of available biomass [36].

actual unused biomass represented here accounts for the subtraction of biomass used for domestic purposes, animal fodder, heating, etc. These values here represent biomass burned by farmers in the field. This biomass is truly unused and serves no other purposes.

The amount of energy available comes from the product of the present supply and the lower heating value (LHV). This is taken through the amount of cultivated land with each of the above-described crops, as well as the reduction due to moisture. The final available amounts are shown in **Table 11**.

With tens of millions of tonnes of unused biomass going to waste, the potential for on-demand electricity generation is immense. In total, there exists over 200 TJ of unused energy from all the biomass over the entire Punjab region. Obviously, this is only one specific instance, and many other areas may come in well below this, but even at a fraction of the potential, the remaining energy is immense. From this point, the major hurdles become collection and storage. Fortunately, at the microgrid scale, the volumes required are not overwhelming.

The capital costs and operation can be handled in a similar way to renewables or diesel generation, but with a few caveats. Whereas diesel would be purchased, stored, and used by the microgrid operator, the biomass is locally sourced. Two prominent incentives exist for engaging the community and collecting the biomass required to operate the generation unit. Either cash can be directly paid out per kilogram of biomass collected by farmers, or discounts/vouchers for electricity can be distributed.

Category	Type of biomass	Name of crop	Cultivated area (km^2)	Moisture content (%)	Total biomass (dry basis) (kt)	Used (%)	Unused biomass (dry basis) (kt)
A1	Straw	Wheat	34,765	9.2	14,317.30	80	2863.45
		Barley	70	–	45.28	20	36.22
		Paddy	35,406	10.6	8774.14	16.45	7417.70
		Seasum	206	–	19.35	20	15.48
		Pulses	274	–	31.92	80	6.88
Total					**23,187.99**	–	**10,339.73**
A2	Stalk	Cotton	6043	12	707.47	31.3	486.03
		Maize	1649	11.5	800.44	24.2	598.70
		Arhar	91	–	39.00	70	11.70
		Rapeseed and Mustard	498	–	142.49	70	42.75
Total					**1689.43**	–	**1139.18**
A3	Bagasse*	Sugar Cane	1441	15	1154.14	40-50	577.07
	Tops and Leaves	Sugar Cane		59.2	940.99	60	376.40
Total					**2095.13**	–	**953.47**
A4	Cobs*	Maize	1649	8.6	207.78	24.2	154.70
	Husks*	Paddy	25,406	9.6	2417.18	49	1152.76
	Shells*	Groundnuts	37	9.87	0.92	36	0.49
Total					**2625.88**	–	**1307.95**

*Indicates processing residue.

Table 11. Available energy stores derived from unused biomass [36].

8. Conclusions

It is often said that no single renewable energy source will be able to entirely replace our dependence on fossil fuels, but instead it will take a combination of solar, wind, hydro, biomass, and others to wean our dependence on pollution-heavy and unsustainable fuels. The idea of a multipart solution is not a new one, and it is certainly not limited to power production. Much in the same way that our energy needs must come from multiple sources, there are multiple levers to pull in order to reduce the cost of rural electrification and bring it within the reach of developing countries.

Just like solar, wind, hydro, and biomass come together to offer a solution to our power production needs, microgrids equally rely not only on multiple generation sources, but also on multiple consumption strategies to increase viability. The microgrid by itself is expensive and clunky and does not utilize new technological developments or advancements to improve on itself. At its core, it is a scaled-down technology that has not changed in over 100 years.

However, when combined with the techniques and technologies discussed in this chapter, the microgrid takes on a new form. It is no longer wasteful—carrying extra capacity which cannot be used—or allowing itself to be stretched and overloaded leading to total failure. This smart microgrid recognizes its own capacity, stores as much excess energy as possible, and recommends the shifting or clipping of noncritical loads for the benefit of the entire community. Tightening down and preventing electrical waste directly translates to a less expensive, more efficient distribution of energy. Cheaper electricity opens the door to rural electrification where it was once too costly to distribute power, and direct access to inexpensive power provides a boost in quality of life unimaginable to those who have been fortunate enough to have access to a near unlimited supply of inexpensive power.

Author details

Andrew Hubble and Taha Selim Ustun*

*Address all correspondence to: ustun@cmu.edu

School of Electrical and Computer Engineering at Carnegie Mellon University, Pittsburgh, Pennsylvania, USA

References

[1] Kanchev H, Colas F, Lazarov V, Francois B. Emission reduction and economical optimization of an urban microgrid operation including dispatched PV-based active generators. IEEE Transactions on Sustainable Energy. 2014;**5**:1397-1405. DOI: 10.1109/TSTE.2014.2331712

[2] Schmitt L, Kumar J, Sun D, Kayal S, Venkata S. Ecocity upon a hill: Microgrids and the future of the European City. IEEE Power and Energy Magazine. 2013;**11**:59-70. DOI: 10.1109/MPE.2013.2258286

[3] Manfren M, Caputo P, Costa G. Paradigm shift in urban energy systems through distributed generation: Methods and models. Applied Energy. 2011;**88**:1032-1048. DOI: 10.1016/j.apenergy.2010.10.018

[4] World Energy Outlook Electricity Access Database. World Energy Outlook [Internet]. 2015. Available from: http://www.worldenergyoutlook.org/resources/energydevelopment/energyaccessdatabase [Accessed: October 16, 2016]

[5] Hubble AH, Ustun TS. Scaling renewable energy based microgrids in underserved communities: Latin America, South Asia, and Sub-Saharan Africa. In: IEEE PES Power Africa Conference; 28 June-03 July 2016. Livingstone: IEEE; 2016. pp. 134-138

[6] CORE International Inc. Zambia Rural Electrification Master Plan. Phase 1: Rapid Resource Assessment. Final Report. Technical Report, Core International, 2005

[7] Holland CW, Derbyshire W. Electrifying 20 million households an Indonesian rural electrification strategy. In: EEA Conference and Exhibition; 19-20 June 2009. Christchurch: EEA; 2009. pp. 19-20

[8] Income Share Held by Lowest 10%. World Bank [Internet]. 2014. Available from: http:// http//data.worldbank.org/indicator/SI.DST.FRST.10 [Accessed: October 01, 2016]

[9] Cruz M, Foster J, Quillin B, Schellekens P. Ending Extreme Poverty and Sharing Prosperity: Progress and Policies. World Bank Group: Policy Research Note [Internet]. 2015. Available from: http://documents.worldbank.org/curated/en/801561468198533428/pdf/101740-NWP-PRN03-Oct2015-TwinGoals-Box393265B-PUBLIC.pdf [Accessed: November 13, 2016]

[10] Murenzi J, Ustun TS. The case for microgrids in electrifying sub-Saharan Africa. In: IEEE Renewable Energy Congress (IREC); 24-26 March 2015. Tunisia: IEEE; 2015. pp. 1-6

[11] Buchana P, Ustun TS. The role of microgrids & renewable energy in addressing sub-Saharan Africa's current and future energy needs. In: IEEE Renewable Energy Congress (IREC); 24-26 March 2015. Tunisia: IEEE; 2015. pp. 1-6

[12] Solar and Wind Energy Resource Assessment. UNEP [Internet]. 2015. Available from: http://en.openei.org/apps/SWERA/ [Accessed: November 03, 2016]

[13] Pump Price for Diesel Fuel (USD per Litre). World Bank [Internet]. 2015. Available from: http://data.worldbank.org/indicator/EP.PMP.DESL.CD [Accessed: September 20, 2016]

[14] Kagimu V, Ustun TS. Novel business models and policy directions based on SE4ALL global framework for minigrids. In: IEEE International Conference on Technological Innovations in ICT for Agriculture and Rural Development (TIAR); 10-12 July 2016. Chennai: IEEE; 2016. pp. 1-6

[15] Bhatia M, Angelou N. Beyond Connections Energy Access Redefined. Energy Sector Management Assistance Program. World Bank; 2015

[16] Modi V, McDade S, Lallement D, Saghir J, editors. Energy Services for the Millennium Development Goals. Washington, DC: World Bank; 2005. p. 116

[17] Barnes DF, Khandker SR, Samad HA. Energy poverty in rural Bangladesh. Energy Policy. 2011;**39**:894-904. DOI: 10.1016/j.enpol.2010.11.014

[18] Nussbaumer P, Nerini FF, Onyeji I, Howells M. Global insights based on the multidimensinal energy poverty index (MEPI). Sustainability. 2013;**5**:2060-2076. DOI: 10.3390/su5052060

[19] Energy for Community Services. Poor People's Energy Outlook. Warwickshire: Practical Action Publishing; 2013. p. 71. ISBN: 978 1 85339 789 9

[20] Tenenbaum B, Greacen C, Siyambalapitiya T, Knuckles J. From the Bottom Up: How Small Producers and Minigrids can Deliver Electrification and Renewable Energy in Africa. World Bank [Internet]. 2104. Available from: http://www.esmap.org/sites/esmap.org/files/ESMAP_From%20the%20Bottom%20Up_Optimizedpdf.pdf [Accessed: October 16, 2016]

[21] Franz M, Peterschmidt N, Rohrer M, Kondev B. Mini-grid Policy Toolkit. Africa-EU Renewable Energy Cooperation Programme [Internet]. 2014. Available from: http://www.ren21.net/Portals/0/documents/Resources/MGT/MinigridPolicy-Toolkit_Sep2014_EN.pdf [Accessed: September 30, 2016]

[22] Groh S, Taylor H. The role of microfinance in energy access: Changing roles, changing paradigms, and future potential. Enterprise Development and Microfinance. 2015;**26**:307-324. DOI: 10.3362/1755-1986.2015.026

[23] Ustun TS. The Importance of Microgrids & Renewable Energy in Meeting Energy Needs of the Brazilian Amazon. Unpublished Manuscript. 2015

[24] G. Roberta, T. Esteves. Energy supply on Amazon Remote Communities in Brazil: A case study for the Communities of Terra Nova, Mourao and Sao Sebastiao Methods, Calculatio Formulae. In: International Conference on Microperspectives for Decentralized Energy Supply; 28 February-01 March 2013. Berlin. 2013. pp. 100-103

[25] SIN Maps. Operador Nacional do Sistema Eletrico [Internet]. 2011. Available from: http://www.ons.org.br/conheca{_}sistema/mapas{_}sin.aspx [Accessed: November 17, 2016]

[26] Boletim Mensal de Geracao Eolica. Operador Nacional do Sistema Eletrico [Internet]. 2015. Available from: http://www.ons.org.br/download/resultados{_}operacao/boletim{_}mensal{_}geracao{_}eolica/Boletim{_}Eolica{_}mai{_}2016.pdf [Accessed: November 17, 2016]

[27] Redes E Fluxos Do Territorio. Instituto Brasilerio de Geografia e Estatistica (IBGE) [Internet]. 2015. Available from: http://biblioteca.ibge.gov.br/visualizacao/livros/liv97260.pdf [Accessed: November 17, 2016]

[28] HidroWeb Sistemas de Informacoes Hidrologicas. National Water Agency of Brazil [Internet]. 2016. Available from: http://www.snirh.gov.br/hidroweb/ [Accessed: November 17, 2016]

[29] Rwanda: Country at a Glance. World Bank [Internet]. 2015. Available from: http://www.worldbank.org/en/country/rwanda [Accessed: November 20, 2016]

[30] Installed Generating Capacity. Rwanda Energy Group [Internet]. 2014. Available from: http://www.reg.rw [Accessed: November 29, 2016]

[31] JRC's Institute for Energy and Transport. RE2nAF - European Commission [Internet]. 2015. Available from: http://re.jrc.ec.europa.eu/re2naf.html [Accessed: November 29, 2016]

[32] Niyigena D, Habineza C, Ustun TS. Computer-based smart energy management system for rural health centers. In: Renewable and Sustainable Energy Conference (IRSEC); 10-13 December 2015. Mococo: IEEE; 2015. pp. 1-6

[33] Thornburg J, Ustun TS, Krogh B. Smart microgrid operation simulator for management and electrification planning. In: IEEE PES Power Africa Conference; 28 June-03 July 2016. Livingstone: IEEE; 2016. pp. 1-5

[34] Agriculture Overview. World Bank [Internet]. 2016. Available from: http://www.worldbank.org/en/topic/agriculture/overview [Accessed: December 15, 2016]

[35] Energy on Demand. Village Industrial Power [Internet]. 2016. Available from: http://villageindustrialpower.com/ [Accessed: December 15, 2016]

[36] Singh J, Panesar B, Sharma S. Energy potential through agricultural biomass using geographical information system - A case study of Punjab. Biomass and Bioenergy. 2017; **32**:301-307. DOI: 10.1016/j.biombioe.2007.10.003

Energy Management in Microgrids: A Combination of Game Theory and Big Data-Based Wind Power Forecasting

Zhenyu Zhou, Fei Xiong, Chen Xu and Runhai Jiao

Additional information is available at the end of the chapter

http://dx.doi.org/10.5772/intechopen.68980

Abstract

Energy internet provides an open framework for integrating every piece of equipment involved in energy generation, transmission, transformation, distribution, and consumption with novel information and communication technologies. In this chapter, the authors adopt a combination of game theory and big data to address the coordinated management of renewable and traditional energy, which is a typical issue on energy interconnections. The authors formulate the energy management problem as a three-stage Stackelberg game and employ the backward induction method to derive the closed-form expressions of the optimal strategies. Next, we study the big data-based power generation forecasting techniques and introduce a scheme of the wind power forecasting, which can assist the microgrid to make strategies. Simulation results show that more accurate prediction results of wind power are conducive to better energy management.

Keywords: energy internet, Stackelberg game, microgrid energy management, wind power forecasting

1. Introduction

Energy internet has been identified as a key enabler of the third industrial revolution [1], which represents a new paradigm shift for both energy industry and consumers. In this new paradigm, the energy provisioning and demand sides are connected more closely and promptly than ever before by implementing distributed and flexible energy production and consumption while hiding the diversity of underlaying technologies through standardized interfaces [2, 3]. In addition, energy consumers with colocated distributed energy sources and distributed energy storage devices within limited areas, such as school, office building, industrial park, and residence community, etc., can form a local energy internet, that is, the microgrid, which provides a promising way of relieving the stress caused by the increasing energy demands and penetrations of renewable energy sources.

Microgrid is, in essence, a flexible and efficient network for interconnecting distributed renewable energy sources, load, and intermediate storage units at consumer premise [4]. It can be treated by the grid as a controllable load or generator and can operate in either islanded or grid-connected mode [5]. However, due to the intermittent and fluctuating characteristics of renewable energy sources and limited generation capacity, the large penetration of uncontrolled and uncoordinated renewable generators into the microgrid especially distribution network will cause a high level of volatility and system disturbances. For instance, the uncertainties brought by renewable energy sources will lead to significant mismatch between generation and load, which results in numerous critical problems such as power imbalance, voltage instability, interarea oscillations, and frequency fluctuations [6]. Hence, novel energy management methodologies are required to harness the full potential of the microgrid to reduce the energy supply-demand imbalance by making the full use of widespread renewable energy resources.

We study a distributed energy management problem in order to efficiently use renewable energy, with the aim of maximizing the individual objective function of each market player while guaranteeing the reliable system operation and satisfying users' electricity demands. Due to the uncertainty and uncontrollability of renewable generation, the authors utilize the big data-based renewable power forecasting techniques to obtain the short-term prediction value [7]. Then, the authors focus on solving the distributed microgrid energy management problem by employing noncooperative game theory [8], which provides an effective mathematical tool for analyzing optimization problems with multiple conflicting objective functions. The major contributions are summarized as follows:

- We adopt a combination of game-theoretical and data-centric approaches to address the microgrid energy management problem in energy internet. To address the uncertainties brought by wind turbine, the authors propose a deep learning-based short-term wind power forecasting algorithm by combining stacked autoencoders (SAE), the back-propagation algorithm, and the genetic algorithm. The authors employ SAE with three hidden layers in the pre-training process to extract the characteristics from the training sequence and the back-propagation algorithm to calculate the weights of the overall neural network in the fine-tuning process. Then, the authors adopt a genetic algorithm to optimize the neuron number of hidden layers and the learning rate of autoencoders.

- We provide thorough introduction and summary of the related works and the state-of-the-art progress in the research direction of energy management in microgrids. The authors have categorized the existing literature based on research motivations and application scenarios. The authors provide in-depth analysis and discussion on the contributions of the surveyed works, common assumptions, application scenarios, advantages, disadvantages, and possible future directions. The extensive review of available works sheds new insights to the underexplored open issues of energy management design in microgrids.

http://dx.doi.org/10.5772/intechopen.68980

- We model the energy management problem as a three-stage Stackelberg game to capture the dynamic interactions and interconnections among electricity users, the microgrid, the utility company, and the energy storage company. In the first stage, both the utility company and the energy storage company issue real-time electricity prices to the microgrid. In the second stage, the microgrid adjusts its electricity price offered to electricity users and the amounts of electricity procured from the utility and the energy storage companies. In the third stage, electricity users adjust their electricity demands based on the price offered by the microgrid. The objective function of each game player is well designed based on multiobjective optimization approaches, and practical constraints such as active power generation limits, power balance, electricity demands, etc., have been taken into consideration.
- Based on the short-term wind power prediction, we employ the backward induction method to analyze the proposed three-stage Stackelberg game and derive the closed-form analytical expressions for optimal energy management solutions. In the simulation, the authors compare the optimal payoff of the microgrid with different prediction errors of wind power forecasting. Numerical results show that accurate prediction results of wind power are conducive to better energy management.

The structure of this chapter is organized as follows. In Section 2, we give a brief review of related works on energy management and prediction technologies. The system model of energy management and problem formulation are provided in Section 3. Section 4 introduces the proposed game-theoretical and data-centric energy management algorithm. The simulation results and analyses are presented in Section 5. Finally, Section 6 gives the conclusion.

2. Related works

The aim of this chapter is to solve the distributed microgrid energy management problem by exploring both game theory and big data analysis in energy internet. The comprehensive summary of the classifications of distributed microgrid energy management is shown in **Table 1**. Some literature studies propose mathematical tools to deal with uncertainties of renewable energy in energy management problems. Two main methods that have been widely applied to handle day-to-day uncertainties of renewable energy are stochastic optimization

Application scenarios	Solution methods	Optimization goals	Literature
Renewable energy generation	Stochastic optimization	Handling date uncertainties of renewable energy	[10–12]
	Robust optimization		[14–17]
Wind power forecasting	Linear methods	Increasing the accuracy of prediction model	[19, 20]
	Nonlinear methods		[24–27]
Microgrid management	Ordinary decision theory	Optimizing energy-scheduling strategies	[28–30]
	Noncooperative games		[33–36]
	Cooperative games		[37–40]

Table 1. A comprehensive summary of distributed microgrid energy management.

and robust optimization [9]. On the one hand, stochastic optimization provides an effective framework to optimize statistical objective functions while the uncertain numerical data are assumed to follow a proverbial probability distribution. In Ref. [10], a multistage framework is presented to minimize the cost of the total energy management system based on stochastic optimization. The authors developed a stochastic dynamic programming method for optimizing the multidimensional energy management problem in Ref. [11]. A stochastic optimization-based real-time energy management approach was adopted to minimize the operational cost of the total energy system in Ref. [12]. However, considering the complex operation details and various practical constraints in practical applications, the precise estimation of the probability distributions of uncertain data can be a tremendous challenge. Hence, the impact of data uncertainties on the optimality performance may not be sufficiently captured in the stochastic optimization-based energy management approaches.

On the other hand, robust optimization, which considers the worst-case operation scenarios, only requires appropriate information and enable a distribution-free model of data uncertainties [13]. Hence, robust energy management can mitigate the negative effect of uncertainty on the optimality performance and thus overcome the aforementioned limitations of stochastic optimization. In Ref. [14], a novel pricing strategy was presented to enable robustness against the uncertainty of power input. The authors proposed a robust energy-scheduling approach for solving the uncertainty brought by electric vehicles in Ref. [15]. Robust energy management methods were proposed to optimize the energy-dispatching problem while the worst-case scenarios of renewable energy integration have been considered [16, 17]. However, due to the fact that the worst-case scenarios of all uncertain factors are assumed to provide the highest protection against uncertainties, the optimality performance is also severely degraded as the price paid for robustness.

With the development of advanced information and communication technologies, the big data-based forecasting approach can learn from these massive amounts of real-world data, and thus adapt conventional energy management design to this new data-centric paradigm by utilizing the historical knowledge. Taking wind power forecasting as an example, the data-centric approaches mine the relationship between historical data and knowledge to build the prediction model through various approaches, such as persistence methods, linear methods, and nonlinear methods. The persistence method is one of the classic methods for wind power forecasting and is usually utilized as a benchmark method while short-term wind speeds are assumed highly correlated [18]. Linear methods have been shown to outperform most persistence methods in short-term forecasting as they can capture the time relevance and probability distribution of wind speed data [19, 20]. Nonlinear methods such as artificial neural networks (ANNs) [21], support vector machines (SVM) [22, 23], etc., are demonstrated to outperform linear methods in nonlinear models. ANN, which is a simplified model of human brain neural processing, has the advantage of fast self-learning capability, easy implementation, and high prediction accuracy [24]. SVM is a machine-learning model of ANNs to analyze data which is used for classification and regression analysis [25]. To efficiently handle the complex, unlabeled and high-dimensional time series data, deep learning has been proposed in Ref. [26]. As an essential deep learning architecture, SAE plays a fundamental role in unsupervised learning and the objective function can be solved efficiently via fast back propagation [27].

http://dx.doi.org/10.5772/intechopen.68980

There already exists some work about energy management design in microgrid. In Ref. [28], a double-layer control model, which consists of a dispatch layer to offer the output power of each unit and a schedule layer to provide the operation optimization, is proposed for microgrid energy management. The authors presented a fair energy-scheduling strategy in Ref. [29] to maximize the total system benefit while providing higher energy utilization priorities to users with larger contributions. In Ref. [30], the authors took demand side management and generation scheduling into consideration for ensuring the real-time operation of energy management system. However, the previous studies mainly focus on the total benefit in the energy management system, and ignore the interactions and interconnections among multiple market players, including utility companies, storage companies, microgrids, customers, and so on.

Game theory has widely been applied in microgrid energy management to provide a distributed self-organizing and self-optimizing solution for optimization problems with conflicting objective functions in Ref. [31]. Games can be classified into two categories based on whether or not binding agreements among players can be enforced externally, that is, noncooperative and cooperative games [32]. Noncooperative games, which offer an analytical framework tailored for characterizing the interactions as well as decision-making process among multiple game players, focus on predicting players' individual strategies and analyzing the competitive decision-making involving players to find the Nash equilibrium. The players will influence the decision-making process despite their partially or even completely conflicting interests upon the result of a decision. In contrast, cooperative games offer mathematical tools to study the interactions of rational cooperative players, and the strategic outcome among those players as well as their utilities can be improved under a common agreement.

For noncooperative game-based microgrid energy management, the authors proposed a multiuser Stackelberg game model for maximizing the benefit of each player in Ref. [33]. In Ref. [34], a new model of electricity market operation was adopted to optimize the objective function of each player. The authors provided a dynamic noncooperative repeated game model to optimize the energy-trading amounts of users with distributed renewable generators [35]. In Ref. [36], a distributed real-time game-theoretical energy management scheme was employed to maximize the total social benefit while minimizing the cost of each player. For microgrid energy management schemes based on cooperative games, the authors proposed a cooperative demand response scheme for reducing the electricity bills of users in Ref. [37]. In Ref. [38], a cooperative energy-trading approach was proposed for the downlink coordinated multipoint transmission powered by smart grids to reduce energy cost. The authors developed a cooperative distributed energy-scheduling algorithm to optimize the energy dispatch problem while considering the integration of renewable generation and energy storage in Ref. [39]. In Ref. [40], the authors provided a multistage market model for minimizing the operational cost of the utility company while maximizing the total benefit of the market. Compared to cooperative games, the noncooperative games have the advantage of a lower communication overhead and do not require a common commitment among various market players. As one kind of noncooperative game models, the Stackelberg game can efficiently model the hierarchy among players, where the leaders have dominant market positions over followers, and can impose their own strategies upon the followers. Considering above two points, the authors propose the noncooperative game-theoretical approach and model the microgrid energy management problem as a three-stage Stackelberg game.

In summary, most of the previous studies have not provided a comprehensive framework for how to utilize the real-world data to improve the energy management performance. The prior statistic knowledge of uncertain renewable power outputs was assumed to be perfectly known and its impact on the energy-trading process among market players has not been fully analyzed. This motivates us to explore the integration of deep learning-based wind power forecasting technique with Stackelberg game-based energy management strategy, so as to make a further step to enable data-centric energy management in future energy internet.

3. System model and problem formulation

3.1. System model

Figure 1 presents a structure of a typical microgrid energy management system with the utility company, the energy storage company, users, and various kinds of renewable energy sources. In this system, without loss of generality, the authors assume that there is a single conventional energy generation company, which is denoted as the utility company, and a renewable sources-based energy storage company, which is denoted as the storage company. The energy storage company which operates independently from the utility company can store and absorb excess

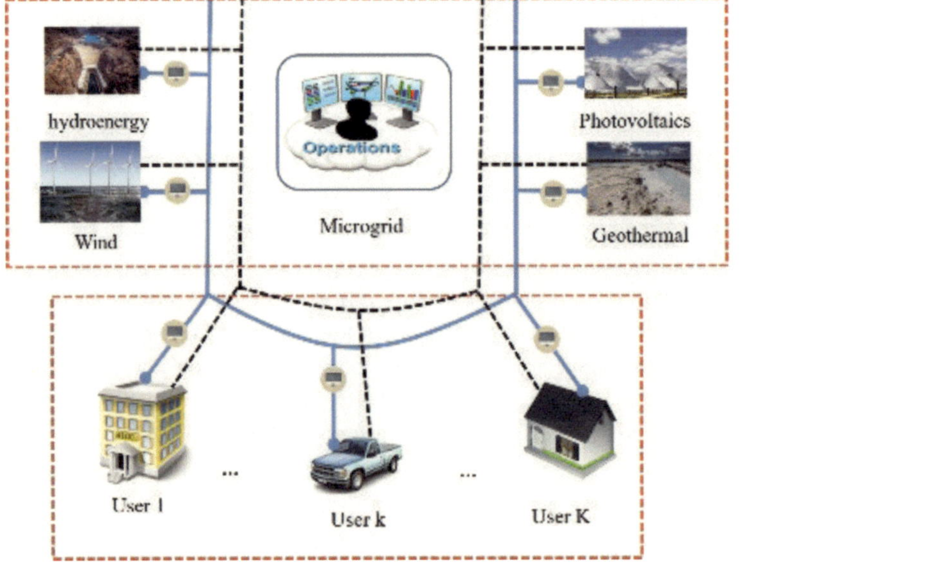

Figure 1. System model of microgrid energy management.

energy during nonpeak periods and deliver it back to the grid during the peak times. Furthermore, the authors assume that there is a single microgrid and there are $\mathcal{K}$ users, denoted as $\mathcal{K} = \{1, \ldots, k, \ldots, K\}$, in this model. The utility company and the storage company are regarded as energy suppliers to meet the electric power demand of the microgrid and ensure the stability of the power system. To implement efficient energy management, the microgrid should be in charge of energy dispatching and be responsible for meeting users' electricity demands based on the forecasting of renewable energy generation. However, due to renewables' uncontrollable fluctuations, variability, intermittent nature, and the capacity limitation of the microgrid, the microgrid may not be able to meet the electricity demand of users by itself and has to purchase electricity from the utility company and the storage company.

3.2. Objective function

3.2.1. Objective function of the utility company

The definition of the utility company's objective function is rather flexible. Generally, the authors consider the cost function consisting of the electricity generation cost denoted as $C(L)$ and the pollutant emission cost denoted as $I(L)$ [41]. Each of them can be modeled as a quadratic function of the electricity demand L. Besides, line loss, which is mainly caused by resistance of the transmission lines, has been taken into consideration to ensure energy supply. Hence, the objective function of the utility company is formulated as

$$U_g(L_{m,g}, p_g) = R_g(L_{m,g}, p_g) - C_g(\varepsilon_g L_{m,g}) - I_g(\varepsilon_g L_{m,g}), \tag{1}$$

where

$$\begin{aligned} R_g(L_{m,g}, p_g) &= L_{m,g} p_g, \\ C_g(\varepsilon_g L_{m,g}) &= a_g(\varepsilon_g L_{m,g})^2 + b_g(\varepsilon_g L_{m,g}) + c_g, \\ I_g(\varepsilon_g L_{m,g}) &= \alpha_g(\varepsilon_g L_{m,g})^2 + \beta_g(\varepsilon_g L_{m,g}). \end{aligned} \tag{2}$$

$R_g(L_{m,g}, p_g)$ denotes the electricity revenue; $C_g(\varepsilon_g L_{m,g})$ and $I_g(\varepsilon_g L_{m,g})$ are the cost functions of the power generation and the pollutant emission, respectively; $L_{m,g}$ denotes the quantity of electricity bought from the utility company by the microgrid; p_g is the unit electricity price of the utility company; and $a_g, b_g, c_g, \alpha_g, \beta_g$ are the cost parameters of $C_g(\varepsilon_g L_{m,g})$ and $I_g(\varepsilon_g L_{m,g})$. Assuming that ρ_g denotes the power loss percentage during power transmission, which is related to voltage, efficiencies of transformers, and resistance of the transmission line. Hence, $\varepsilon_g L_{m,g}$ is the actually generated electricity to satisfy the microgrid demand $L_{m,g}$, where $\varepsilon_g = 1/(1 - \rho_g)$.

3.2.2. Objective function of the storage company

The authors considered the power loss inefficiency during the battery charging and discharging processes, as well as line loss, and the objective function of the storage company is formulated as

$$U_s(L_{m,s}, p_s) = R_s(L_{m,s}, p_s) - C_s(\varepsilon_s L_{m,s}), \tag{3}$$

where

$$\begin{aligned} R_s(L_{m,s}, p_s) &= L_{m,s} p_s, \\ C_s(\varepsilon_s L_{m,s}) &= \frac{c_s \varepsilon_s L_{m,s}}{\eta_c \eta_d}. \end{aligned} \tag{4}$$

$R_g(L_{m,s}, p_s)$ denotes the electricity revenue; $C_s(\varepsilon_s L_{m,s})$ is the cost function of energy storage; $L_{m,s}$ denotes the quantity of electricity bought from the storage company by the microgrid; p_s is the unit electricity price of the storage company; η_c and η_d are the charging and discharging efficiencies of storage equipment, respectively; and c_s denotes the unit cost of operation and maintenance. The meaning of ε_s is the same as ε_g introduced above.

3.2.3. Objective function of the microgrid

The authors focus on renewable energy which is the main source of the microgrid and consider the satisfaction function based on quality of service of the electricity provided by the utility and storage companies [42]. Hence, the objective function of the microgrid is formulated as

$$\begin{aligned} &U_m(L_{m,g}, L_{m,s}, p_m) = R_{m,g}(L_{m,g}) + R_{m,s}(L_{m,s}) \\ &- C_{m,g}(L_{m,g}, p_g) - C_{m,s}(L_{m,s}, p_s) + R_m(L_{k,m}, p_m) \\ &- C_m(\hat{L}_r + \Delta) - I_m(\hat{L}_r + \Delta) + F|\Delta|, \end{aligned} \tag{5}$$

where

$$\begin{aligned} &R_{m,g}(L_{m,g}) = X_{m,g} L_{m,g} - \frac{d_{m,g}}{2}(L_{m,g})^2, \\ &R_{m,s}(L_{m,s}) = X_{m,s} L_{m,s} - \frac{d_{m,s}}{2}(L_{m,s})^2, \\ &R_m(L_{k,m}, p_m) = \sum_{k=1}^{K} L_{k,m} p_m, \\ &C_{m,g}(L_{m,g}, p_g) = L_{m,g} p_g, \\ &C_{m,s}(L_{m,s}, p_s) = L_{m,s} p_s, \\ &C_m(\hat{L}_r + \Delta) = a_m(\hat{L}_r + \Delta)^2 + b_m(\hat{L}_r + \Delta) + c_m, \\ &I_m(\hat{L}_r + \Delta) = \alpha_m(\hat{L}_r + \Delta)^2 + \beta_m(\hat{L}_r + \Delta). \end{aligned} \tag{6}$$

$R_{m,g}(L_{m,g})$ denotes the satisfaction value; $C_{m,g}(L_{m,g}, p_g)$ denotes the payment of the microgrid for electricity bought from the utility company; and $X_{m,g}$ denotes the satisfaction parameter for the utility company. As the satisfaction parameters depend on various factors, such as electricity demands, electricity prices, preferences in different energy sources, weather conditions, etc., it is hard to model the satisfaction parameters accurately. Thus, the authors assume that these parameters are predefined. Analogously, $d_{c,m}$ denotes predefined satisfaction parameters of the microgrid for the utility company. The definitions of $R_{m,s}(L_{m,s})$ and $C_{m,s}(L_{m,s}, p_s)$ are similar to those of $R_{m,g}(L_{m,g})$ and $C_{m,g}(L_{m,g}, p_g)$ as introduced above; $R_m(L_{k,m}, p_m)$ denotes the electricity revenue acquired from users while $L_{k,m}$ is the quantity of electricity bought by the kth user and p_m is the unit electricity price of the microgrid; $C_m(\hat{L}_r + \Delta)$ and $I_m(\hat{L}_r + \Delta)$ are the cost functions of wind power generation and wind power pollutant emission, respectively; $a_m, b_m, c_m, \alpha_m, \beta_m$ are the cost parameters of $C_m(\hat{L}_r + \Delta)$ and $I_m(\hat{L}_r + \Delta)$. $\hat{L}_r + \Delta$ denotes the prediction result of wind power while $\hat{L}_r$ is the real wind power and Δ is the prediction error. F denotes the penalty factor of the prediction error Δ that satisfies $F < 0$. That is, the payoff of the microgrid will decrease when the result of wind power forecasting is not accurate, which reflects the restriction of the power purchase agreement in the market.

3.2.4. *Objective function of users*

In a similar way, the authors also take the satisfaction function into consideration. Hence, the objective function of the kth user is given by

$$U_k(L_{k,m}, p_m) = R_{k,m}(L_{k,m}) - C_{k,m}(L_{k,m}, p_m), \tag{7}$$

where

$$\begin{aligned} R_{k,m}(L_{k,m}) &= X_{k,m}L_{k,m} - \frac{d_{k,m}}{2}(L_{k,m})^2, \\ C_{k,m}(L_{k,m}, p_m) &= L_{k,m}p_m. \end{aligned} \tag{8}$$

$R_{k,m}(L_{k,m})$ denotes the satisfaction value and $C_{k,m}(L_{k,m}, p_m)$ denotes the payment that the kth user pays for electricity bought from the microgrid. The meanings of $X_{k,m}$ and $d_{k,m}$ are similar to $X_{m,g}$ and $d_{m,g}$.

3.3. Problem formulation

The authors propose a three-stage Stackelberg game, which consists of leaders and followers to describe the interconnection of each stage and model the energy management process. The three-stage Stackelberg game is described in a distributed manner in **Figure 2**:

- Stage I: The utility and the storage companies, as leaders of the game, announce the unit electricity price p_g and p_s to the microgrid. By setting reasonable prices, the companies hope to maximize their own payoffs. Thus, the authors can describe the optimization problem for the utility and storage companies as

$$\max_{p_g} U_g(p_g), \tag{9}$$

$$\max_{p_s} U_s(p_s). \tag{10}$$

- Stage II: The microgrid can be assumed as the follower of the utility and the storage companies as well as the leader of users. On the one hand, the microgrid determines

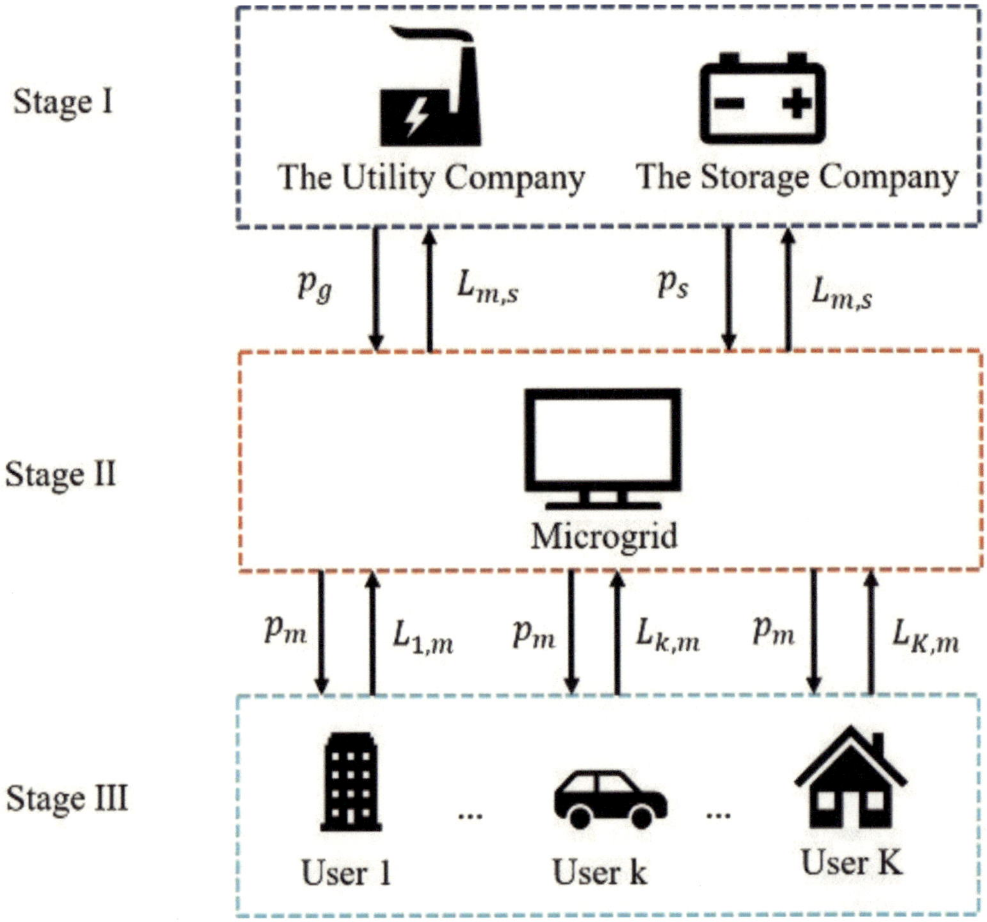

Figure 2. The diagram of the three-stage Stackelberg game.

electricity demand $L_{m,g}$ and $L_{m,s}$ based on the prediction result of the wind power and the unit prices p_g, p_s. On the other hand, it announces electricity price p_m to users. The objective of the microgrid is also to maximize its payoff by adjusting $L_{m,g}$, $L_{m,s}$, and p_m. We describe the optimization problem for the microgrid as

$$\begin{aligned} &\max_{L_{m,g},\, L_{m,s},\, p_m} \quad U_m(L_{m,g}, L_{m,s}, p_m), \\ s.t. \quad & C_1 : 0 \le \varepsilon_g L_{m,g} \le L_{g,\max}, \\ & C_2 : 0 \le \varepsilon_s L_{m,s} \le L_{s,\max}, \\ & C_3 : 0 \le p_m \le p_{m,\max}, \\ & C_4 : L_{m,s} + L_{m,g} = \sum_{k=1}^{K} L_{k,m} - \hat{L}_r - \Delta > 0, \end{aligned} \tag{11}$$

where $L_{g,\max}$, $L_{s,\max}$, and $p_{m,\max}$ denote the capacity and pricing constraints.

- Stage III: The kth user ($\forall k \in \{1, 2, \ldots, K\}$), as the follower of the microgrid, determines electricity amount $L_{k,m}$ purchased from the microgrid based on p_m to maximize its payoff. We can describe the optimization problem for the kth user as

$$\max_{L_{k,m}} U_k(L_{k,m}), \tag{12}$$

$$s.t. \quad C_5 : L_{k,m} \geq L_{k,b}, \tag{13}$$

where $L_{k,b}$ is the basic electricity demand of the kth user.

4. Algorithms and analysis

In this section, we first propose a distributed energy management algorithm based on the three-stage Stackelberg game. Then, the big data analysis-based wind power forecasting algorithm is derived by combining SAE, the back-propagation algorithm, and the genetic algorithm.

4.1. Distributed energy management algorithm

We propose a three-stage Stackelberg game to describe the interconnections of each stage and use the backward induction to capture the interrelation of the decision-making process in each stage.

4.1.1. Analysis of the third-stage user game

The optimization objective of the kth user is defined in Eq. (12), which is a standard concave function. Hence, the authors can use the Karush-Kuhn-Tucker (KKT) conditions to solve the optimization problem. The optimal solution of the kth user is given by

$$\begin{cases} \hat{L}_{k,m1} = \dfrac{X_{k,m} - p_m}{d_{k,m}}, \\ \hat{L}_{k,m2} = L_{k,b}, \end{cases} \tag{14}$$

where $\hat{L}_{k,m1}$ denotes the optimal electricity procurement quantities; $\hat{L}_{k,m2}$ denotes the scenario where the optimal electricity procurement quantity lines on the boundary of the inequality constraint.

4.1.2. Analysis of the second-stage microgrid game

In stage II, the authors assume user $k' \in \mathcal{K}' = \{1, \ldots, i, \ldots, K'\}$ purchases electricity $L_{k,m1}$ and user $k'' \in \mathcal{K}'' = \{1, \ldots, i, \ldots, K''\}$ purchases electricity $L_{k,m2}$. While $\mathcal{K} = \mathcal{K}' \bigcup \mathcal{K}''$, the authors can obtain

$$\sum_{k=1}^{K} L_{k,m} = \sum_{K'=1}^{K'} \frac{X_{k,m} - p_m}{d_{k,m}} + \sum_{K''=1}^{K''} L_{k,b}. \tag{15}$$

Based on KKT conditions, the optimal amount of electricity procured from the utility company is given by

$$\begin{cases} \hat{L}_{m,g1} = 0, \\ \hat{L}_{m,g2} = \dfrac{X_{m,g} - p_g - \mu_{m,1}}{d_{m,g}}, \\ \hat{L}_{m,g3} = \dfrac{L_{g,\max}}{\varepsilon_g}, \end{cases} \tag{16}$$

In a similar way, based on KKT conditions, the optimal amount of electricity procured from the storage company is given by

$$\begin{cases} \hat{L}_{m,s1} = 0, \\ \hat{L}_{m,s2} = \dfrac{X_{m,s} - p_s - \mu_{m,1}}{d_{m,s}}, \\ \hat{L}_{m,s3} = \dfrac{L_{s,\max}}{\varepsilon_s}, \end{cases} \tag{17}$$

The optimal price is given by

$$\begin{cases} \hat{p}_{m1} = 0, \\ \hat{p}_{m2} = \dfrac{\sum_{K'=1}^{K'} \frac{X_{k,m}}{d_{k,m}} + \sum_{K''=1}^{K''} L_{k,b} - \mu_{m,1} \sum_{K'=1}^{K'} \frac{1}{d_{k,m}}}{\sum_{K'=1}^{K'} \frac{2}{d_{k,m}}}, \\ \hat{p}_{m3} = p_{m,\max}, \end{cases} \tag{18}$$

$\hat{L}_{m,g1}$, $\hat{L}_{m,g3}$, $\hat{L}_{m,s1}$, $\hat{L}_{m,s3}$, $\hat{p}_{m1}$, and $\hat{p}_{m3}$ denote the scenarios that where the optimal solutions line on the boundaries of the inequality constraints. $\hat{L}_{m,g2}$, $\hat{L}_{m,s2}$, and $\hat{p}_{m2}$ denote the interior solutions. When $L_{m,g} = 0$ or $L_{m,g} = \frac{L_{g,\max}}{\varepsilon_g}$ and $L_{m,s} = 0$ or $L_{m,s} = \frac{L_{s,\max}}{\varepsilon_s}$, there is no price competition between the utility and storage companies. Thus, the analysis of the corresponding p_g and p_s is omitted here. Considering the price competition game between the utility company and the storage company, p_m can be viewed as a function of p_g and p_s based on Eq. (18), which is given by

$$p_m = A_{m,1} p_g + A_{m,2} p_s + A_{m,3}, \tag{19}$$

where

$$A_{m,1}=\frac{\frac{1}{d_{m,g}}}{\left(1+\frac{\sum_{k'=1}^{K'}\frac{1}{d_{k,m}}}{\frac{2}{d_{m,g}}+\frac{2}{d_{m,s}}}\right)\left(\frac{2}{d_{m,g}}+\frac{2}{d_{m,s}}\right)},$$

$$A_{m,2}=\frac{\frac{1}{d_{m,s}}}{\left(1+\frac{\sum_{k'=1}^{K'}\frac{1}{d_{k,m}}}{\frac{2}{d_{m,g}}+\frac{2}{d_{m,s}}}\right)\left(\frac{2}{d_{m,g}}+\frac{2}{d_{m,s}}\right)}, \tag{20}$$

$$A_{m,3}=\frac{\sum_{k'=1}^{K'}\frac{X_{k,m}}{d_{k,m}}+\sum_{k''=1}^{K''}L_{k,b}}{\left(1+\frac{\sum_{k'=1}^{K'}\frac{1}{d_{k,m}}}{\frac{2}{d_{m,g}}+\frac{2}{d_{m,s}}}\right)\sum_{k'=1}^{K'}\frac{2}{d_{k,m}}}-\frac{\frac{X_{m,g}}{d_{m,g}}+\frac{X_{m,s}}{d_{m,s}}-\left(\sum_{k'=1}^{K'}\frac{X_{k,m}}{d_{k,m}}+\sum_{k''=1}^{K''}L_{k,b}\right)+\hat{L}_r+\Delta}{\left(1+\frac{\sum_{k'=1}^{K'}\frac{1}{d_{k,m}}}{\frac{2}{d_{m,g}}+\frac{2}{d_{m,s}}}\right)\left(\frac{2}{d_{m,g}}+\frac{2}{d_{m,s}}\right)}.$$

4.2.3. Analysis of the first-stage utility and storage company game

In this case, defining $L_{m,g}$ as a function of p_g, we have

$$\hat{L}_{m,g}(p_g)=A_{g,1}p_g+A_{g,2}, \tag{21}$$

where

$$A_{g,1}=-\frac{1}{d_{m,g}}+\frac{\frac{1}{d_{m,g}}-\sum_{k'=1}^{K'}\frac{A_{m,1}}{d_{k,m}}}{1+\frac{d_{m,g}}{d_{m,s}}},$$

$$A_{g,2}=\frac{X_{m,g}}{d_{m,g}}-\frac{\frac{X_{m,g}}{d_{m,g}}+\frac{X_{m,s}-p_s}{d_{m,s}}\sum_{k'=1}^{K'}\frac{X_{k,m}-A_{m,2}p_s-A_{m,3}}{d_{k,m}}+\sum_{k''=1}^{K''}L_{k,b}-\hat{L}_r-\Delta}{1+\frac{d_{m,g}}{d_{m,s}}}. \tag{22}$$

Hence, U_g can be written as a quadratic function of p_g, which is given by

$$U_g(p_g)=A_{g,3}(p_g)^2+A_{g,4}p_g+A_{g,5}, \tag{23}$$

where

$$
\begin{aligned}
A_{g,3} &= A_{g,1} - \varepsilon_g^2(a_g + \alpha_g)A_{g,1}^2, \\
A_{g,4} &= A_{g,2}[1 - 2\varepsilon_g^2(a_g + \alpha_g)A_{g,1}] - \varepsilon_g(b_g + \beta_g)A_{g,1}, \\
A_{g,5} &= -\varepsilon_g^2(a_g + \alpha_g)A_{g,2}^2 - \varepsilon_g(b_g + \beta_g)A_{g,2} - c_g.
\end{aligned} \tag{24}
$$

Since U_g is a convex function of p_g based on Eq. (22), the authors can obtain $\hat{p}_g$ by solving the convex function that

$$
\hat{p}_g = -\frac{A_{g,4}}{2A_{g,3}}. \tag{25}
$$

In the same way, $\hat{p}_s$ can be obtained similarly as above since $\hat{p}_s$ has the same solution structure with $\hat{p}_g$. The detailed process is omitted here due to space limitations.

4.2. Algorithm of wind power forecasting

We propose a deep learning-based short-term wind power forecasting algorithm by combining SAE, the back-propagation algorithm, and the genetic algorithm. It is noted that the proposed forecasting model can also be applied for other distributed renewable energy sources such as solar energy, hydroenergy, etc. The reason why the authors study the wind power forecasting in this chapter is mainly due to the illustration purpose and the availability of the wind big data. The core of the algorithm is to establish a forecasting model through training on the historical data. Exploiting the statistical relationship among the historical time series data can be divided into two processes: the pre-training process and the fine-tuning process. In the pre-training process, three stacked AEs, which consist of one visible layer, one hidden layer, and one output layer form a neural network. In the fine-tuning process, one more layer is added to the end of the neural network and back-propagation algorithm is applied to obtain more appropriate initial weights of the whole network. Furthermore, for improving the forecasting accuracy, we adopt genetic algorithm to optimize the learning rate of each AE and the number of neurons of each layer.

4.2.1. Training process of the proposed genetic SAE forecasting model

As shown in **Figure 3(a)**, SAE consists of one input layer x, the first hidden layer h_1, and one output layer $\hat{x}$. We adopt encoder function $f_{\theta 1}$ to transform x to a low or a high-dimensional code h_1 and adopt decoder function $g_{\theta 1}$ to reconstruct the original data as $\hat{x}$. We can obtain the values of parameters $\theta_j = \{w_j, b_j, w_j^T, d_j\}$, $j \in \{1, 2, \ldots, J\}$ (J denotes the number of layers in SAE) through back propagation, where w_j and w_j^T are weight matrices of the encoder and the decoder, b_j and d_j are biases of the encoder and the decoder, respectively.

We add a new hidden layer h_2 to the whole network, new layer and the original layers are stacked into the existing AE in **Figure 3(b)**. There is a new AE illustrated since h_1 and h_2 are combined as the input layers. Hence, the authors can stack more auto encoders by removing the last layer h_1 and add one more layer. Considering computation complexity, three auto coders are stacked together in this section. The pre-training process is shown as **Figure 3(a)** and **(b)**, which consists of two hidden layers h_1, h_2 and trains the initial weights of the whole network.

http://dx.doi.org/10.5772/intechopen.68980

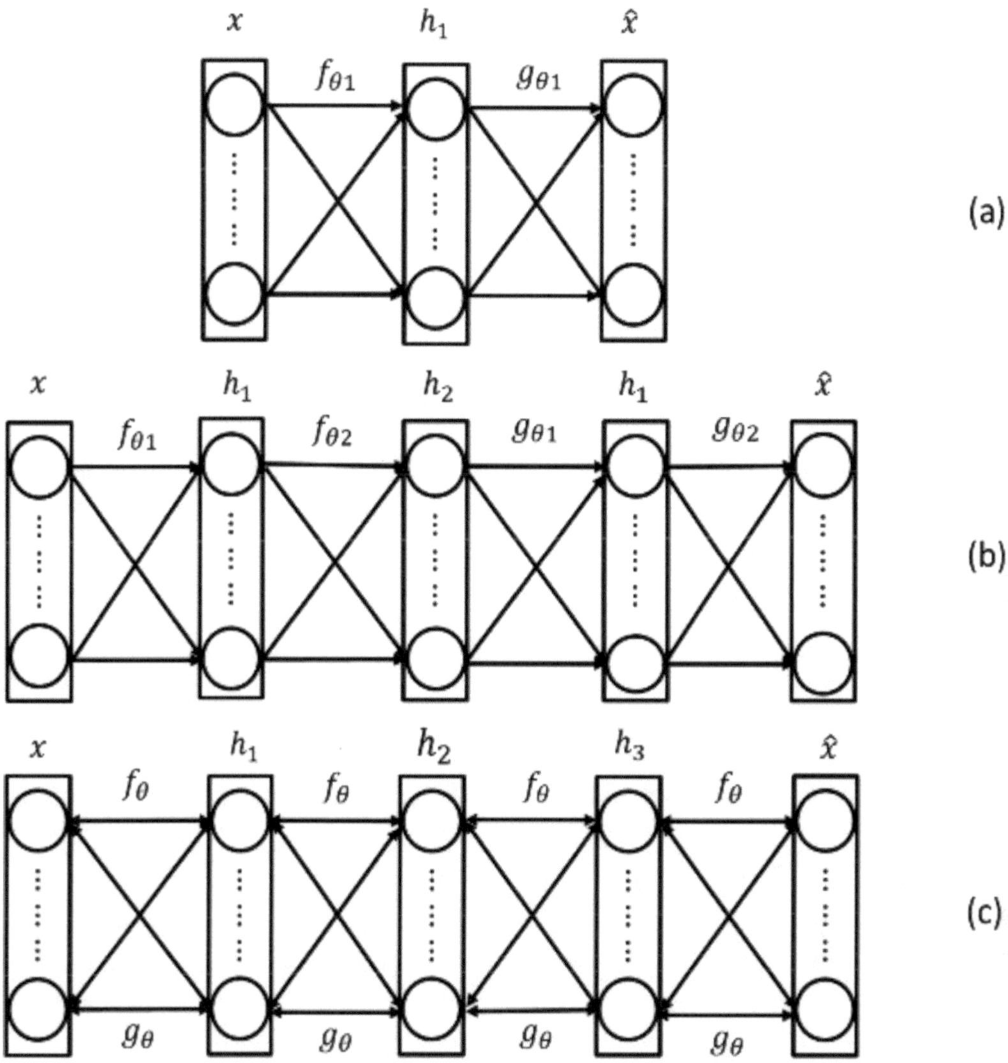

Figure 3. The pre-training and fine-tuning process of genetic SAE.

In **Figure 3(c)**, to form the whole genetic SAE neural network, we add an output layer and initialize the set of parameter w_4, b_4 between the last hidden layer and the output layer. The process which we adopt back-propagation algorithm to train all the weights and biases of the whole network is called the fine-tuning process. Hence, a deep network with three hidden layers can be trained to converge to a global minimum by the process we proposed.

4.2.2. Optimization of the proposed model

The learning rate of the network and the number of neurons in hidden layer are the key parameters which have a significant impact on the final prediction performance. Hence, we adopt the genetic algorithm to optimize the parameters of the SAE and the whole network for improving the performance of the models. We regard the historical time series data **x** as the

Algorithm 1 The proposed genetic SAE Algorithm

Procedure: Genetic Algorithm
Begin
Initialize: $\mathbf{P}(\mathbf{0}, \mathbf{0})$
Set $t = 0$
while $t < T$ **do**
  for $d \in \mathcal{D}$ **do**
    The Pre-training Process:
    Evaluate fitness of $\mathbf{P}(\mathbf{d}, \mathbf{t})$: Calculate and store the best, worst, and average objective value for current individuals.
    Select operation to $\mathbf{P}(\mathbf{d}, \mathbf{t})$: Select optimal individual for the next generation.
    The Fine-turning Process:
    Crossover operation to $\mathbf{P}(\mathbf{d}, \mathbf{t})$: Do crossover operation on the selected individuals and obtain better individuals.
    Mutation operation to $\mathbf{P}(\mathbf{d}, \mathbf{t})$: Do mutation operation to $\mathbf{P}(\mathbf{d}, \mathbf{t})$ based on a certain mutation probability.
  end for
  Update: $t = t + 1$
end while

individuals of population in genetic algorithm and obtain a multidimensional vector $\mathbf{P}(\mathbf{d}, \mathbf{t})$, where there are d individuals in the population denoted as $d \in \mathcal{D} = \{1, \ldots, d, \ldots, D\}$ and $t \in \mathcal{T} = \{1, \ldots, t, \ldots, T\}$ is the number of evolution. We assume that the size of the population is D and the maximum of evolution is T. First, we set the initial population as $\mathbf{P}(\mathbf{0}, \mathbf{0})$. Then, we calculate the objective value and the fitness value to select optimal individual for the next generation. After crossover and mutation, we can obtain optimal individual $\mathbf{P}(\mathbf{d}, \mathbf{T})$. Algorithm 1 shows the optimization process of the proposed model. To make a fair comparison, we optimize the parameters of the BP algorithm and the SVM algorithm in the similar way. The mean absolute percentage error (MAPE) provides a statistical measure of prediction accuracy of a forecasting method, which is expressed in percentage. It measures how much forecasts can differ from the actual data, which is summed for every evaluation points and divided by the total number of points. Since MAPE has been widely adopted in wind power forecasting, the authors also adopt it to evaluate the accuracy of the prediction model.

5. Simulation results

In order to evaluate the prediction accuracy of the proposed wind-forecasting model, real data of wind turbines, which were collected form a local micorgrid in Hebei Province, China, are employed to perform the training and forecasting processes. By excluding unnecessary information, the 1-year data samples of active power, which spans from September 2015 to October 2016, are utilized for simulations. The proposed game-theoretical energy management algorithm with big data-based wind power forecasting is implemented based on Matlab. Simulation results are performed for a scenario which consists of the utility company, the energy storage company, the microgrid, and the users. The simulation parameters are summarized in **Table 2**. **Figure 4** shows the optimal electricity prices of the utility company, the energy storage company, and the microgrid, that is, $\hat{p}_g$, $\hat{p}_s$, and $\hat{p}_m$, versus the basic electricity demands of users $L_{k,b}$. $L_{k,b}$ is increased from 10 to 100 kW with a step of10, and the corresponding $\hat{p}_g$, $\hat{p}_s$, and $\hat{p}_m$ are obtained by the proposed algorithm. The simulation results demonstrate that $\hat{p}_g$, $\hat{p}_s$, and $\hat{p}_m$ increase monotonically as $L_{k,b}$ increases, which is reasonable since the electricity generation cost also increase dramatically as $L_{k,b}$ increases. $\hat{p}_s > \hat{p}_g$ is due to the preference of the microgrid to use clean renewable energy stored by the energy storage company. In addition, we have $\hat{p}_m > \hat{p}_g$ and $\hat{p}_m > \hat{p}_s$. Since only one microgrid has been considered in the second stage, the microgrid is always able to make more profits by announcing higher prices toward users than those of the utility and the energy storage companies.

Figures 5 and **6** show the optimal payoff of the microgrid $U_m(\hat{p}_m, \hat{L}_{m,g}, \hat{L}_{m,s})$ versus the prediction error of wind power forecasting Δ for the two scenarios $\Delta > 0$ and $\Delta < 0$, respectively. Here, $\Delta > 0$ represents that the actual wind power output is less than the predicted amount, and the microgrid has to procure more electricity from both the utility and the energy storage companies. In comparison, $\Delta < 0$ represents that the actual wind power output is more than the predicted amount, and the microgrid will not procure the specified amount of electricity from both the utility and the energy storage companies. Three cases where $L_{k,b} = 40,\ 60,$ and 80 kW

Parameter	Value
Power generation cost parameter of utility company a_g	0.03
Pollutant emission cost parameter of utility company α_g	0.08
The unit cost of operation and maintenance c_s	1.5
Charging efficiencies of storage equipment η_c	0.5
Discharging efficiencies of storage equipment η_d	0.5
Power generation cost parameter of microgrid a_m	0.05
Pollutant emission cost parameter of microgrid α_m	0.05
Satisfaction parameter for utility company $X_{m,g}$	5
Satisfaction parameter for utility company $d_{m,g}$	0.21
Satisfaction parameter for storage company $X_{m,s}$	10
Satisfaction parameter for storage company $d_{m,s}$	0.21
Satisfaction parameter for microgrid $X_{c,m}$	50
Satisfaction parameter for microgrid $d_{m,s}$	0.15
Capacity of utility company $L_{g,\max}$	200 kW
Capacity of storage company $L_{m,\max}$	100 kW
The highest price users can afford $p_{m,\max}$	50 cents/kWh
The real wind power $\hat{L}_r$	20 kW
The penalty factor F	−50

Table 2. Simulation parameters.

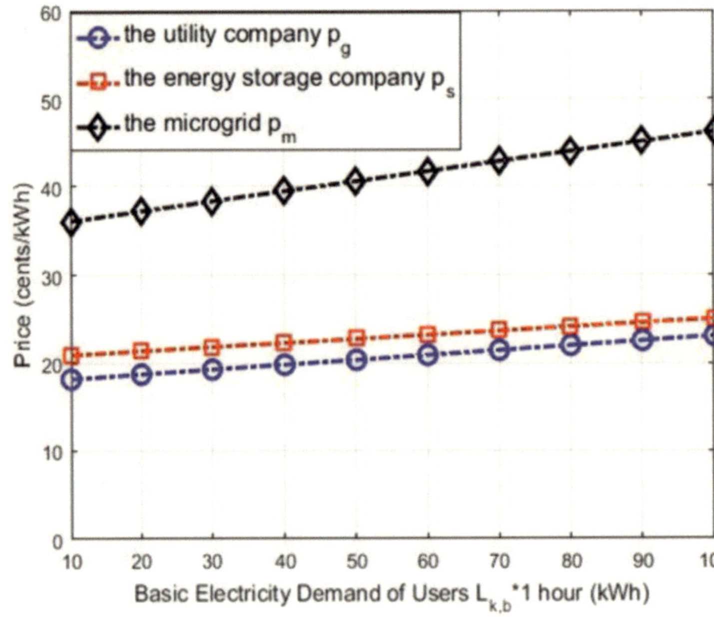

Figure 4. The optimal electricity prices of the utility company $\hat{p}_g$, the energy storage company $\hat{p}_s$, and the microgrid $\hat{p}_m$ versus the basic electricity demands of user $L_{k,b}$.

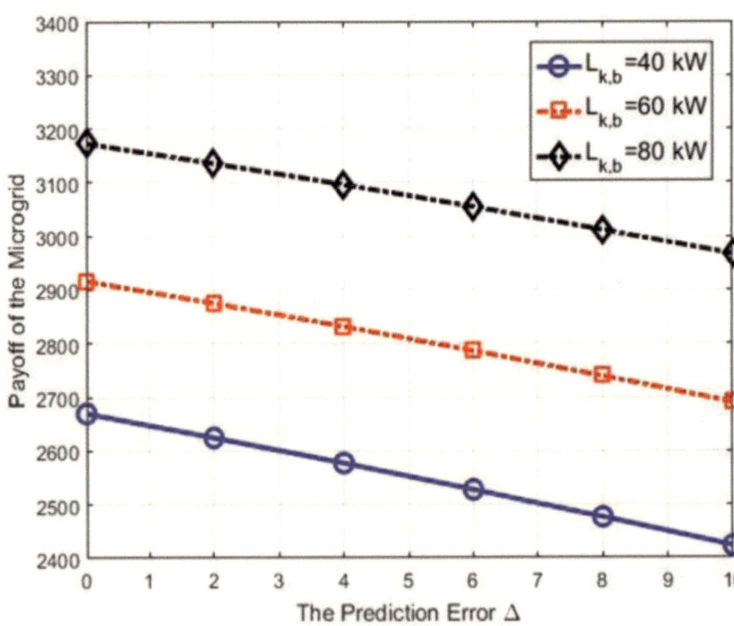

Figure 5. The optimal payoff of the microgrid U_m versus the prediction error of wind power forecasting $\Delta > 0$.

have been considered. Both **Figures 5** and **6** show that the optimal payoff of the microgrid decreases monotonically as $|\Delta|$ increases. For example, if Δ is increased from 0 to 10 kW or decreased from 0 to -10 kW, the optimal payoff will be decreased by 9.2 and 22.1% when

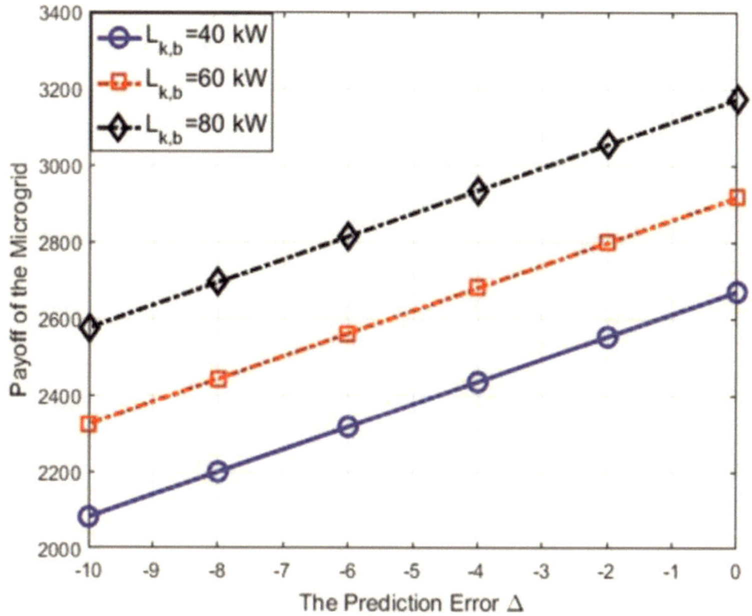

Figure 6. The optimal payoff of the microgrid U_m versus the prediction error of wind power forecasting $\Delta < 0$.

$L_{k,b} = 40$ kW, respectively. The reason is that the microgrid will be charged for the difference between the predicted and actual electricity procurement quantities, due to the restriction of power purchase agreement. It is also clear that the optimal payoff is degraded more severely

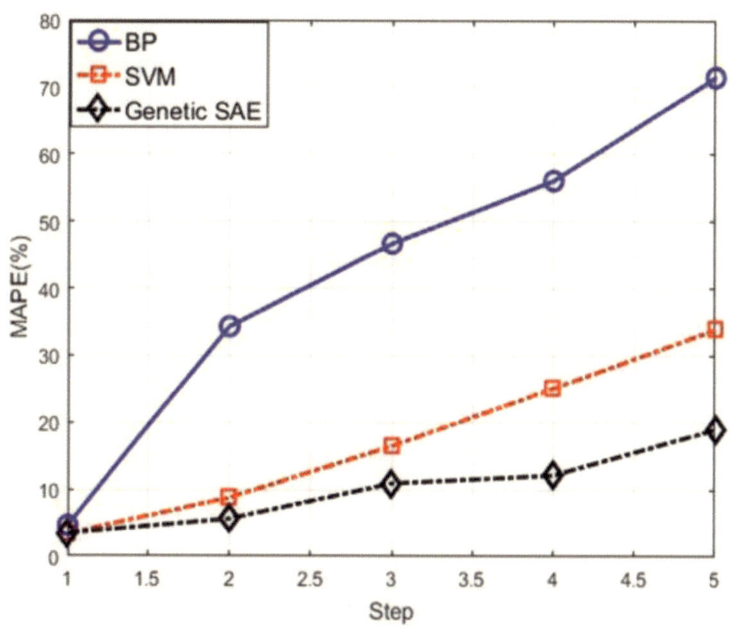

Figure 7. MAPE of three different models with wind power forecasting step varies.

when $\Delta < 0$ compared to $\Delta > 0$. The reason is that the electricity prices of the utility and the energy storage companies are higher when $\Delta < 0$ compared to the case of $\Delta > 0$.

Figure 7 shows the MAPE value of three different algorithms including BP, SVM, and genetic SAE versus wind power forecasting step. The process of wind power forecasting based on historical data in current time is called step 1. By adding the prediction result to the historical data, the authors can obtain a new prediction result in next hour and the process is called step 2, and so on. A higher step means longer period of prediction, which presents lead to less precise predictions and high MAPE. From the simulation results, the authors found that MAPE increases as prediction step increases. Thus, we can come to the conclusion that the result becomes inaccurate as the step increases. Furthermore, the simulation results demonstrate the authors obtain a minimum prediction error by genetic SAE algorithm compared to the other two algorithms. More concretely, the predicted absolute error decreases by 7.3% compared with the SVM algorithm and 32.4% compared with the BP algorithm when step 5.

6. Conclusions

In this chapter, the authors proposed to utilize the big data-based power generation forecasting techniques to obtain the short-term wind power forecasting results that assist the microgrid to implement energy management strategies. Simulation results validated the proposed algorithm and demonstrated that the optimal payoff of the microgrid is decreased due to the prediction error. The proposed genetic SAE algorithm is demonstrated to provide the most accurate predictions, which is helpful for energy management. In future work, we will emphasize on cooperative energy management among multiple microgrids based on the predictions of renewable power and electricity consumption.

Author details

Zhenyu Zhou*, Fei Xiong, Chen Xu and Runhai Jiao

*Address all correspondence to: zhenyu_zhou@ncepu.edu.cn

The State Key Laboratory of Alternate Electrical Power System with Renewable Energy Sources, School of Electrical and Electronic Engineering, North China Electric Power University, Beijing, China

References

[1] Rifkin J. The Third Industrial Revolution: How Lateral Power is Transforming Energy, The Economy, and The World. New York: Palgrave Macmillan Trade; 2011

[2] Katz R, Culler D, Sanders S, Lutz K. An information-centric energy infrastructure: The Berkeley view. Sustainable Computing: Informatics and Systems. 2011;**1**(1):7–22

[3] Huang A, Crow M, Heydt G, Dale S. The future renewable electric energy delivery and management (FREEDM) system: The energy internet. Proceedings of the IEEE. 2011;**99**(1):133–148

[4] Conti S, Nicolosi R, Rizzo S. Generalized systematic approach to assess distribution system reliability with renewable distributed generators and microgrids. IEEE Transactions on Power Delivery. 2012;**27**(1):261–270

[5] Zhang Y, Asr N, Duan J, Chow M. Day-ahead smart grid cooperative distributed energy scheduling with renewable and storage integration. IEEE Transactions on Sustainable Energy. 2016;**7**(4):1739–1748

[6] Clement K, Haesen E, Driesen J. The impact of charging plug-in hybrid electric vehicles on a residential distribution grid. IEEE Transactions on Power Systems. 2010;**25**(1):371–380.

[7] Haupt S, Kosovic B. Variable generation power forecasting as a big data problem. IEEE Transactions on Sustainable Energy. 2017;**8**(2): 725–732

[8] Tadelis S. Game Theory: An Introduction. Princeton, NJ: Princeton University Press; 2013

[9] Wei W, Liu F, Mei S, Hou Y. Robust energy and reserve dispatch under variable renewable generation. IEEE Transactions on Smart Grid. 2015;**6**(1):369–380

[10] Darivianakis G, Georghiou A, Smith R, Lygeros J. A stochastic optimization approach to cooperative building energy management via an energy hub. In: Proc. IEEE CDC'15; Osaka, Japan, 2015. pp. 6212–6215

[11] Li L, Yan B, Yang C, Zhang Y, Chen Z, Jiang G. Application-oriented stochastic energy management for plug-in hybrid electric bus with AMT. IEEE Transactions on Vehicular Technology. 2016;**65**(6):4459–4470

[12] Chen T, Wang X, Giannakis G. Cooling-aware energy and workload management in data centers via stochastic optimization. IEEE Journal of Selected Topics in Signal Processing. 2016;**10**(2):402–415

[13] Valencia F, Collado J, Saez D, Marin L. Robust energy management system for a microgrid based on a fuzzy prediction interval model. IEEE Transactions on Smart Grid. 2016;**7**(3):1486–1494

[14] Chiu W, Sun H, Poor H. Energy imbalance management using a robust pricing scheme. IEEE Transactions on Smart Grid. 2013;**4**(2):896–904

[15] Morales J, Cervantes I, Castillo U. On the design of robust energy management strategies for FCHEV. IEEE Transactions on Vehicular Technology. 2015;**64**(5):1716–1728

[16] Zhang Y, Gatsis N, Giannakis G. Robust energy management for microgrids with high-penetration renewables. IEEE Transactions on Sustainable Energy. 2013;**4**(4):944–953

[17] Xiang Y, Liu J, Liu Y. Robust energy management of microgrid with uncertain renewable generation and load. IEEE Transactions on Smart Grid. 2016;**7**(2):1034–1043

[18] Khalid M, Savkin A. A method for short-term wind power prediction with multiple observation points. IEEE Transactions on Power Systems. 2012;**27**(2):579–586

[19] Zhang J, Wang C. Application of ARMA model in ultra-short term prediction of wind power. In: Proc. CSA'13; Wuhan China; 2013. pp. 361–364

[20] Yunus K, Thiringer T, Chen P. ARIMA-based frequency-decomposed modeling of wind speed time series. IEEE Transactions on Power Systems. 2016;**31**(4):2546–2556

[21] Kurian S, Krishnan S, Cheriyan E. Real time implementation of artificial neural networks-based controller for battery storage supported wind electric generation. IET Generation, Transmission and Distribution. 2015;**9**(10):937–946

[22] Gu B, Sheng V. A robust regularization path algorithm for v-support vector classification. IEEE Transactions on Neural Networks and Learning Systems. 2017;**28**(5): 1241–1248

[23] Gu B, Sheng V, Tay K, Romano W, Li S. Incremental support vector learning for ordinal regression. IEEE Transactions on Neural Networks and Learning Systems. 2015;**26**(7):2546–2556

[24] Methaprayoon K, Yingvivatanapong C, Lee W, Liao J. An integration of ANN wind power estimation into unit commitment considering the forecasting uncertainty. IEEE Transactions on Industry Applications. 2007;**43**(6):1441–1448

[25] Yang L, He M, Zhang J, Vittal V. Support-vector-machine-enhanced Markov model for short-term wind power forecast. IEEE Transactions on Sustainable Energy. 2015;**6**(3):791–799

[26] Hosseini-Asl E, Zurada J, Nasraoui O. Deep learning of part-based representation of data using sparse autoencoders with nonnegativity constraints. IEEE Transactions on Neural Networks and Learning Systems. 2016;**27**(12):2486–2498

[27] Thirukovalluru R, Dixit S, Sevakula R, Verma N, Salour A. Generating feature sets for fault diagnosis using denoising stacked auto-encoder. In: Proc. IEEE ICPHM'16; Ottawa, Canada; 2016. pp. 1–7

[28] Jiang Q, Xue M, Geng G. Energy management of microgrid in grid-connected and standalone modes. IEEE Transactions on Power Systems. 2013;**28**(3):3380–3389

[29] Xie S, Zhong W, Xie K, Yu R, Zhang Y. Fair energy scheduling for vehicle-to-grid networks using adaptive dynamic programming. IEEE Transactions on Neural Networks and Learning Systems. 2016;**27**(8):1697–1707

[30] Yu R, Zhong W, Xie S, Yuen C, Gjessing S, Zhang Y. Balancing power demand through EV mobility in vehicle-to-grid mobile energy networks. IEEE Transactions on Industrial Informatics. 2016;**12**(1):77–90

[31] Saad W, Han Z, Poor H, Basar T. Game-theoretic methods for the smart grid: An overview of microgrid systems, demand-side management, and smart grid communications. IEEE Signal Processing Magazine. 2012;**29**(5):86–105

[32] Maschler M, Solan E, Zamir S. Game Theory. Cambridge, UK: Cambridge University Press; 2013

[33] Maharjan S, Zhu Q, Zhang Y, Gjessing S, Basar T. Dependable demand response management in the smart grid: A Stackelberg game approach. IEEE Transactions on Smart Grid. 2013;**4**(1):120–132

[34] Khoussi S, Bilil H, Aniba G. Optimal time of use of renewable electricity pricing: Three-player games model. In: Proc. IEEE SmartGridComm'15; Miami, FL; 2015. pp. 199–204

[35] Mediwaththe C, Stephens E, Smith D, Mahanti A. A dynamic game for electricity load management in neighborhood area networks. IEEE Transactions on Smart Grid. 2016;7 (3):1329–1336

[36] Tushar M, Assi C, Maier M. Distributed real-time electricity allocation mechanism for large residential microgrid. IEEE Transactions on Smart Grid. 2014;**6**(3):1353–1363

[37] Ma K, Hu G, Spanos C. A cooperative demand response scheme using punishment mechanism and application to industrial refrigerated warehouses. IEEE Transactions on Industrial Informatics. 2015;11(6):1520–1531

[38] Xu J, Zhang R. Cooperative energy trading in CoMP systems powered by smart grids. IEEE Transactions on Vehicular Technology. 2016;**65**(4):2142–2153

[39] Zhang Y, Rahbari-Asr N, Duan J, Chow M. Day-ahead smart grid cooperative distributed energy scheduling with renewable and storage integration. IEEE Transactions on Sustainable Energy. 2016;**7**(4):1739–1748

[40] Gkatzikis L, Koutsopoulos I, Salonidis T. The role of aggregators in smart grid demand response markets. IEEE Journal on Selected Areas in Communications. 2013;**31**(7):1247–1257

[41] Mohsenian-Rad A, Wong V, Jatskevich J, Schober R, Leon-Garcia A. Autonomous demand-side management based on game-theoretic energy consumption scheduling for the future smart grid. IEEE Transactions on Smart Grid. 2010;**1**(3):320–331

[42] Machowski J, Bialek J, Bumby D. Power System Dynamics: Stability and Control. New York: Wiley; 2008

Chapter 10

Operation of Plug-In Electric Vehicles for Voltage Balancing in Unbalanced Microgrids

Guido Carpinelli, Fabio Mottola, Daniela Proto and Angela Russo

Additional information is available at the end of the chapter

http://dx.doi.org/10.5772/intechopen.68894

Abstract

The widespread use of distributed energy resources in the future electric distribution systems represents both a challenge and an opportunity for all the Smart Grid operators. Among these resources, plug-in electric vehicles are expected to play a significant role not only for the economic and environmental benefits they involve but also for the ancillary services they can provide to the supplying grid. This chapter deals with real-time operation of unbalanced microgrids including plug-in electric vehicles. The operation is achieved by means of an optimal control strategy aimed at minimizing the costs sustained for the energy provision while meeting various technical constraints. Among the technical constraints, the optimal control allows guaranteeing the satisfaction of power quality requirements such as the containment of slow voltage variations and the unbalance factors. Case studies are investigated in order to show the feasibility and the effectiveness of the proposed approach.

Keywords: microgrids, slow voltage variations, unbalances, optimal control strategy, plug-in vehicles

1. Introduction

Smart grids (SGs) are playing a vital role in the future of power systems mainly due to their potential to minimize costs and environmental impacts while maximizing reliability, power quality (PQ), resilience, and stability. These goals can be attained by means of a proper exploitation of the SG's distributed energy resources (DERs). However, the integration of DERs (i.e., generation units, controllable loads, and energy storage-based devices) is a challenging task in the view of the optimal planning and operation of the SGs. Thus, new research contributions aimed to guarantee the DER correct behavior are of high interest for all the involved operators (from System Operators to Electric Utilities and Energy Traders, Independent Power Producers and Consumers). Among the energy storage-based devices, plug-in electric vehicles (PEVs) are achieving particular attention in the relevant literature. PEVs, in fact, are expected to have a wide diffusion in the next future especially due to the current attention paid worldwide to the reduction of greenhouse gas emissions mainly caused by the use of internal combustion engines. However, PEVs require to charge from the power grid, then their massive spread could pose critical issues. The charge of a large number of PEVs is a huge load demand that the grid is called to face and a number of problems can occur in terms of PQ degradation, instability of electrical networks, and degradation of operation efficiency [1]. On the other hand, PEVs connection to the grid could represent also an opportunity due to the presence of on-board batteries that could have an important role in grid operation. The electrical energy storage devices, in fact, are a particular typology of loads that are expected to be key components in modern power systems. When opportunely coordinated, these devices can become distributed resources for the grid so providing benefits rather than criticisms to it. The benefits achievable by their use are mainly related to voltage support, price arbitrage, PQ, reliability improvement, reduction of the cost for energy provision, and renewable energy fostering. In order to gain these benefits, the storage systems need specific charge/discharge control strategies able to account for the requirements of the storage itself and of the grid they are connected to, in addition to adequate communication infrastructures. In particular, when the storage systems are included in a microgrid (μG), the control strategies should guarantee the correct operation of the whole μG taking into account also the presence of other controllable sources, such as DG units and dispatchable loads. In these cases, the adopted control tasks should take into account also the possibility of load shedding or generation curtailment in case of increased demand or generation, respectively.

A proper PEV control strategy would take into account targets such as cost minimization and PQ improvement.

Regarding the first target, in the frame of demand response paradigms, PEVs have a great potential. More specifically, in the case of dynamic pricing schemes, an appropriate control of PEVs charging would allow reducing costs by shifting the peak load in response to changes in the price of electricity over time. Different time-based rate programs are available: time-of-use (TOU) pricing refers to different prices applying to determined periods of the day (block of hours of the day). In these periods, the tariff is constant. Real-time pricing (RTP) is characterized by pricing rates typically applying on an hourly basis. Variable peak pricing (VPP) is a hybrid of time-of-use and real-time pricing not only with different periods of the day for pricing but also with a price for the on-peak period that depends on market conditions. With the critical peak pricing (CPP), utilities identify critical events during a specified time period when the price for electricity may substantially increase (e.g., very hot hours in a summer afternoon). Finally, critical peak rebate (CPR) is a pricing structure similar to the CPP that allows customers to be paid for cutting back on electricity during critical events relative to the amount they normally use [2]. In this variegated pricing scenario, the optimal operation of PEVs allows substantial reduction of the operation costs for the μG that could optimize the expense for the energy provision. Of course, an optimized control should be performed by taking into account the preferences and the comfort levels required by the vehicle owners.

Regarding the PQ improvement target, several benefits can be achieved by a proper control of the PEV charging. In the case of low voltage (LV) μGs, typically characterized by the presence of unbalances, the benefits PEVs can provide are even more significant. LV μGs, in fact, are characterized by significant unbalances due to their structure, to the presence of single-phase loads and renewable generation units (i.e., single-phase photovoltaic systems). For this reason, voltage unbalances can be excessive and represent a burdensome PQ issue [3]. Consequences of unbalances in LV μGs can be the increase of losses and heating effects affecting power electronic converters and induction motors [4], overloading of distribution feeders and transformers [5], excessive mechanical stress and noise due to a double system frequency in the synchronous generators equipping microgeneration units [6].

This chapter is framed in this multifaceted scenario of μG real-time operation and deals with advanced strategies for the control of μG's distributed energy sources (DERs) aimed at minimizing operation costs and at ensuring an optimal grid and storage systems operation with reference to various technical constraints. More specifically, the DERs to be optimally controlled in the considered μG are: the batteries on board the electric PEVs connected to the grid for charging and the DG units connected to the grid through power converters. The optimal operation of μG refers to the cost minimization under real-time pricing while grid constraints on bus voltages and line currents are satisfied. Among the voltage constraints, a maximum allowed voltage unbalance factor is imposed on the basis of the current standards [7]. Specific requirements for the storages (e.g., a vehicle's battery has to be fully charged at a specified departure time) are also guaranteed.

The proposed control strategy refers to real-time operation and helps the μG's operator better control the operational processes and plan appropriate energy market bidding, due to the most proficient use of electrical storage systems. It is formulated as a multi-period non-linear constrained optimization problem (OP) which aims at minimizing the energy costs for the whole μG while satisfying technical constraints related to the correct operation of the components and of the grid as well as to the needs of the vehicle owners.

The technical strength of the proposed operation strategy is related to the proper control of DERs which allows

- minimizing the costs for energy provision;
- complying with the standard limits for the slow voltage variations and voltage unbalances, which are among the most severe PQ phenomena in LV μGs;
- mitigating the impact of very large loads, by containing line currents and bus voltages within admissible ranges; and
- increasing grid capability in terms of power injection from distributed energy sources.

This chapter is structured as follows. The state of the art of the PEV charging devices and the challenges and benefits related to their inclusion in μGs are described in Section 2. Section 3 shows the analytical formulation of the proposed control strategy. The results of numerical applications are reported in Section 4. Finally, Section 5 draws conclusions and opens problems for future research.

2. Plug-in vehicles in the frame of low voltage microgrids: State of the art

The relevant literature on plug-in electric vehicles is very wide and rich. In this section, review papers are mentioned to a large extent and, in some cases, research papers are shown on particular topics.

2.1. Classification of charging systems

One of the key elements of a successful electric vehicle diffusion is the charging equipment. Some common features are required for the electric vehicle charger (EVC), for example, battery performance, high reliability of the charger, high power efficiency, and minimum impact on the PQ [8]. In addition, since a large-scale diffusion of PEVs can also pose new challenges for the power system operators (e.g., congestions, voltage profiles, etc.), the "smart" charging solutions may transform PEVs in a resource for the system [9–11].

With specific reference to PEVs, in Ref. [1], the battery charging schemes are classified as follows: uncontrolled, indirectly controlled, smart, and bidirectional. In the uncontrolled charging, the vehicle immediately begins recharging as soon as it is connected to the grid. Indirectly controlled, smart, and bidirectional charging schemes are used to control energy prices; in the smart charging schemes, the battery charging is subject to some measure of intelligent control by the utility or system operator.

Regarding the charger device, an AC/DC converter is required since the batteries charge by absorbing DC power. The main function of this converter is to rectify the AC power from the electrical grid. Chargers can be installed on board and off board of the vehicles. On-board charger is often designed in small size to reduce the weight burden for PEV. Off-board chargers are particularly convenient in fast charging service [12].

Different charging methods can be implemented. Ref. [12] classifies the charging methods as constant current, constant voltage, constant power, taper charging, trickle charging, advanced charging that involves combination of the above methods, pulse-charging and negative pulse-charging.

2.2. Adverse effects and services provided by plug-in electric vehicles

PEVs can be operated in two configurations when connected to the electric grid:

- Grid-to-vehicle, usually referred to as V1G (also G2V); and
- Vehicle-to-grid, usually referred to as V2G.

In the first configuration, the flow of the electric power is uni-directional, that is, the electric power flows from the grid to the PEVs. In the second configuration, the flow of the electric power is bi-directional, that is, the PEV can absorb electric power from the grid and inject electric power to the grid. In this case, PEVs can be assimilated to a distributed energy resource (distributed generator or storage system).

A large number of studies indicates some adverse effects of PEVs on electric distribution systems (for instance, see the review Ref. [13]). Among them, we can mention voltage instability, increased

http://dx.doi.org/10.5772/intechopen.68894

peak demand, PQ problems (e.g., voltage profile, voltage unbalance, harmonics, etc.), increased system energy losses, transformer heating, and overloading [12–14]. It should be noted that most of them occur when the charging of vehicles is uncoordinated.

In spite of the impact on the distribution systems, PEVs are considered valuable resources for the electric power grids because they can provide a large number of services to the local distribution system and also to the power system. When V1G configuration is adopted, the charging method assumes an important role. For instance, shifting the charging to off-peak periods can be exploited to better integrate a large diffusion of PEVs into the electricity grid. Another contribution [15] proposes an integration of PEVs able to improve the PQ levels of the distribution system. An autonomous control of PEVs charging systems (in V1G configuration) is proposed in order to reduce voltage unbalances in a distribution network, in which local voltage measurements drive the modulation of the charging current.

However, the most promising configuration is the V2G.[1] V2G power is an interesting concept that was first proposed in Ref. [16]. Operating in V2G mode, a PEV can provide services to the grid when it is parked. Considering that PEVs are parked much of the time (e.g., Ref. [17] estimates a parking time equal to 96% of the time), the contribution can be important. Typical services that PEVs are able to provide in the V2G operating mode are [12, 18]:

- baseload power and peak power provision; and
- ancillary services (frequency control, load balance, and spinning reserve services).

Other services provided in V2G mode [18, 13, 9] are reactive power support (for instance, Ref. [19] proposes a coupled energy and reactive power market considering the contribution of PEVs), power leveling (i.e., valley filling and peak shaving), and PQ improvement (for instance, harmonic filtering). In addition, in the presence of high penetration of intermittent renewable power generation, the V2G mode can be exploited to use PEVs as storage systems, backup systems, and to balance power fluctuations.

As a consequence of V2G diffusion, an increase of stability and reliability of the distribution system and a reduction of distribution costs are expected [20].

2.3. Plug-in electric vehicles modeling in μGs

μG operation is significantly affected by the wide diffusion of PEVs, in both configuration V1G and V2G. In the relevant literature, a lot of scientific contributions inquire into the possible exploitation of PEVs to provide numerous services at μG level and to strengthen the operation of a μG.

An early contribution [21] analyzes a large exploitation of PEV batteries in islanded operated μGs and proposes several control strategies to improve μG islanding.

One issue considered in the relevant literature is the day-ahead scheduling when PEVs are present in the μGs. Many contributions are available [22–29]. The proposed models may take

---

[1] A thorough comparison between V1G and V2G configurations is provided in Ref. [13].

into account the uncertainties of the problem variables, characteristics of different intermittent energy resources, and different optimization objectives. The integration of intermittent generation and PEVs into scheduling problems is taken into account in [23, 24, 26–29].

A two-stage energy scheduling for a μG with renewable generation is proposed in Ref. [23] and the impact of PEVs on the energy scheduling is evaluated considering different charging schemes. A two-stage framework for a μG, including PEVs and solar energy generation, is proposed also in Ref. [26]; at the first stage, a stochastic energy schedule problem is formulated for the next 24-h operation; then at the second stage, a predictive online charging control method is implemented to account for uncertainties. Also in Ref. [28], the uncertainty of the load demand and of PEVs along with the intermittency of the renewable energy resources are taken into account in the μG energy scheduling. In particular, peculiarities of wind generation and coordination among wind power plants and PEVs are considered in Refs. [22, 30, 31]. In Ref. [22], a solution of the coordinated wind and PEV (in V2G configuration) energy dispatch problem in a stochastic framework is proposed; uncertainties of wind power and of PEV driving patterns are taken into consideration. In Ref. [30], the problem of power balance in a μG including PEVs and wind turbines is addressed and a hierarchical stochastic control scheme for the coordination of PEV charging and wind power is proposed.

The presence of responsive loads together with PEVs and intermittent generation is considered in Ref. [27], where the PEVs are used to reduce peak power and modify load curves.

In Ref. [25], a regional energy management strategy based on the use of PEVs and battery swapping stations whose charging is coordinated by a price-incentive model is proposed.

The optimization models formulated for the μG energy scheduling take into account several objectives, such as the minimization of the expected total operation costs [28], the minimization of the total cost of PEVs and maximization of the profit of battery swapping stations [25], the minimization of the total cost of the network including the cost of power supply for loads and PEVs as well as the cost of energy not supplied [29].

Particular case studies are presented in Refs. [32–34]. Ref. [32] proposes to jointly control the electricity consumption of home appliances and PEVs in a μG context; in Ref. [33], an office building μG, and in Ref. [34], a commercial μG are considered.

Further topics related to the presence of PEVs in μGs are considered in the relevant literature. Refs. [35, 36] address the problem of the unit commitment for μGs including also PEVs. In particular, Ref. [36] proposes a probabilistic model able to account for the uncertainties on wind power generation, loads, and PEVs operation.

V2G strategies are developed to allow PEVs to contribute to congestion management to help maintain a secure state of the grid supporting the system in critical conditions [37]. The presence of PEVs can also be used to actively participate in μG service restoration, as shown in Ref. [38]. In addition, PEVs have been demonstrated to be able to support the system also in smoothing frequency fluctuations due to insufficient load frequency control capacity. In Ref. [39], the power control of PEVs in V2G configuration is applied to compensate for the inadequate load frequency control capacity and to improve the frequency stability of μGs, especially

http://dx.doi.org/10.5772/intechopen.68894

in island operation mode. Ref. [40] proposes a new controller able to take instantaneous power available from the fleet of PEVs in order to provide a frequency support service.

Finally, a day-ahead scheduling strategy is proposed in Ref. [3] which refers to unbalanced μGs including in addition to PEVs, other distributed resources. The operation strategy is based on a multi-objective approach, whose objective functions refer to cost minimization, PQ improvements, and energy savings.

3. Plug-in vehicles in the frame of low voltage microgrids: The proposed optimal operation strategy

SGs need to optimally operate its DERs. PEVs are particular types of DERs whose benefits can be achieved by adopting appropriate charge/discharge control strategies able to account for the requirements of the on-board storage units and of the grid they are connected to. These control strategies should take into account also the presence of other DERs and should guarantee the correct operation of the whole μG during its operation.

In this section, we formulate a control strategy based on the solution of a multi-period, non-linear, constrained OP which aims at minimizing the energy costs for the whole μG while satisfying technical constraints related to the correct operation of the components and of the grid as well as to the needs of the vehicle owners. In particular, the technical constraints include limits on some power quality disturbances such as slow voltage variations and unbalances.

The strategy is applied to a three-phase LV μG that includes loads and distributed resources, such as DG units and batteries on board PEVs, and is connected to the MV distribution grid through an MV/LV transformer.

The loads of the μG are supposed to be non-responsive, single-phase and three-phase loads.

The DG units are assumed non-dispatchable, single-phase PV systems which are the most extensively used type of DG units in LV systems. The connection to the distribution network of the PV units through power converters allows the control of their reactive power.

The PEVs' batteries are connected to the μG using EVCs which consist of single-phase devices connected to the grid through power converters that allow the control of active power and reactive power in both charging and discharging stages.

The proposed control strategy is used to evaluate the reference signals for the converters which connect the DERs to the grid. To obtain the above signals for the kth control interval, the non-linear, constrained OP: (i) is formulated over a time horizon Ω_k, composed of elementary time intervals, each of duration Δt, and (ii) is solved within one elementary time interval ahead the considered control interval (e.g., the control signals for the kth elementary time interval are obtained within the (k-1)th elementary time interval).

More specifically, the dispatching of the DERs must guarantee the desired energy level to be stored in all of the connected PEVs at their departure time; thus, the procedure for the kth control interval takes into account all of the time intervals included within the longest among the plug-in times of the connected vehicles (time horizon Ω_k). Based on this consideration, the time horizon Ω_k to be considered will change along with the arrivals of PEVs, and, at each new arrival, it has to be properly defined once more. To explain how to select the adequate time horizon Ω_k for the *k*th control interval, let us refer to the example shown in **Figure 1**, in which the time is discretized in time intervals of duration Δt, as mentioned above. In **Figure 1**, N_{EV} is the total number of EVCs in the μG.

In the figure, for each EVC, the time intervals between the arrival and the departure of the vehicle are highlighted. For instance, the vehicle charged at the *ith* bus is already connected at the *k*th time interval and its departure is scheduled at the end of the $(k+n_{k,i})$th time interval. The set of time intervals $\{k,\ldots, k+n_{k,i}\}$ is referred to as $\omega_{k,i}$. The time horizon of the optimization has to account for all the vehicles connected at the *k*th control interval. Generally speaking, since the control procedure would account for the desired energy stored into the batteries of all of the plugged vehicles, the time horizon of the optimization at the *k*th control interval is:

$$\Omega_k = \{k, \ldots, k + n_k\}, \tag{1}$$

being

$$n_k = max\{n_{k,1}, \ldots, n_{k,N_{EV}}\} \tag{2}$$

that is the last interval of the time horizon Ω_k that has to be considered when the procedure is applied to the *k*th time control interval.

Due to the presence of constraints referring to different time intervals, the OP involved in the desired control strategy is multi-period and solved for all the time intervals of Ω_k at the same time.

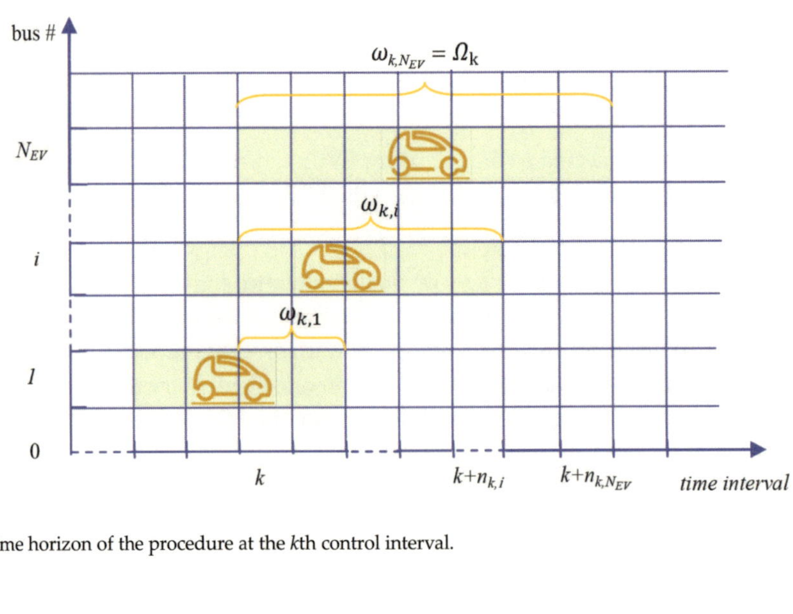

Figure 1. Time horizon of the procedure at the *k*th control interval.

http://dx.doi.org/10.5772/intechopen.68894

The OP is a single-objective (SO) OP that uses as inputs the values, for all of the time intervals included in Ω_k, of (i) forecasted load demand, (ii) forecasted DG production, and for all of the vehicles that are plugged in at the time interval k, (iii) initial state of charge (SOC) of the on-board batteries, (iv) departure time, and (v) desired value of SOC at the departure time.

Regarding the outputs of the solution of the OP, they are control actions and/or data useful for the monitoring of μG operation. More in detail, in the case of control action, the outputs of the procedure are reference signals that the CCS sends to the interfacing converters of the distributed resources that are the active and reactive phase powers that the EVC exchange with the grid when the vehicles are plugged in, and the reactive phase power requested to the DG units. Outputs useful for the monitoring are (i) the voltage profile at each bus, (ii) the unbalance factor at each bus, and (iii) the currents flowing in the lines. The active power exchanged with the upstream grid at the Point of Common Coupling (PCC) is another output of the procedure that is useful for the knowledge of the μG's energy request and the corresponding cost. **Figure 2** shows a schematic of the proposed control strategy evidencing its inputs and outputs.

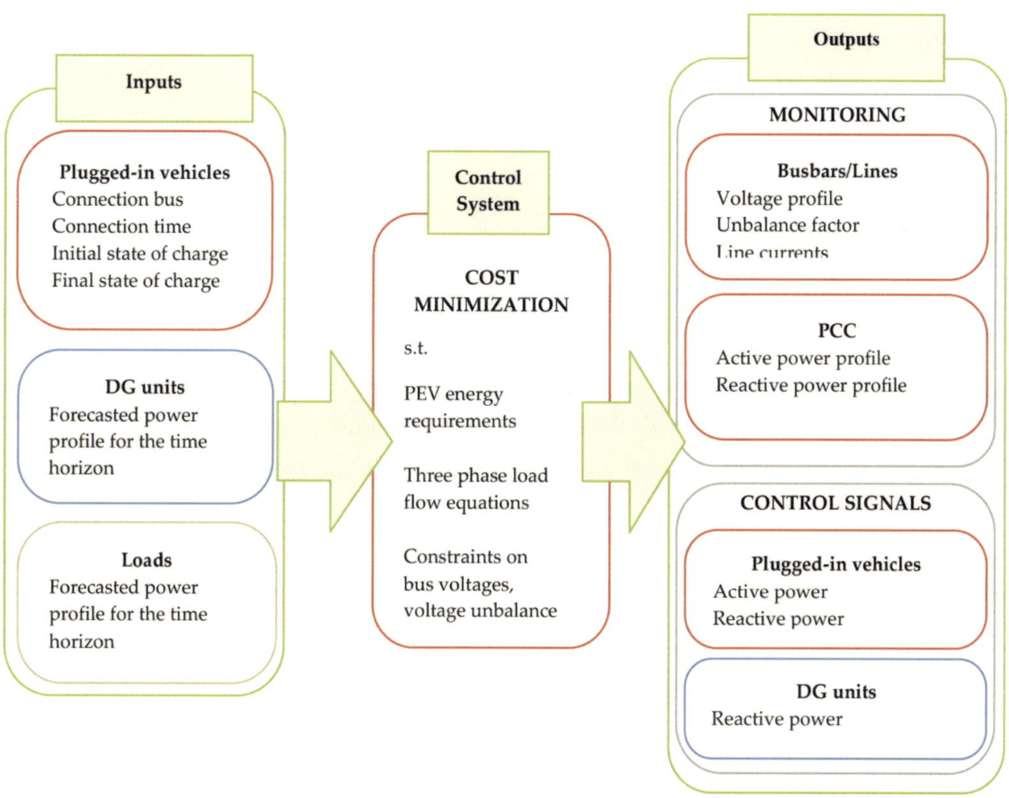

Figure 2. Schematic of the proposed control strategy.

The OP to be solved is formulated as:

$$min\ F(\boldsymbol{x}) \tag{3}$$

subject to:

$$\Psi_h(\boldsymbol{x}) = 0, \quad h = 1,\dots, n_{eq} \tag{4}$$

$$\eta_m(\boldsymbol{x}) \le 0, \quad m = 1,\dots, n_{in} \tag{5}$$

where Ψ_h is the hth equality constraint, η_m is the mth inequality constraint, $\boldsymbol{x}$ is the vector of the optimization variables, $F(\boldsymbol{x})$ is the objective function.

The objective function refers to the cost of the energy purchased from the upstream grid, that is:

$$CE_k = \sum_{\tau \in \Omega_k} \sigma_\tau (P^1_{1,\tau} + P^2_{1,\tau} + P^3_{1,\tau})\Delta t \tag{6}$$

where $P^1_{1,\tau}$, $P^2_{1,\tau}$, and $P^3_{1,\tau}$ are the phase active powers imported from the upstream grid at the τth time interval and σ_τ is the price of energy during the τth time interval. Regarding this last, a real-time tariff is considered whose variation is on an hourly basis. It has to be noted that, in this formulation, it is assumed that the μG is not allowed to sell energy to the upstream grid.

The equality and inequality constraints can be classified on the basis of the part of the system they refer to. Thus, they are grouped into constraints imposed on:

i. the whole μG;

ii. the slack bus;

iii. the busses where loads are connected;

iv. the busses where DG units are connected, and

v. the busses where plug-in vehicles are connected.

For the sake of clarity, in what follows, busses are supposed to be pure load or pure generation or EVC busses; it is trivial to derive the case of mixed typologies.

3.1. Constraints on the whole μG

Since we deal with unbalanced μGs, the three-phase grid, loads, and generations systems were modeled in phase coordinates [41, 42], and then the three-phase load flow equations are included in the optimization model:

$$P^p_{i,\tau} = V^p_{i,\tau} \sum_{j=1}^{ng} \sum_{q=1}^{3} V^q_{j,\tau} [G^{pq}_{ij} \cos(\vartheta^p_{i,\tau} - \vartheta^q_{j,\tau}) + B^{pq}_{ij} \sin(\vartheta^p_{i,\tau} - \vartheta^q_{j,\tau})] \tag{7}$$

$$\begin{gathered} Q^p_{i,\tau} = V^p_{i,\tau} \sum_{j=1}^{ng} \sum_{q=1}^{3} V^q_{j,\tau} [G^{pq}_{ij} \sin(\vartheta^p_{i,\tau} - \vartheta^q_{j,\tau}) - B^{pq}_{ij} \cos(\vartheta^p_{i,\tau} - \vartheta^q_{j,\tau})], \\ \tau \in \Omega_k, i \in \Omega_G, p = 1, 2, 3 \end{gathered} \tag{8}$$

http://dx.doi.org/10.5772/intechopen.68894

where ng is the number of μG busses, B_{ij}^{pq} and G_{ij}^{pq} are the terms of the susceptance matrix and of the conductance matrix, respectively, corresponding to the bus i with phase p and the bus j with phase q; $V_{i,\tau}^{p}$ ($V_{j,\tau}^{q}$) is the magnitude of the ith (jth) bus voltage with phase p (q) during the τth time interval; $\vartheta_{i,\tau}^{p}$ ($\vartheta_{j,\tau}^{q}$) is the phase-voltage argument of the ith (jth) bus voltage with phase p (q) during the τth time interval; $P_{i,\tau}^{p}$ and $Q_{i,\tau}^{p}$ are the active and reactive powers injected at the phase p of the ith bus during the τth time interval; Ω_G is the set of μG busses.

Inequality constraints are imposed on the phase-voltage magnitudes that must be within admissible ranges and on the unbalance factor at all busbars that cannot exceed a maximum value:

$$V_{min} \le V_{i,\tau}^{p} \le V_{max}, \quad \tau \in \Omega_k, \ i \in \Omega_G, \ p = 1,2,3 \tag{9}$$

$$kd_{i,\tau} \le kd_{max}, \quad \tau \in \Omega_k, \ i \in \Omega_G \tag{10}$$

where V_{min} and V_{max} are minimum and maximum values for voltage magnitudes, $kd_{i,\tau}$ is the unbalance factor at the ith bus during the τth time interval and kd_{max} is the maximum value imposed to the unbalance factor. The unbalance factor is given by the ratio of the positive and negative components of voltages that are expressed as functions of the bus phase voltages.

The line phase currents cannot exceed the lines' ampacities:

$$I_{l,\tau} \le I_l^r, \quad \tau \in \Omega_k, l \in \Omega_l \tag{11}$$

where $I_{l,\tau}$ is the phase current flowing through line l during the τth time interval expressed as function of bus voltages, I_l^r is the ampacity of the line l, and Ω_l is the set of the μG lines.

3.2. Constraints on the slack bus

Constraints (ii) refer to the magnitude and phase of the voltage at the slack bus ($V_{1,\tau}^{p}$), ($\vartheta_{1,\tau}^{p}$) and on the apparent three-phase power $(P_{1,\tau}{}^2 + Q_{1,\tau}{}^2)^{1/2}$ that can be exchanged with the upstream grid and that must be within the size of the interconnecting transformer (S_{tr}):

$$V_{1,\tau}^{p} = V^{slack}, \quad \tau \in \Omega_k, \ p = 1,2,3 \tag{12}$$

$$\vartheta_{1,\tau}^{p} = \frac{2}{3}\pi(1-p), \quad \tau \in \Omega_k, \ p = 1,2,3 \tag{13}$$

$$(P_{1,\tau}{}^2 + Q_{1,\tau}{}^2)^{1/2} \le S_{tr}, \quad \tau \in \Omega_k \tag{14}$$

with V^{slack} the specified value of the voltage magnitude at slack bus, $P_{1,\tau}$ and $Q_{1,\tau}$ are the three-phase active and reactive powers at the slack bus expressed as functions of the active and reactive phase powers.

3.3. Constraints on the load busses

They refer to the load busses (whose set is Ω_L) where active and reactive powers are equal to specified values ($P_{i,\tau}^{p,sp}$ and $Q_{i,\tau}^{p,sp}$):

$$P_{i,\tau}^{p} = P_{i,\tau}^{p,sp}, \qquad \tau \in \Omega_k, \quad i \in \Omega_L, \quad p = 1,2,3 \tag{15}$$

$$Q_{i,\tau}^{p} = Q_{i,\tau}^{p,sp}, \qquad \tau \in \Omega_k, \quad i \in \Omega_L, \quad p = 1,2,3 \tag{16}$$

3.4. Constraints on DG busses

They refer to the active power ($P_{i,\tau}^{\xi_i}$) at the ith bus DG busses (whose set is Ω_{DG}) with phase ξ_i that must be equal to a specified value ($P_{i,\tau}^{\xi,sp}$) and to the maximum DG unit's apparent power that is constrained by the rating of the interfacing converter at the ith bus with phase ξ_i ($S_{DG,i}^{\xi_i}$):

$$P_{i,\tau}^{\xi_i} = P_{i,\tau}^{\xi_i,sp}, \qquad \tau \in \Omega_k, \quad i \in \Omega_{DG}, \tag{17}$$

$$(P_{i,\tau}^{\xi_i,sp^2} + Q_{i,\tau}^{\xi_i^2})^{\frac{1}{2}} \leq S_{DG,i}^{\xi_i}, \qquad \tau \in \Omega_k, \qquad i \in \Omega_{DG} \tag{18}$$

Since we deal with single-phase DG units, the active and reactive powers at the other phases are imposed equal to zero.

3.5. Constraints on EVC busses

Finally, constraints on the EVC bus refer to the limits on the active power at ith bus with the specified phase ξ_i during the τth time interval ($P_{i,\tau}^{\xi_i}$) that is limited by the maximum power that the EVC can absorb from ($P_{EVC_{i,\tau}}^{\xi_i,ch}$) or supply to ($P_{EVC_{i,\tau}}^{\xi_i,dch}$):

$$-P_{EVC_{i,\tau}}^{\xi_i,ch} \leq P_{i,\tau}^{\xi_i} \leq P_{EVC_{i,\tau}}^{\xi_i,dch}, \qquad \tau \in \omega_{k,i}, \qquad i \in \Omega_{EVk} \tag{19}$$

with Ω_{EVk} the set of nodes where the vehicles are plugged in at the time interval k and $\omega_{k,\,i}$ the set of time intervals within Ω_k in which the vehicle at the ith bus is connected (see **Figure 1**). Obviously, when the ith vehicle is not connected to the grid, its active power ($P_{i,\tau}^{\xi_i}$) and reactive powers ($Q_{i,\tau}^{\xi_i}$) must be equal to zero:

$$P_{i,\tau}^{\xi_i} = 0, \qquad \tau \notin \omega_{k,i}, \quad i \in \Omega_{EVk} \tag{20}$$

$$Q_{i,\tau}^{\xi_i} = 0, \qquad \tau \notin \omega_{k,i}, \quad i \in \Omega_{EVk} \tag{21}$$

Again, the apparent power at the ith bus, phase ξ_i, must be limited by rating of the charging equipment ($SE_{EVC,i}^{\xi_i}$):

$$(P_{i,\tau}^{\xi_i\,2} + Q_{i,\tau}^{\xi_i\,2})^{\frac{1}{2}} \leq SE_{EVC,i}^{\xi_i}, \qquad \tau \in \omega_{k,i}, \quad i \in \Omega_{EVk} \tag{22}$$

Finally, constraints on the SOC of the on-board batteries impose that it is limited by the size of the battery ($SE_i^{\xi_i,max}$) and by a minimum value corresponding to the admissible depth of discharge ($SE_i^{\xi_i,min}$):

http://dx.doi.org/10.5772/intechopen.68894

$$SE_{i,k}^{\xi_i,0} - \sum_{m=k}^{\tau}(\delta_{i,m}^{\xi_i} P_{i,m}^{\xi_i} \Delta t) \leq SE_i^{\xi_i,\max}, \qquad \tau \in \omega_{k,i}, \quad i \in \Omega_{EVk} \tag{23}$$

$$SE_{i,k}^{\xi_i,0} - \sum_{m=k}^{\tau}(\delta_{i,m}^{\xi_i} P_{i,m}^{\xi_i} \Delta t) \geq SE_i^{\xi_i,\min}, \qquad \tau \in \omega_{k,i}, \quad i \in \Omega_{EVk} \tag{24}$$

where $SE_{i,k}^{\xi_i,0}$ is the initial value of the energy stored in the battery on board the PEV plugged into the EVC connected to the ith bus with phase ξ_i, and where

$$\delta_{i,m}^{\xi_i} = \begin{cases} \dfrac{1}{\eta_{EVC}} & \text{if } P_{i,m}^{\xi_i} > 0 \\ \eta_{EVC} & \text{if } P_{i,m}^{\xi_i} \leq 0 \end{cases} \tag{25}$$

with η_{EVC} the charging efficiency related to the EVC. A specified SOC value is also imposed to the vehicle's battery at the time of departure:

$$SE_{i,k}^{\xi_i,0} - \sum_{m=k}^{k+n_k}(\delta_{i,m}^{\xi_i} P_{i,m}^{\xi_i} \Delta t) = SE_i^{\xi_i,\exp}, \qquad \tau \in \omega_{k,i}, \quad i \in \Omega_{EVk} \tag{26}$$

where, $SE_i^{\xi_i,\exp}$ is the specified value of the energy stored in the battery of the PEV plugged-into the EVC connected to the ith bus with phase ξ_i, at the departure time.

4. Numerical applications

In this section, some results of numerical applications of the proposed control strategy are reported. The test μG considered in this application is the three-phase unbalanced μG reported in **Figure 3**, whose lines' parameters are reported in **Tables 1** and **2**.

The μG is connected to the upstream medium voltage (MV) distribution network through a 250 kVA, 20/0.4 kV transformer. Both single-phase and three-phase loads are connected to the

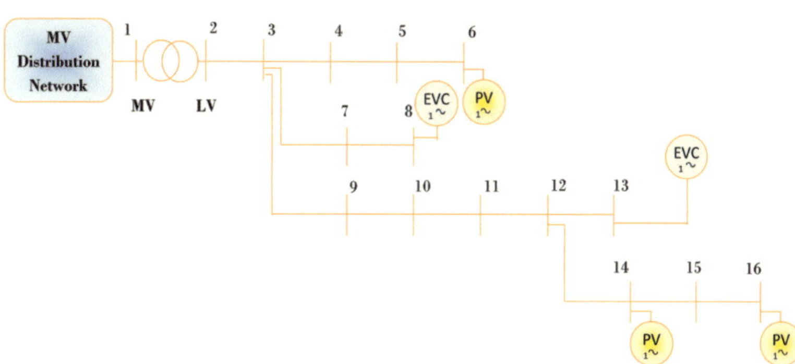

Figure 3. Test network.

Bus		Type	Length [m]	Bus		Type	Length [m]
From	To			From	To		
2	3	T1	100	9	10	T2	81
3	4	T2	137	10	11	T2	70
4	5	T2	168	11	12	T2	93
5	6	T2	10	12	13	T4	174
3	7	T4	107	12	14	T4	66
7	8	T4	102	14	15	T4	86
3	9	T3	162	15	16	T4	173

Table 1. Network data.

Type	Material	$n \times mm^2$	Diam. [mm]	Ampacity [A]
T1	Copper	3×150+95N	53	311
T2	Copper	4×25	28.3	112
T3	Aluminium	3×70+54.6N	37	180
T4	Copper	4×16	24	85

Table 2. Line type.

μG: their nominal active values are reported in **Table 3** ($cos\ \phi$= 0.9 is assumed for all of the loads).

In **Table 3**, bus locations and rated active powers of the photovoltaic systems connected to the μG and of the EVCs are also reported. In this application, it is assumed that all the considered DG units are able to control their reactive power according to the control signals that are outputs of the proposed strategy performed by the μG operator. The rated power values of the converters interfacing DG units correspond to the rated powers of the DG units.

PEVs can be connected to different busses of the μG through EVCs. As an example, the numerical application was performed with reference to the case of two PEVs connected to the μG according to the locations and time schedule of **Figure 4**. The batteries on board the PEVs are assumed to have capacities of 24 kWh; it is assumed that the full SOC is requested at their departure time. The charging (discharging) efficiency is 90% (93%).

With reference to the case of **Figure 4**, the control strategy was applied at each elementary time interval (Δt = 10 min) of the following two time horizons:

- T_1: one PEV is connected to bus i = 13 with phase p = 1: it corresponds to the plugged-in time of the first PEV (from 6:20 to 8:40 a.m.);
- T_2: two PEVs are connected to the μG; the first is connected to bus i = 13 with phase p = 1 (from 8:00 to 8:40 a.m.) and the second is connected to bus i = 8, phase p = 2 (from 8:00 to 11:00 a.m.): it corresponds to the plugged-in time of the PEV connected at 8:00 a.m.

Type	Bus #	Phase	Active power [kW]	Type	Bus #	Phase	Active power [kW]
PV	6	1	5	Load	7	2	6
PV	14	2	5	Load	7	3	3
PV	16	3	5	Load	8	1	3
EVC	8	2	7.4	Load	8	2	3
EVC	13	1	7.4	Load	8	3	3
Load	4	1,2,3	12	Load	13	1	6
Load	10	1,2,3	12	Load	13	2	3
Load	5	1	3	Load	13	3	3
Load	5	2	3	Load	14	1	3
Load	5	3	6	Load	14	2	3
Load	6	1	9	Load	14	3	3
Load	6	2	3	Load	16	1	3
Load	6	3	3	Load	16	2	6
Load	7	1	3	Load	16	3	6

Table 3. Allocation nodes and rated power of loads and DG.

The energy tariff used in this application is an actual real-time tariff [43] whose values in the two considered time horizons are reported in **Figure 5**. This price is assumed to be known 24 h ahead. Loads and DG units' active powers are derived by short-term forecasting procedures. **Figure 6** shows the active power injected at bus $i = 6$ (phase $p = 1$) where both the load and the PV unit are connected. In the figure, two forecasted profiles are shown: the first forecast (**Figure 6a**) refers to all of the intervals of T_1 and is known one time interval ahead the first interval of T_1; the second forecast (**Figure 6b**) refers to all of the intervals of T_2 and is known one time interval ahead the first interval of ahead the first interval of T_2.

Figures 7 and **8** show the active and reactive power profiles of the PEVs connected in the two considered time frames. They are outputs of the procedure and correspond to control signals for the EVCs. **Figure 9** shows the energy stored by the PEVs during the two time horizons.

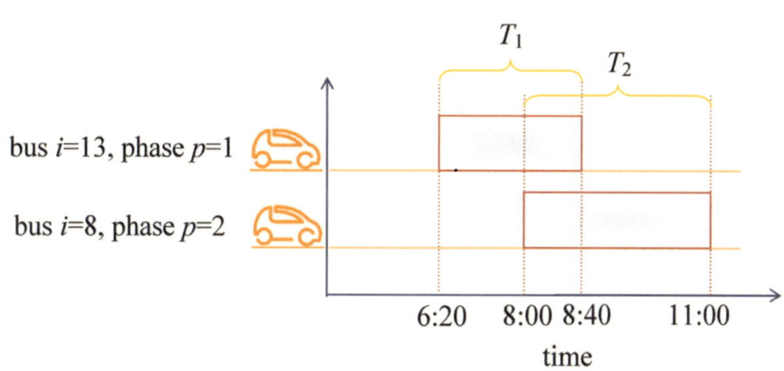

Figure 4. Case study: plug-in time of two plug-in electric vehicles (PEVs).

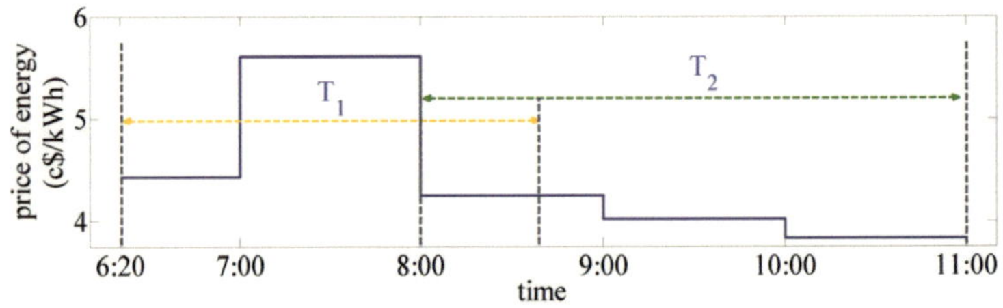

Figure 5. Price of energy.

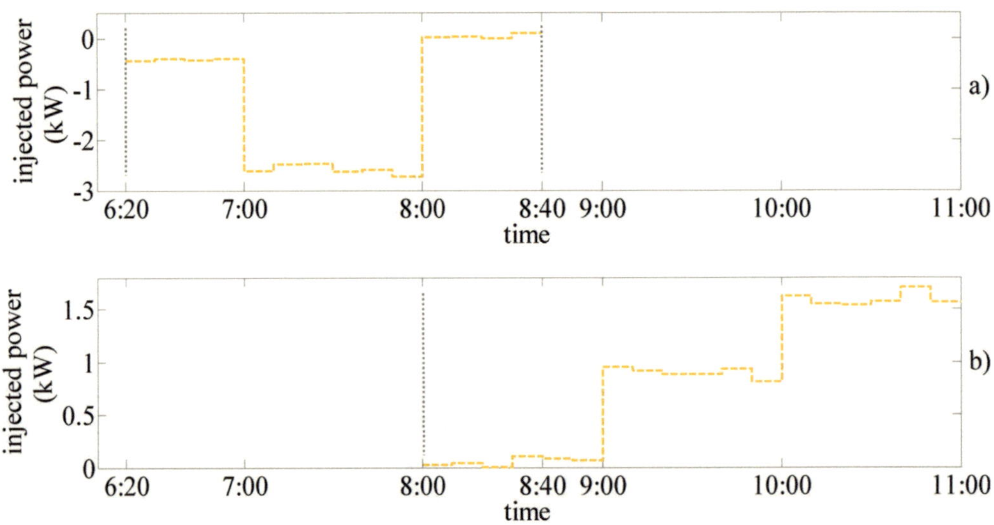

Figure 6. Forecasted value of the injected active power at bus $i = 6$ (phase $p = 1$).

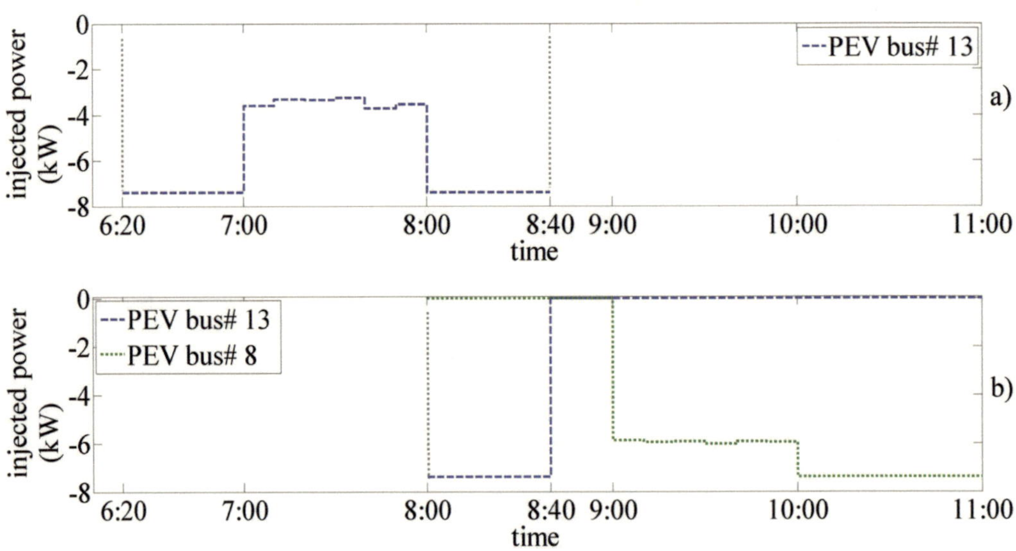

Figure 7. Active power of the PEVs.

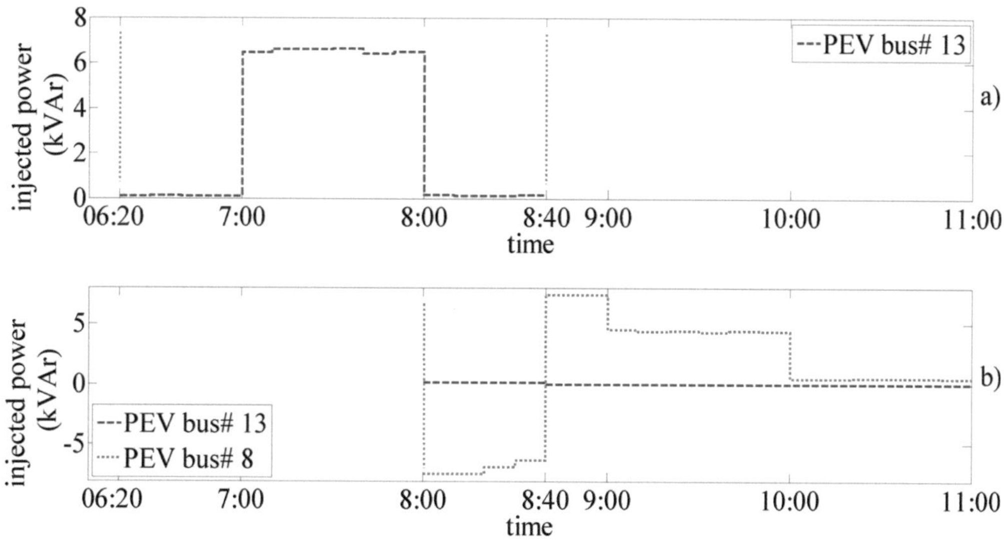

Figure 8. Reactive power of the PEVs during T_1 (a) and T_2 (b).

By the analysis of **Figure 7a**, it clearly appears that in T_1 the EVC reduces its charging in the time interval from 7:00 to 8:00 a.m., that is, the time interval characterized by a higher value of the energy price. In this time interval, the EVC's reactive power increases, in order to reduce both losses and the power requested at the PCC, thus reducing the overall cost for the energy supply sustained by the μG. The same consideration applies to the case of **Figure 7b**. Due to the shortness of the PEV connection time, it can also be noted that the EVCs never discharge.

Figure 10 shows the active power imported at the PCC during T_1 and T_2 in both the uncontrolled and controlled cases. The uncontrolled case corresponds to PEVs charged at the maximum EVC's rated power until the full charge is reached and neither the DG units nor the EVCs provide reactive power support. The figure shows that in the time interval characterized

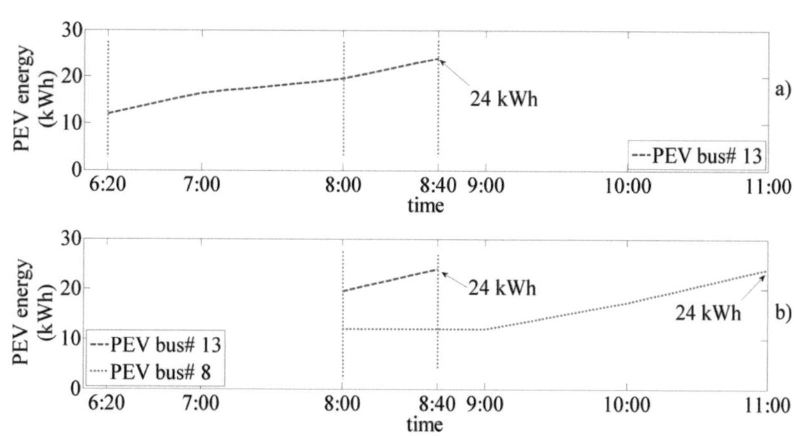

Figure 9. Stored energy of the PEVs during T_1 (a) and T_2 (b).

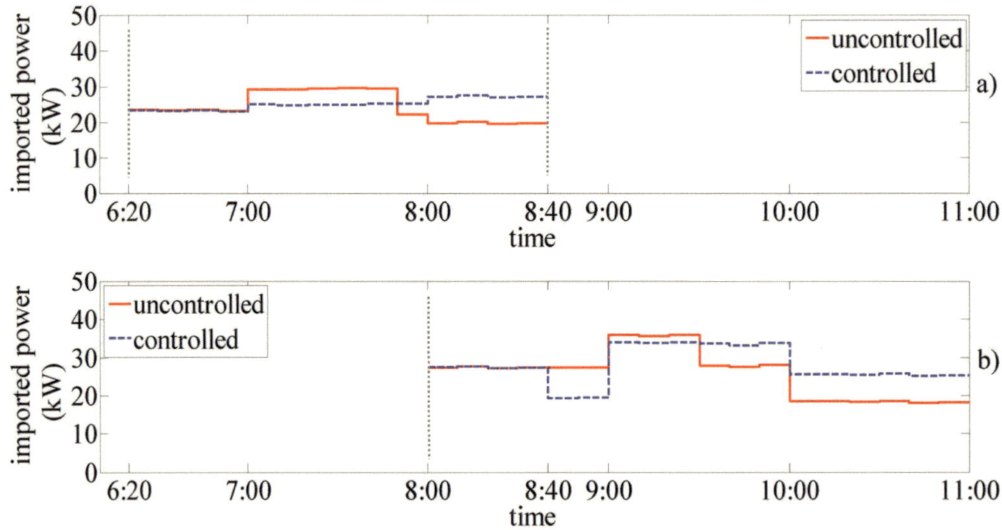

Figure 10. Active power imported at the PCC during T_1 (a) and T_2 (b).

by higher energy price (i.e., 7:00–8:00 a.m.), the energy imported at the PCC in the uncontrolled case is higher than that imported in the controlled case.

Finally, **Figures 11** and **12** show the voltage profile and the voltage unbalance factor, both at the first time interval of T_1.

Figure 11 demonstrates that the bus voltages are always within the limits imposed by regulations (i.e., 0.9–1.1 p.u.). **Figure 12** shows that, in the controlled case, the unbalance factor is

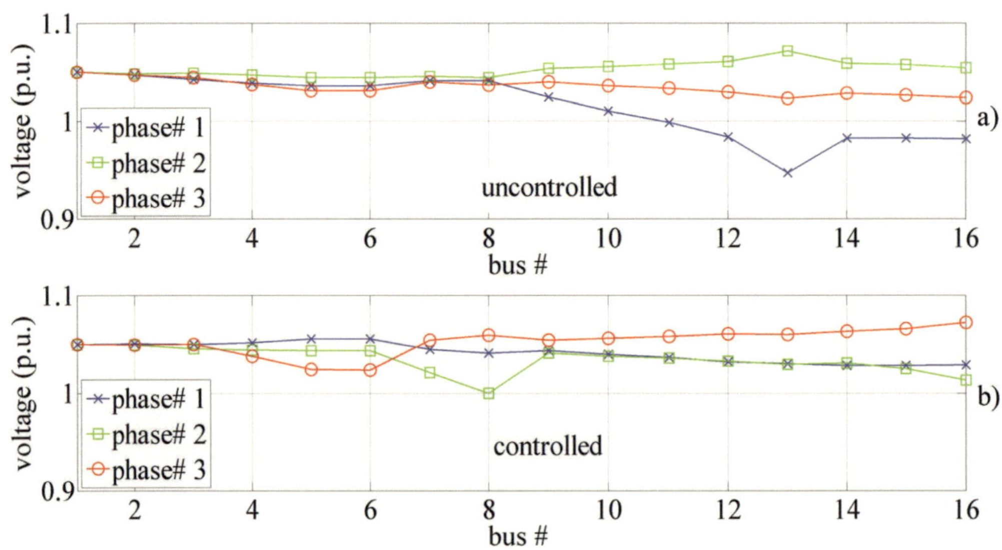

Figure 11. Voltage profile at the first time interval of T_1 in the uncontrolled case (a) and in the controlled case (b).

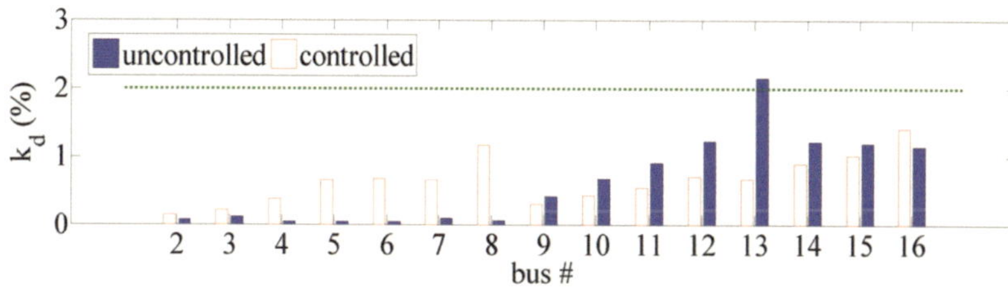

Figure 12. Voltage unbalance factor at the first time interval of T_1.

always lower than the limit of 2%. The proposed control strategy is able to satisfy the requirements in terms of unbalance limits even in particularly critical cases such as the case of bus i = 13, when in the uncontrolled case, the unbalance factor limit is exceeded. In this case, the use of reactive power provided by the DG units (here not reported for the sake of conciseness) and the PEVs is able to reduce the unbalances.

5. Conclusions

In this chapter, an optimal operation strategy was proposed that allowed managing the charging/discharging patterns of the plug-in vehicles connected to an unbalanced LV μG. Aim of the proposed procedure was the minimization of the costs sustained by the μG for the energy provision. The operating strategy allowed complying also with the standard limits for voltage unbalances and slow voltage variations.

Author details

Guido Carpinelli[1]*, Fabio Mottola[1], Daniela Proto[1] and Angela Russo[2]

*Address all correspondence to: guido.carpinelli@unina.it

1 University of Naples Federico II, Naples, Italy

2 Politecnico di Torino, Torino, Italy

References

[1] Kong PY, Karagiannidis GK. Charging schemes for plug-in hybrid electric vehicles in smart grid: A survey. IEEE Access. 2016;**4**:6846–6875

[2] Badtke-Berkow M, Centore M, Mohlin K, Spiller B. Making the Most of Time-Variant Electricity Pricing. Environmental Defense Fund Report; 2015

[3] Carpinelli G, Mottola F, Proto D, Varilone P. Minimizing unbalances in low-voltage microgrids: Optimal scheduling of distributed resources. Applied Energy. 2017;**191**:170–182

[4] Alam MJE, Muttaqi KM, Sutanto D. Alleviation of neutral-to-ground potential rise under unbalanced allocation of rooftop PV using distributed energy storage. IEEE Transactions on Sustainable Energy. 2015;**6**(3):889–898

[5] Li YW, Vilathgamuwa DM, Loh PC. Microgrid power quality enhancement using a three-phase four-wire grid-interfacing compensator. IEEE Transactions on Industry Applications. 2005;**41**(6):1707–1719

[6] Oe SP, Christopher E, Sumner M, Pholboon S, Johnson M, Norman SA. Microgrid Unbalance Compensator: Mitigating the Negative Effects of Unbalanced Microgrid Operation. Lyngby: IEEE/PES Innovative Smart Grid Technologies Europe; 2013

[7] Std EN 50160, Voltage characteristics in public distribution systems; 1999

[8] Gallardo-Lozano J, Milanés-Montero MI, Guerrero-Martínez MA, Romero-Cadaval E. Electric vehicle battery charger for smart grids. Electric Power Systems Research. 2012;**90**:18–29

[9] Hu J, Morais H, Sousa T, Lind M. Electric vehicle fleet management in smart grids: A review of services, optimization and control aspects. Renewable and Sustainable Energy Reviews. 2016;**56**:1207–1226

[10] Richardson DB. Electric vehicles and the electric grid: A review of modeling approaches, impacts, and renewable energy integration. Renewable and Sustainable Energy Reviews. 2013;**19**:247–254

[11] García-Villalobos J, Zamora I, San Martín JI, Asensio FJ, Aperribay V. Plug-in electric vehicles in electric distribution networks: A review of smart charging approaches. Renewable and Sustainable Energy Reviews. 2014;**38**:717–731

[12] Yong JY, Ramachandaramurthy VK, Tan KM, Mithulananthan N. A review on the state-of-the-art technologies of electric vehicle, its impacts and prospects. Renewable and Sustainable Energy Reviews. 2015;**49**:365–385

[13] Shareef H, Islam M, Mohamed A. A review of the stage-of-the-art charging technologies, placement methodologies, and impacts of electric vehicles. Renewable and Sustainable Energy Reviews. 2016;**64**:403–420

[14] Sharma I, Cañizares C, Bhattacharya K. Smart charging of PEVs penetrating into residential distribution systems. IEEE Transactions on Smart Grid. 2014;**5**(3):1196–1209

[15] Martinenas S, Knezović K, Marinelli M. "Management of Power Quality Issues in Low Voltage Networks Using Electric Vehicles: Experimental Validation," in IEEE Transactions on Power Delivery, vol. 32, no. 2, pp. 971–979, April 2017

[16] Kempton W, Letendre SE. Electric vehicles as a new power source for electric utilities. Transportation Research Part D: Transport and Environment. 1997;**2**(3):157–175

[17] Tomić J, Kempton W. Vehicle-to-grid power implementation: From stabilizing the grid to supporting large-scale renewable energy. Journal of Power Sources. 2005;**144**:280–294

[18] Habib S, Kamran M, Rashid U. Impact analysis of vehicle-to-grid technology and charging strategies of electric vehicles on distribution networks: A review. Journal of Power Sources. 2015;**277**(1):205–214

[19] Rabiee A, Feshki Farahani H, Khalili M, Aghaei J, Muttaqi KM. Integration of plug-in electric vehicles into microgrids as energy and reactive power providers in market environment. IEEE Transactions on Industrial Informatics. 2016;**12**(4):1312–1320

[20] Tomić J, Kempton W. Using fleets of electric-drive vehicles for grid support. Journal of Power Sources. 2007;**168**(2):459–468

[21] Peças Lopes JA, Polenz SA, Moreira CL, Cherkaoui R. Identification of control and management strategies for LV unbalanced microgrids with plugged-in electric vehicles. Electric Power Systems Research. 2010;**80**(8):898–906

[22] Wu T, Yang Q, Bao Z, Yan W. Coordinated energy dispatching in microgrid with wind power generation and plug-in electric vehicles. IEEE Transactions on Smart Grid. 2013;**4** (3):1453–1463

[23] Su W, Wang J, Roh J. Stochastic energy scheduling in microgrids with intermittent renewable energy resources. IEEE Transactions on Smart Grid. 2014;**5**(4):1876–1883

[24] Honarmand M, Zakariazadeh A, Jadid S. Integrated scheduling of renewable generation and electric vehicles parking lot in a smart microgrid. Energy Conversion and Management. 2014;**86**:745–755

[25] Zhang M, Chen J. The energy management and optimized operation of electric vehicles based on microgrid. IEEE Transactions on Power Delivery. 2014;**29**(3):1427–1435

[26] Guo Y, Xiong J, Xu S, Su W. Two-stage economic operation of microgrid-like electric vehicle parking deck. IEEE Transactions on Smart Grid. 2016;**7**(3):1703–1712

[27] Rabiee A, Sadeghi M, Aghaeic J, Heidari A. Optimal operation of microgrids through simultaneous scheduling of electrical vehicles and responsive loads considering wind and PV units uncertainties. Renewable and Sustainable Energy Reviews. 2016;**57**:721–739

[28] Mortaz E, Valenzuela J. Microgrid energy scheduling using storage from electric vehicles. Electric Power Systems Research. 2017;**143**:554–562

[29] Tabatabaee S, Mortazavi SS, Niknam T. Stochastic scheduling of local distribution systems considering high penetration of plug-in electric vehicles and renewable energy sources. Energy. 2017;**121**:480–490

[30] Kou P, Liang D, Gao L, Gao F. Stochastic coordination of plug-in electric vehicles and wind turbines in microgrid: A model predictive control approach. IEEE Transactions on Smart Grid. 2016;**7**(3):1537–1551

[31] Yang H, Pan H, Luo F, Qiu J, Deng Y, Lai M, Dong ZY. "Operational Planning of Electric Vehicles for Balancing Wind Power and Load Fluctuations in a Microgrid," in IEEE Transactions on Sustainable Energy, vol. 8, no. 2, pp. 592–604, April 2017.

[32] Tushar M, Assi C, Maier M, Uddin M. Smart microgrids: Optimal joint scheduling for electric vehicles and home appliances. IEEE Transactions on Smart Grid. 2014;**5**(1):239–250

[33] Van Roy J, Leemput N, Geth F, Büscher J, Salenbien R, Driesen J. Electric vehicle charging in an office building microgrid with distributed energy resources. IEEE Transactions on Sustainable Energy. 2014;**5**(4):1389–1396

[34] Beer S, Gómez T, Dallinger D, Marnay C, Stadler M, Lai J. An economic analysis of used electric vehicle batteries integrated into commercial building microgrids. 2012;**3**(1):517–525

[35] Deckmyn C, Van de Vyver J, Vandoorn TL, Meersman B, Desmet J, Vandevelde L. Day-ahead unit commitment model for microgrids. IET Generation, Transmission & Distribution. 2017;**11**(1):1–9

[36] Tafreshi S, Ranjbarzadeh H, Jafari M, Khayyam H. A probabilistic unit commitment model for optimal operation of plug-in electric vehicles in microgrid. Renewable and Sustainable Energy Reviews. 2016;**66**:934–947

[37] López MA, Martín S, Aguado JA, de la Torre S. V2G strategies for congestion management in microgrids with high penetration of electric vehicles. Electric Power Systems Research. 2013;**104**:28–34

[38] Gouveia C, Moreira CL, Lopes JAP, Varajao D, Araujo RE. Microgrid service restoration: The role of plugged-in electric vehicles. IEEE Industrial Electronics Magazine. 2013;**7**(4):26–41

[39] Khooban MH, Niknam T, Blaabjerg F, Dragičević T. A new load frequency control strategy for micro-grids with considering electrical vehicles. Electric Power Systems Research. 2017;**143**:585–598

[40] Rana R, Singh M, Mishra S. Design of modified droop controller for frequency support in microgrid using fleet of electric vehicles. IEEE Transactions on Power Systems; in press, IEEE Early Access Article (doi: 10.1109/TPWRS.2017.2651906)

[41] Arrillaga J, Arnold CP. Computer Analysis of Power System. John Wiley & Sons; Chichester, West Sussex, England; 1990

[42] Laughton MA. Analysis of unbalanced polyphase networks by the method of phase co-ordinates. Part 1: System representation in phase frame of reference. IEE Proceedings. 1968;**115**(8):1163–1172

[43] Comed Residential Real-Time Pricing Program [Internet]. Available from: https://rrtp.comed.com/live-prices/?date=20150318

Chapter 11

Hybrid AC/DC Microgrid Mode-Adaptive Controls

Eneko Unamuno and Jon Andoni Barrena

Additional information is available at the end of the chapter

http://dx.doi.org/10.5772/intechopen.69026

Abstract

The lack of inertial response at microgrids is usually compensated by configuring primary controllers of converter-interfaced devices to contribute in the transient response under power disturbances. The main purpose of this chapter is to study the modes of operation of primary level techniques of generation, storage, loads, and other devices attached to hybrid ac/dc microgrids. Although the chapter includes an analysis of the modes of operation of lower-level regulators, the focus is on upper-level or primary controllers. In this context, we analyze mode-adaptive controls based on voltage and frequency levels and we evaluate their behavior by simulation in the Matlab/Simulink® environment. The results demonstrate that mode-adaptive techniques are adequate for maximizing the energy extracted by distributed generation (DG) systems and limit demand side management actuations while ensuring an adequate regulation of the microgrid.

Keywords: demand-response, demand side management, distributed generation, energy storage systems, inertia emulation, microgrids, mode-adaptive control, primary control

1. Introduction

The electric grid is undergoing various changes in its structure, among other reasons, caused by a high dependence on fossil fuels, a constant increment in the power consumption, and the associated environmental problems. The classical top-down structure of the grid is shifting to a more decentralized topology where generation systems are located near consumption points—also known as distributed generation (DG) systems. This shift brings about several challenges, as the electric grid was not originally designed to handle the distributed and intermittent generating systems. In this context, microgrids are arising as one of the most suitable alternatives [1, 2], as they can efficiently integrate different types of DG systems thanks to the energy storage systems (ESSs) and advanced control strategies they include.

Microgrids are used at a wide variety of applications—distribution grids, electric transportation such as vessels or aircraft, isolated grids, etc.—and therefore can have very different features. However, one of the most common approaches is to classify them depending on the nature of their current [3]: ac, dc, or hybrid ac/dc.

Typically, researchers have focused most part of the research activity related to microgrids on ac systems, as they are the most straightforward solution based on the current infrastructure. The knowledge gained over the years with the electric grid can be directly applied for the development of ac microgrids. Therefore, these systems are characterized by efficient modification of voltage levels with transformers and by advanced fault management capabilities with optimally designed devices.

However, during the last two decades, dc microgrids are arising as an interesting alternative due to the inherent advantages they provide over ac ones. The increment of dc-based systems or devices that require a dc stage to operate, the lack of reactive power circulation or the fact that there is no need for synchronization, among other features, is clearing the path toward dc operated distribution networks. The main drawback resides in the fact that a complete substitution of the ac infrastructure would be unfeasible.

In this context, hybrid ac/dc microgrids are an interesting alternative, as they would enable the integration of dc-based systems through a dc network while maintaining the ac infrastructure [4]. This way the advantages of ac as well as dc grids can be combined, facilitating the shift to a more distributed electric grid composed by DG and ESSs.

As shown in **Figure 1**, hybrid microgrids are composed of ac and dc subgrids, which are linked by one or more interface converter. The integration of these converters not only

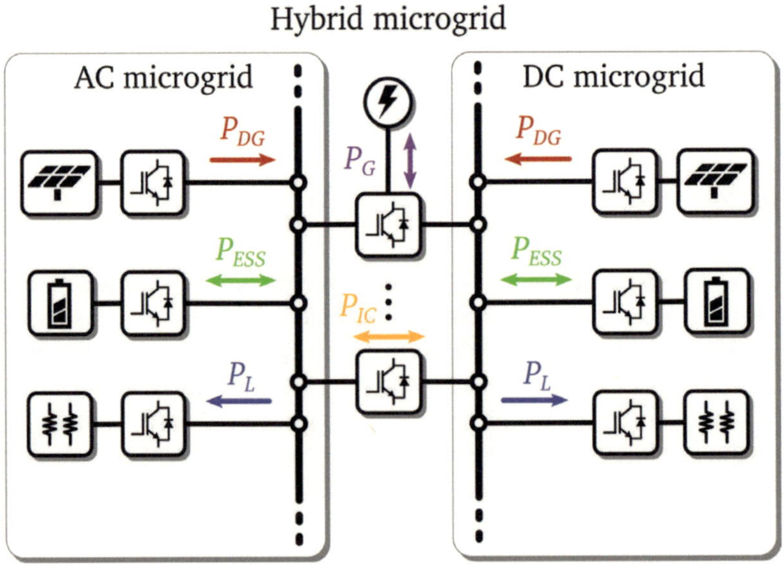

Figure 1. Hybrid ac/dc microgrid.

enables the power exchange between the ac and dc subgrids but they also increase the degrees of freedom regarding the management of the grid, as they can provide other ancillary services [5].

Although these microgrids are gaining a lot of interest especially during the last decades, most of the challenges still reside in their control and management, especially when they operate islanded from the main grid. Even if the control techniques employed at ac, dc, and hybrid ac/dc microgrids can be considerably different, their concept of operation is usually very similar. Inspired by the classical ac grid (**Figure 2**), the management of microgrids is most of the times carried out by employing a hierarchical structure. Each control layer is responsible for certain functions, such as the voltage/frequency control or the management of the islanding/reconnection process.

The main difference is that, at microgrids, conventional synchronous generators are replaced by converter-interfaced generation and storage systems, and therefore, the inertia of the grid is drastically reduced. This is one of the main challenging tasks in the management of microgrids, as lower inertia in the grid means that their voltage and/or frequency is significantly deteriorated under power variations. Consequently, microgrids become more susceptible to failures—especially when operating in the islanded mode—and hence more advanced control strategies need to be adopted to replace the lack of inertial behavior and ensure a stable operation.

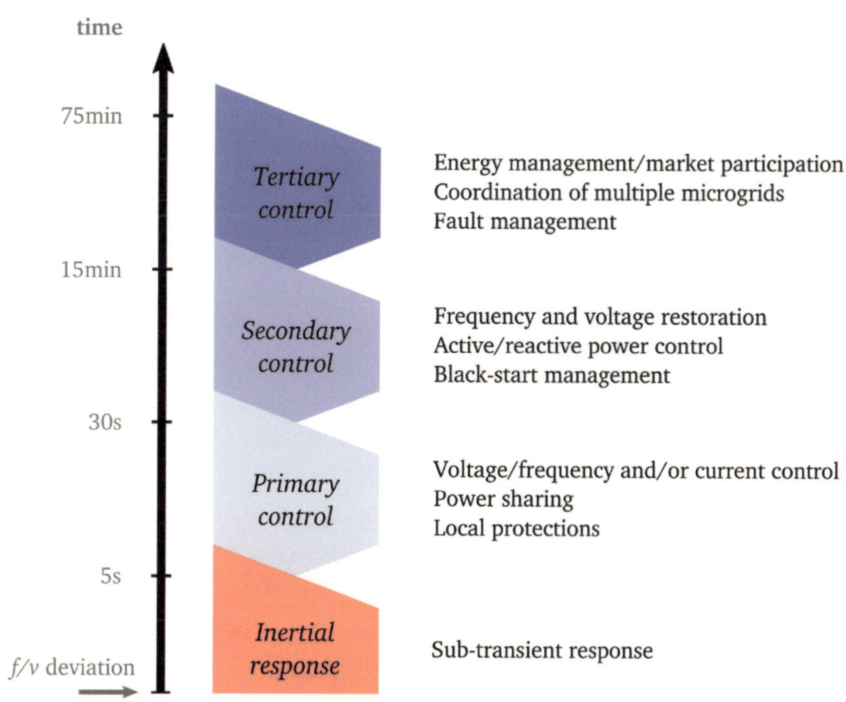

Figure 2. Main ac grid hierarchical control layer functions [6].

One of the most common solutions is to configure the converters associated to DG, ESSs, and even loads to contribute in the voltage and frequency regulation of microgrids. This is usually carried out by primary controllers, which are most of the times integrated locally in each device. **Figure 3** shows the dynamics of each control layer in a microgrid after a power variation.

As shown in the figure, the lack of inertial response is partially replaced by the primary control of converter-interfaced devices connected to the microgrid. In the literature, there is a wide collection of this type of primary techniques that can be integrated at ac, dc, and hybrid microgrids. A review of some of the most relevant strategies can be seen for instance in Refs. [3, 7].

Primary control techniques employed at microgrids are usually composed of two main stages (**Figure 4**): the lower-level stage usually contains the faster regulation loops, which are responsible for the current and/or voltage regulation of converters; similarly, the upper level of the primary control, which is slower than the previous one, determines the reference value where the converter should be controlled, e.g., the active or reactive power.

When there is a variation in the grid, the lower-level control stage primarily defines the transient response of the converter. In addition, the upper-level stage determines the steady-state operation point of the converter. However, depending on how are these stages designed, their effect in the transient as well as steady state response is coupled.

The aim of this chapter is to analyze the operation modes of primary control strategies employed at hybrid ac/dc microgrids, focusing on their lower- and upper-level (primary)

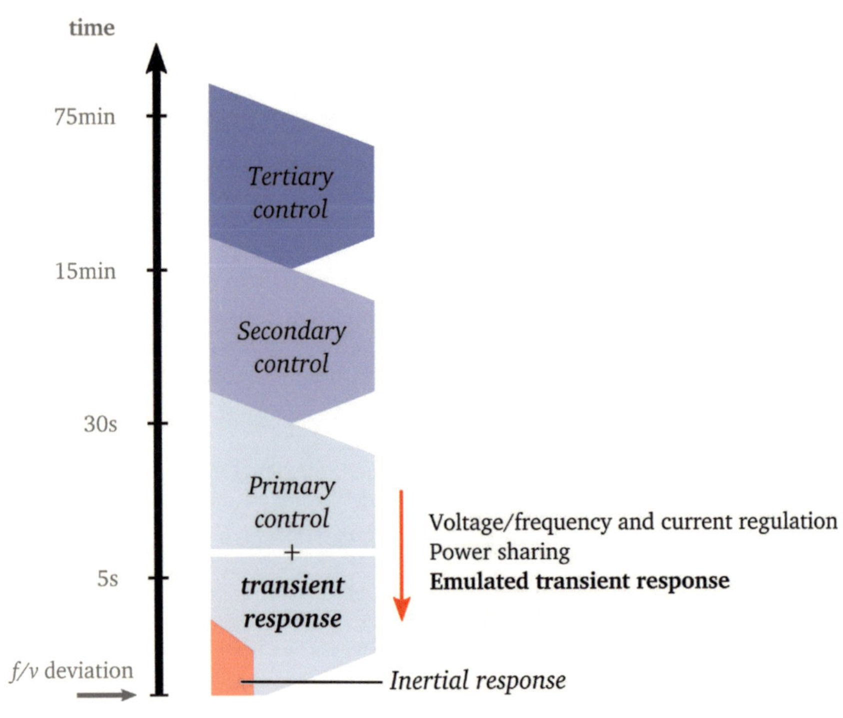

Figure 3. Microgrid hierarchical control layer functions.

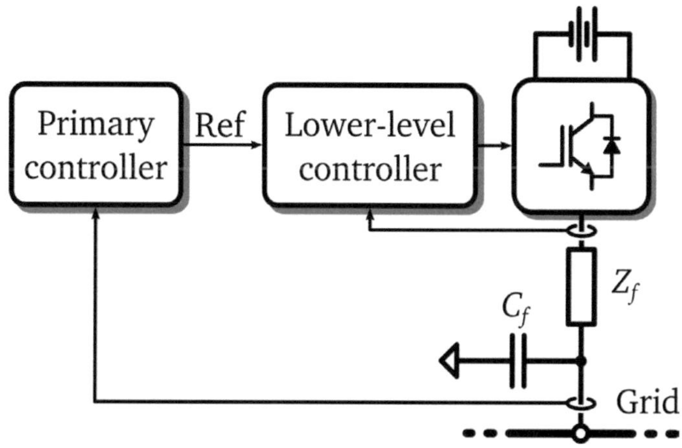

Figure 4. Primary controller configuration.

control stages. Taking into account that ac and dc microgrids are a particular case of hybrid microgrids, the study carried out in this chapter is also appropriate for these systems.

2. Lower-level control operation modes

When designing the lower-level control stages of primary regulators, we can usually follow two main approaches.

In the classical approach, this control stage is composed by one or more cascaded PI regulators that are tuned to follow the reference value provided by upper-level controllers, e.g., a voltage or a power reference. In this case, the regulators do not provide any inherent response over variations in the grid and are mainly designed to control the system so that it reaches the reference value as fast as possible.

On the other hand, in the last decades, a different approach has been proposed where the lower-level regulators are designed to participate in the transient regulation of the voltage and frequency of the network. These regulators are designed to emulate the behavior of classical synchronous generators with power converters associated to DG, ESSs, and loads connected to microgrids. In the literature, these techniques have been widely employed for different applications and are also known as virtual synchronous machines (VSMs) or synchronverters [8–15]. Following the main ac grid configuration—where synchronous generators are directly connected—most of these techniques are usually employed for devices connected to ac microgrids. However, recently, similar approaches have been developed to reproduce an analogous behavior at dc systems. For instance, the study carried out in Ref. [16] shows that a similar response can be emulated at dc systems by employing virtual-impedances in the control strategy, for example, as virtual-capacitors. In this case, instead of controlling the frequency as in VSMs, the variable controlled is the bus voltage.

3. Upper-level (primary) control operation modes

Due to the dispersed nature of microgrids, primary controllers are usually autonomous and operate based on local measurements of the device they are controlling. Whether the device contributes in the frequency/voltage regulation or not, their upper-level regulator is responsible for defining the steady-state point of operation of the converter.

Similar to lower-level regulators, upper-level ones are mainly classified into two different types.

On the one hand, there are certain devices that do not contribute in the frequency/voltage regulation of the microgrid and operate based on the reference provided by another control level (e.g., the secondary) or based on a reference internally calculated to, for example, extract as much energy as possible from the energy source they are connected to—a maximum power point tracking (MPPT) technique. These units are also named grid-following devices, as they do not regulate the bus but rather they "follow" their frequency and/or voltage [5, 7].

On the other hand, systems that contribute in the regulation of the bus are known as grid-forming or grid-supporting systems. These devices share the power variations occurring in the microgrid to decrease the variations of the bus frequency/voltage [5, 7].

We can design generation, storage systems, and loads connected via a power converter to operate differently for example depending on the bus voltage or frequency level. These systems usually include both types of upper-level controllers, and change their mode of operation based on some external condition. This approach is also known as mode-adaptive control, and one of the most interesting methods is to adapt the behavior of each controller based on the level of the bus frequency or voltage.

The authors in Refs. [17–22], for example, propose different types of mode adaptive control strategies for microgrids inspired by this concept, where the devices connected to the microgrid adapt their characteristics based on the frequency or voltage of the grid.

Figure 5 shows a possible configuration for the primary upper-level control of generation, storage systems, and loads connected to a hybrid microgrid. In addition, we have also depicted the connection to the mains grid, in case the microgrid is attached to it.

As we can observe, the primary control of each type of device is configured to behave differently depending on the value of the voltage or frequency. The solid thick lines illustrate an example of these and the shades show that the curves can be modified depending on the requirements. In a

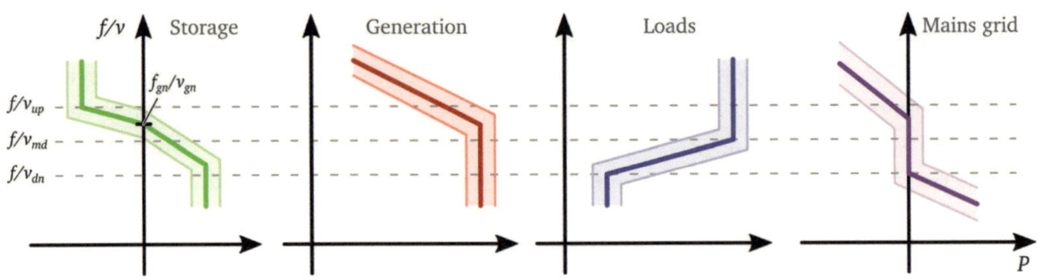

Figure 5. Example of adaptive primary controllers for systems connected to a microgrid.

real application, the ideal approach would be to configure all converter-interfaced devices to contribute in the power regulation. Each system could be designed to participate more or less for certain conditions, but the system would become more robust and stable because power variations would be handled by a higher number of devices with different characteristics and dynamics.

In the following sections, we describe more details of the modes of each type of system.

3.1. Energy storage systems

ESSs are one of the most important agents in the microgrid regulation because they serve as an energy "buffer" to compensate generation and demand deviations during normal operation. We must design and size these systems in order to cope with the most severe conditions of the system; otherwise, a poor regulation would cause the malfunction or disconnection of devices.

Figure 6 shows the most important modes of operation of energy storage systems connected to a microgrid. Under a balanced operation (mode 1), the ESS does not exchange any power with the microgrid. This hysteresis range must be carefully determined in order to avoid an excessive cycling and therefore ageing of the ESS but also to prevent the system from entering into an unstable point of operation.

When the generated power is higher than the demanded one the voltage or frequency of the microgrid increases over the hysteresis upper value and therefore the ESS absorbs power according to the charging droop slope (mode 2). This slope depends not only on the charging capabilities of the ESS but also on the sizing of its power converter.

When the voltage/frequency of the microgrid increases above the preestablished value or the state of charge (SOC) of the ESS is too high, the device turns to mode 3, where the power absorbed from the microgrid is kept constant although the *v*/*f* deviation keeps increasing. As shown in **Figure 6**, if the SOC of the ESS continues to increase while in mode 3, the maximum power level value is decreased. The limit of this saturation would be when the SOC is on its upper limit and hence the ESS would not absorb power anymore.

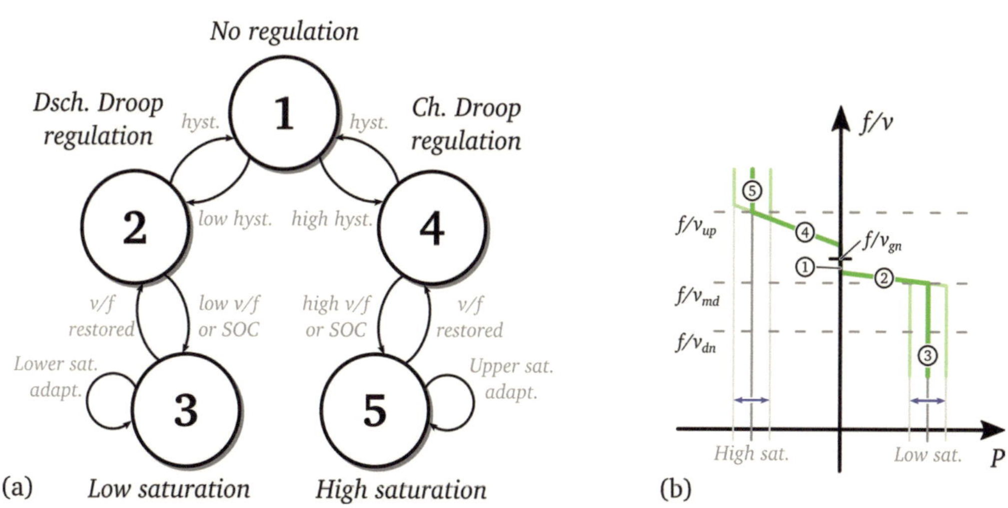

Figure 6. ESS primary controller operation modes.

A similar behavior is reproduced when there is an excessive demand in the microgrid and the voltage or frequency decrease below the hysteresis lower value (mode 4). At this point, the ESS supplies power to the microgrid according to its discharging droop slope. We must note that we could design the droop slopes differently for the ESS charging or discharging process, which means they do not have to be symmetric.

The ESS remains in mode 4 until the voltage or frequency reach their minimum value or the SOC of the ESS is too low. When any of these conditions takes place, the ESS jumps to mode 5 and supplies a constant power regardless of the *v*/*f* negative deviation. As in mode 3, the maximum power provided by the ESS will be decreased if the SOC decreases below the predefined levels. At some point, when the minimum SOC is reached, the ESS will not supply more power to the microgrid.

3.2. Generation systems

Generation systems mainly operate in two different modes (**Figure 7**): maximum energy extraction/constant power operation and droop regulation. During normal operation in mode 1, as most DG systems are based on renewable energy sources (RESs), the converters attached to generation systems are controlled to absorb as much energy as possible from energy sources. In the case of other types of DG systems such as diesel generators, secondary level controllers determine their constant power reference.

When the voltage or frequency of the microgrid increases above the preestablished level, DG systems shift out of their MPP to reduce the power amount they supply to the system. In this mode 2, DG systems contribute in the *v*/*f* regulation of the microgrid through a droop slope (**Figure 7**). As we can see in **Figure 7b**, the droop slope and saturation point depend not only on the value of the voltage or frequency, but also on the maximum power point (MPP) of the DG system. This means that depending on the available power that can be absorbed from the RES, the controllers will have to adapt their operation characteristics to meet the grid codes predefined by the system operator.

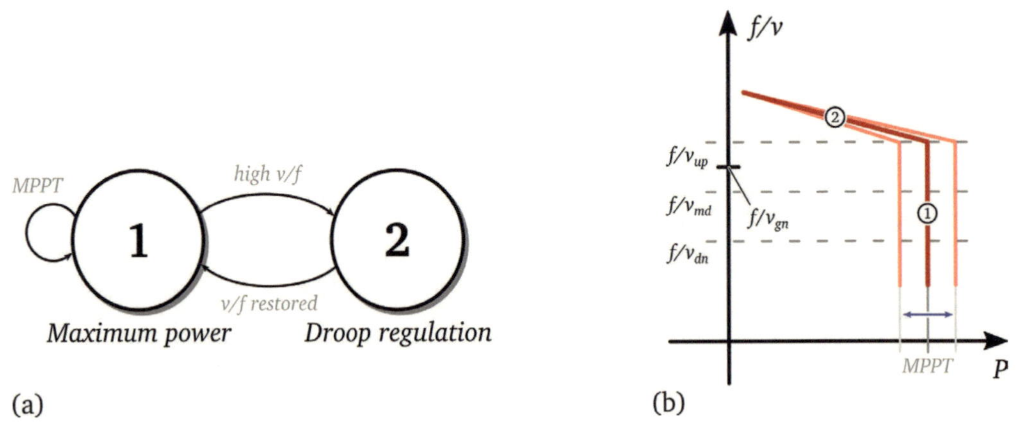

Figure 7. DG primary controller operation modes.

Classically, most converters associated to DG systems have been configured to exclusively operate on the MPP. However, the transition toward a more decentralized electrical system requires the participation of these generators in the regulation of the grid [13].

3.3. Demand-response

Similar to DG systems, loads operate normally in mode 1, absorbing the power required by the attached device. When generation systems are producing all the power they can, no power can be absorbed from the main grid, and energy storage systems are not able to provide more power, the voltage or frequency decreases below the predefined level and the power consumed by loads is consequently decreased (mode 2). In this case, as with DG systems, the droop slopes as well as saturation values are dependent on the instantaneous load (**Figure 8b**).

In the literature, this type of operation is a part of the so-called demand side management, as the loads actively participate in the regulation of the microgrid by reducing their consumed power when required. A high research activity has been carried out in the last years highlighting the importance of the participation of loads in the management of different types of electric systems [23].

3.4. Connection to the main grid

Depending on the topology and type of microgrid, we can follow different approaches with respect to the connection to the main grid. On a classical approach, the connection to the main grid can be employed to contribute in the regulation of the microgrid for the entire voltage range [3]. Another solution would be to use the link to the main grid at specific cases where the voltage or frequency levels are above the maximum or below minimum levels, avoiding the malfunction or disconnection of other devices.

Figure 9 shows the three main operation states of this last approach; as it can be seen, this link only operates when the *v/f* levels are out of some predefined levels. If the voltage or frequency goes above or below boundaries, we employ the link to the main grid to support the regulation

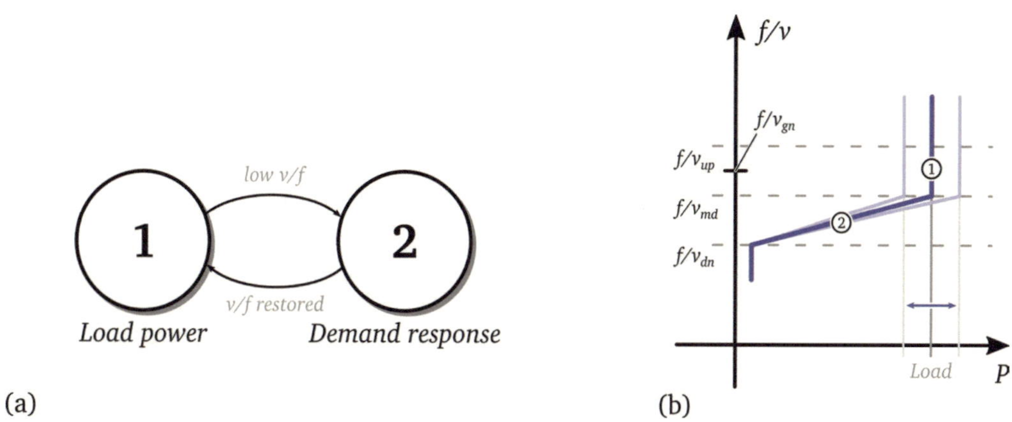

Figure 8. Load primary controller operation modes.

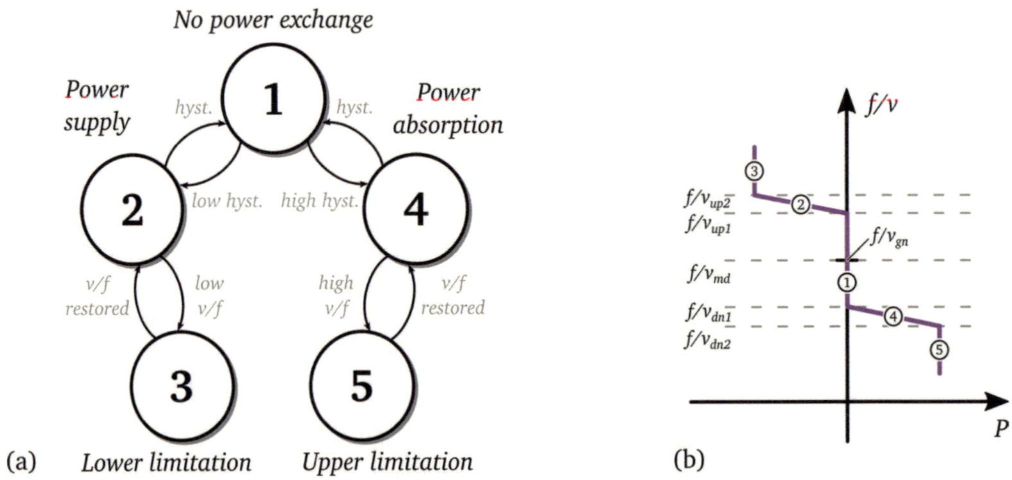

Figure 9. Grid connection primary controller operation modes.

of the system. An example of this situation is an excess of generated energy that causes an increment in the voltage or frequency; as our purpose is to extract as much energy as possible from RESs, we could employ the connection to the mains grid to send this extra power to the grid.

3.5. Interface converters

Interface ac-dc converters located along the microgrid provide an extra degree of freedom in the management of microgrids, as we can control them to perform diverse operations. The most typical approach would be to employ these converters to compensate the *v/f* deviations in the ac and dc subgrid of the hybrid microgrid [5, 22, 24]. The converters would transfer power from one subgrid to the other in order to equalize the excess of generated or demanded power at both systems.

As shown in **Figure 10**, interface converters carry out the power transfer between different subgrids with a droop controller. Unlike classical approaches, this droop is based on the difference between the frequency deviation in the ac subgrid and the voltage deviation in the dc part.

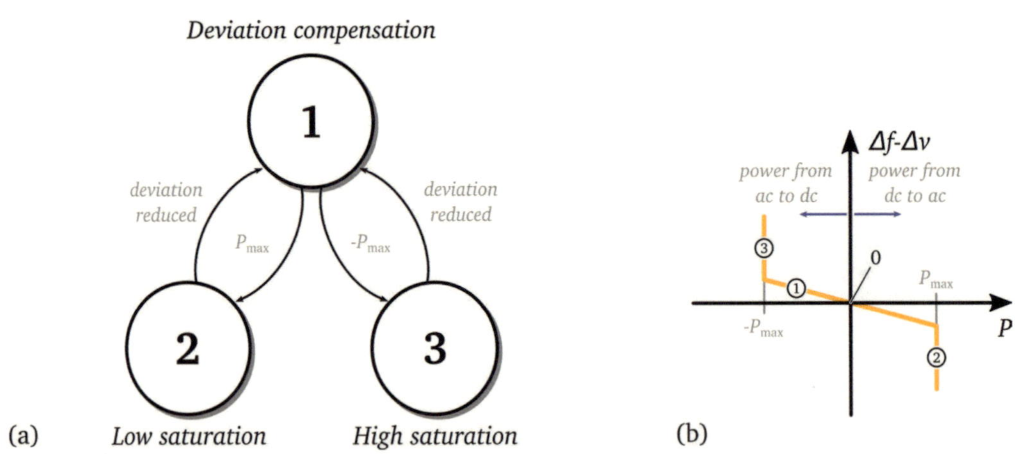

Figure 10. Interface converter primary controller operation modes.

Whenever there is a mismatch between these deviations, ac-dc converters located in the microgrid will transfer power to balance them. Modes 2 and 3 correspond to the maximum power that ac-dc systems can transfer in both directions.

According to Ref. [5], other techniques can be employed in the control of interface converters integrated at hybrid ac/dc microgrids such as the state of charge balancing of ESSs located in the ac and dc subgrid of the system.

4. Simulation results

We have carried out different simulations in the Matlab/Simulink® environment in order to observe the operation modes of primary controllers of a hybrid ac/dc microgrid.

4.1. Simulation scenario

A generation system, a storage system, and a load in the ac as well as dc subgrid of the microgrid compose the simulation scenario, and an interface ac/dc converter links these subgrids (**Figure 11**). We must mention that in this case, the hybrid microgrid also includes a connection point to the main ac grid in order to observe its behavior. As shown in the following sections, the islanded operation is a particular case of microgrids connected to the main grid; the difference is that islanded microgrids do not exchange any power with this grid.

In order to evaluate the behavior of the different agents connected to the hybrid ac/dc microgrid, we introduce a power disturbance in the ac subgrid of the microgrid. The results would be equivalent if we introduce the disturbance in the dc part, so for the sake of simplicity we do not contemplate this case in this chapter.

Figure 12 shows the primary curves employed for ESSs, DG systems, loads, and the connection to the main grid. The purpose is to validate the operation modes of the different agents participating in the regulation of the microgrid, so we have made no distinction in the configuration of DG, ESSs, and loads of the ac and dc part.

In the proposed scenario, if the demanded and generated power is balanced, all systems remain in an equilibrium state and ac and dc buses remain at their rated values (**State 0**). When the voltage or frequency are out of their rated values, ESSs carry out the regulation of the bus based on a predefined droop curve as shown in **Figure 12** (**State I**).

When storage systems reach their boundary operation, the microgrid exchanges power with the main ac grid to carry out the regulation (**State II**). For instance, when the voltage or frequency reaches a certain upper boundary, instead of taking DG systems out of their maximum power point, the exceeding power is supplied to the main ac grid. This is a key aspect of systems connected to the main grid because energy from RESs is not wasted and could be beneficial from the point of view of the electric market.

A similar situation occurs when the voltage or frequency fall below the limit; instead of reducing the power consumed by loads—that could lead to a malfunction of the load—this lack of power is covered by absorbing power from the main grid (**State II**).

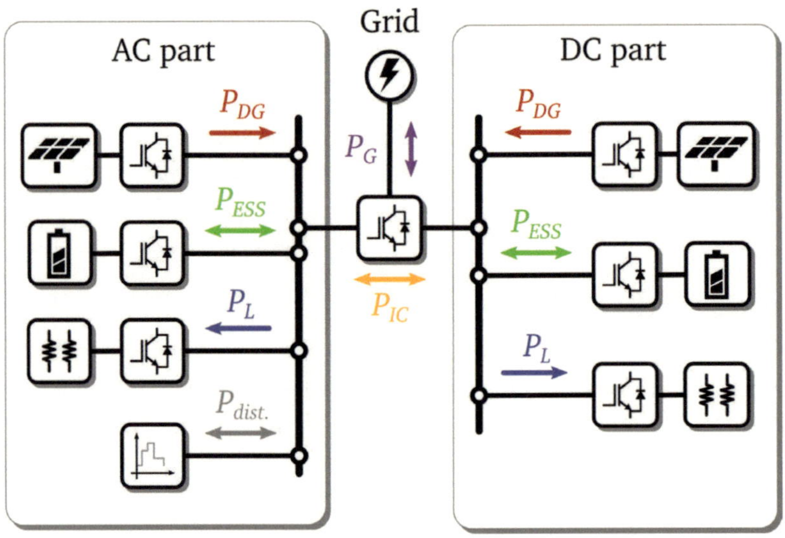

Figure 11. Simulated hybrid ac/dc microgrid scenario.

Usually the exchange of power with the main grid is limited—either by technical or by economical limitations—so when the voltage or frequency reaches the next upper or lower limit, generation systems, and loads are employed to contribute in the power regulation. Therefore, **State III** occurs when DG systems are out of their MPP and are controlled by a droop curve to regulate the power they generate. **State IV**, on the other hand, relates to the situation where loads reduce their consumed power, also known as demand-response.

On an islanded system, where no power can be exchanged with the main grid, if storage systems reach their upper or lower boundaries of operation, DG systems and loads directly carry out the regulation. This means there would not be any State II, and the system would directly enter into the State III or IV.

4.2. Modes of operation under disturbances

In order to reproduce the states mentioned in Section 4.1, we have simulated two different power disturbances, a positive and a negative step-shaped profile. We must mention that all

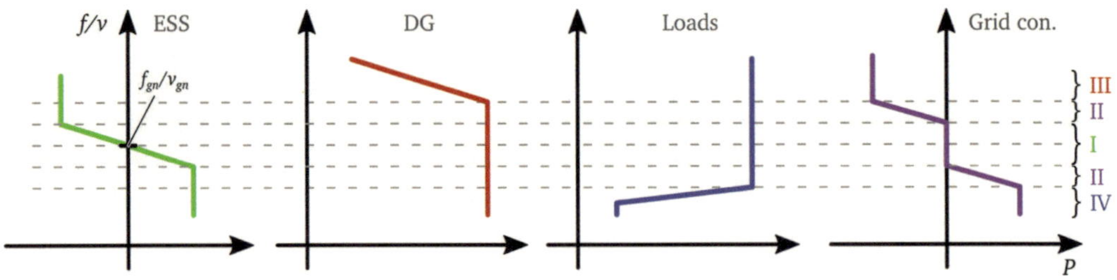

Figure 12. Primary controllers of the simulated scenario.

http://dx.doi.org/10.5772/intechopen.69026

the variables of the following simulations are normalized and shown in per unit (p.u.) values to facilitate the analysis.

4.2.1. Positive power disturbance

The aim of this simulation is to observe the modes of operation of ESS, DG systems, and the grid connection for positive power disturbances or deviations occurring in the microgrid.

Figure 13 shows the most relevant variables of the hybrid ac/dc microgrid during the simulation. We resume these parameters below:

- AC subgrid: we show the frequency and the powers of the DG, ESS, and load. The sign of powers is determined based on whether the device supplies (positive sign) or absorbs (negative sign) power from the ac bus. Therefore, the power of DG systems has a positive sign whereas we represent load power with negative sign.
- DC subgrid: regarding the dc subgrid of the microgrid, we represent the voltage and powers in the figure. The sign of these powers are determined following the same rule as in the ac part.
- Disturbance power: we illustrate the power profile separately and its sign is dependent on whether the power is supplied or absorbed from the ac bus.
- Interface converter: we also illustrate the power transferred between the ac and dc subgrids separately in the graph. Positive power means the power is transferred from the ac subgrid to the dc one, and vice versa.

As we can observe, in the figure we can distinguish different stages during the simulation, based on the values of the frequency and voltage of the hybrid ac/dc microgrid.

The system begins at an equilibrium state with no power disturbance, which in this case is named State 0. At this stage, the demanded and generated power are balanced so the ac and dc buses remain at their rated values.

At the instant $t = 1$ s, a positive power step is introduced and therefore ESSs start regulating the bus by absorbing power, entering into State I. The interface converter transfers the power from the ESS located in the dc subgrid to the ac one in order to compensate v/f deviations in the system, as previously explained in Section 3.5.

At $t = 2$ s, a higher power disturbance is introduced in the system and ESSs continue regulating the bus in State I until they reach their maximum power value. When this boundary is exceeded ($v/f = 1.03$ p.u.), the system enters State II and the extra power is supplied to the main ac grid.

Similarly, at $t = 3$ s, the power disturbance is increased and the microgrid continues supplying power to the main grid in State II until the maximum exchangeable power is reached. After exceeding this limit ($v/f = 1.06$ p.u.), the power supplied to the main grid is fixed and DG systems begin regulating the system.

After the instant $t = 4$ s, the disturbance is reverted and the system recovers its original equilibrium state.

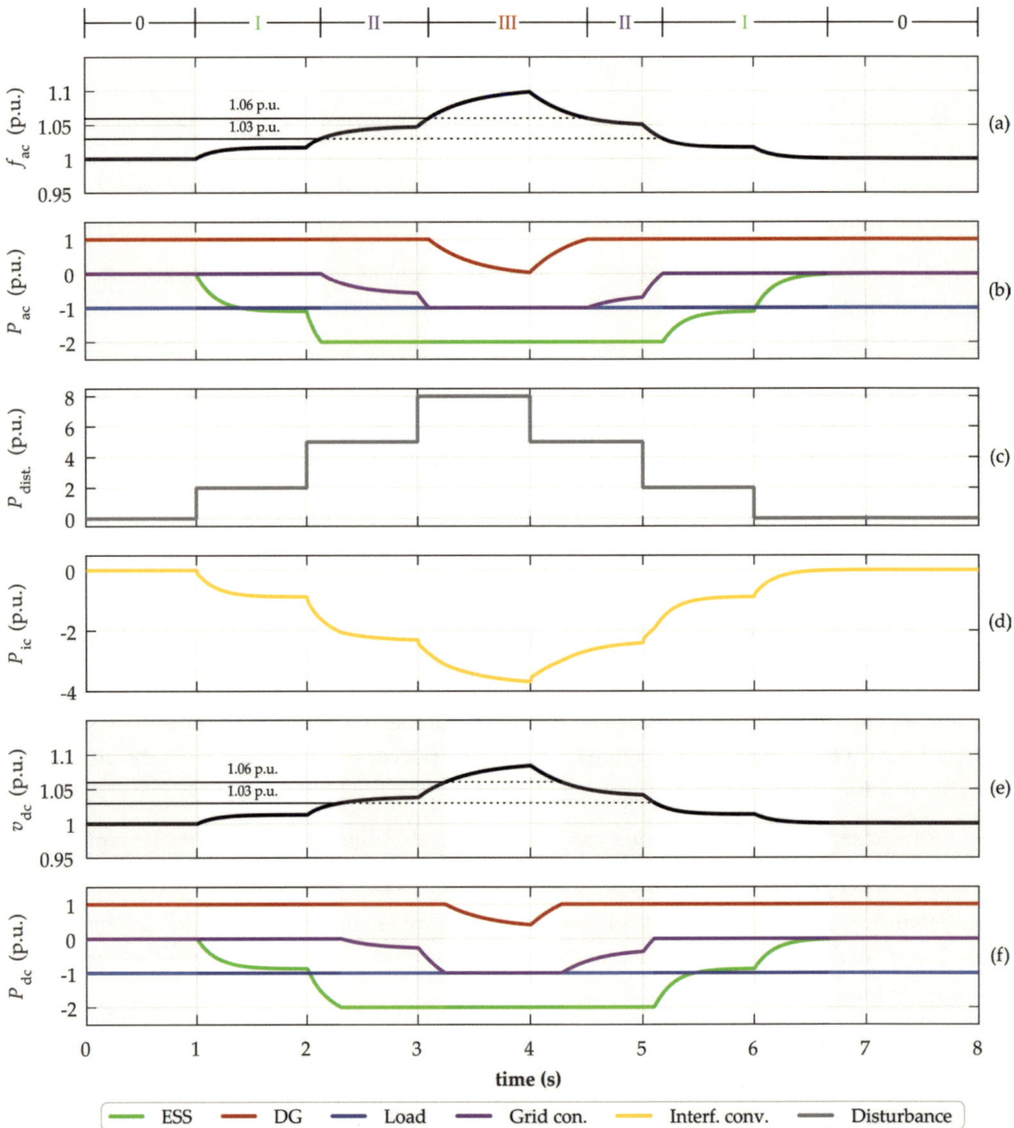

Figure 13. Microgrid response under positive power disturbances: (a) ac bus frequency, (b) ac subgrid power values, (c) disturbance power, (d) interface converter power, (e) dc bus voltage, and (f) dc subgrid power values.

In this simulation, we can see that the interface converter transfers power from the dc subgrid to the ac one in order to compensate the power deviations in both systems. However, it can be also noticed that the frequency in the ac grid and the voltage in the dc one do not reach their upper values at the same time, making the grids to change of state at different instants.

4.2.2. Negative power disturbance

In the following simulation, the main purpose is to observe the mode of operation of not only ESSs and the connection to the main grid but also the behavior of loads when lower limits of *v*/*f* are reached. Therefore, in this case, the power disturbance of Section 4.2.1 has been inversed, emulating different values of demanded power.

We show the most relevant parameters of the hybrid ac/dc microgrid for this simulation in **Figure 14**. As we can observe, the system goes through a very similar process as in the previous case. The difference is that the voltage as well as frequency decrease instead of increasing due to the negative power disturbance.

In this context, the system begins in State 0 at an equilibrium point and enters into State I when the first negative power step is introduced. Afterwards, with the second power step the

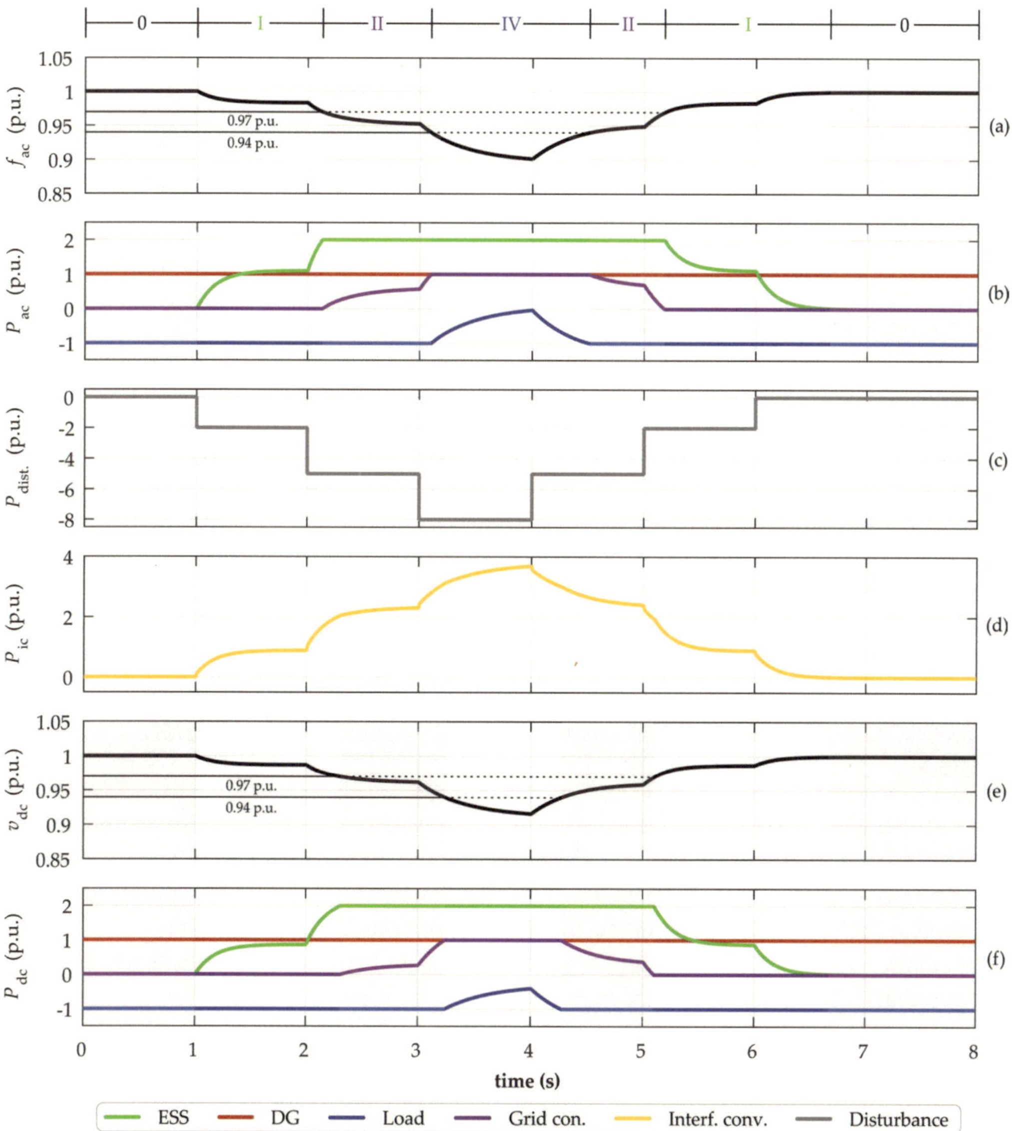

Figure 14. Microgrid response under negative power disturbances: (a) ac bus frequency, (b) ac subgrid power values, (c) disturbance power, (d) interface converter power, (e) dc bus voltage, and (f) dc subgrid power values.

microgrid reaches the $v/f = 0.97$ p.u. limit and enters into State II, where the lack of power is handled by absorbing power from the main grid. Finally, with the third negative power step the system reaches the $v/f = 0.94$ p.u. limit and, as no more power can be absorbed from the grid, loads begin reducing their demanded power (State IV).

In the proposed configuration, all the systems attached to the hybrid microgrid participate in the regulation of the system at different conditions. This structure facilitates the analysis of the mode-adaptive control techniques, but in a more realistic environment the most optimal approach would be to take advantage of the disperse characteristics of ESSs, DG systems, loads, and converters to design a bus regulating strategy composed by a mix of these systems for the entire v/f range.

5. Conclusions

In this chapter, the hierarchical control of microgrids has been initially revised, showing that this operation concept is very similar to the one employed in the conventional ac grid. One of the main differences is that, at microgrids, classical synchronous generators are replaced by converter-interfaced DG and ESSs, drastically reducing the inertial response of the system. In order to cope with this problem, the controllers of converter-interfaced devices are equipped with more advanced primary level techniques that contribute in the regulation of the microgrid.

Throughout the chapter, we have studied the different modes of operation of primary techniques and lower-level controllers of converter-interfaced generation, storage, loads, and other devices. Regarding lower-level techniques, we have shown that we can configure regulators to respond differently under power disturbances, providing more or less "inertial behavior." In the case of upper-level/primary controllers, we have carried out a thorough analysis of their most relevant modes of operation. We can design these controllers to behave differently—by contributing in the bus regulation or not—depending on certain conditions such as the voltage or frequency level in the microgrid. This strategy is also known as a mode-adaptive control, as generation, storage systems or loads adapt their mode of operation depending on some external condition.

The simulations carried out in this chapter demonstrate that a mode-adaptive control is very useful to take advantage of DG systems and loads. In the proposed case, DG systems—which are usually based on RES—are only taken out of their MPP when the voltage or frequency of the bus reach a very high boundary, maximizing the energy produced. Similarly, loads only contribute in the regulation of the system when the voltage or frequency reaches a very low level, which avoids any possible malfunction of these devices. Although the proposed configuration is adequate to observe the mode-adaptive control operation of each device attached to the microgrid, throughout the chapter we highlight that the most optimal approach would be to design a bus-regulating strategy for the entire operation range composed by a combination of ESSs, DG systems, loads, and converters, taking advantage of their different dynamic characteristics.

http://dx.doi.org/10.5772/intechopen.69026

Acknowledgements

This work has been partially funded by a predoctoral grant of the Basque Government (PRE_2016_2_0241).

Author details

Eneko Unamuno* and Jon Andoni Barrena

*Address all correspondence to: eunamuno@mondragon.edu

Electronics and Computing Department, Mondragon Unibertsitatea, Arrasate-Mondragón, Spain

References

[1] Nordman B, Christensen K, Meier A. Think globally, distribute power locally: The promise of nanogrids. Computer (Long Beach Calif). 2012;**45**:89-91. doi:10.1109/MC.2012.32

[2] Farhangi H. The path of the smart grid. IEEE Power Energy Magazine. 2010;**8**:18-28. doi:10.1109/MPE.2009.934876

[3] Unamuno E, Barrena JA. Hybrid ac/dc microgrids—Part I: Review and classification of topologies. Renewable and Sustainable Energy Review. 2015;**52**:1251-1259. doi:10.1016/j.rser.2015.07.194

[4] Unamuno E, Barrena JA. Hybrid ac/dc microgrids—Part II: Review and classification of control strategies. Renewable and Sustainable Energy Review. 2015;**52**:1123-1134. doi:10.1016/j.rser.2015.07.186

[5] Unamuno E, Barrena JA. Primary control operation modes in islanded hybrid ac/dc microgrids. Proceedings of EUROCON 2015, Salamanca: IEEE; 2015, pp. 1-6. doi:10.1109/EUROCON.2015.7313681

[6] Ulbig A, Rinke T, Chatzivasileiadis S, Andersson G. Predictive control for real-time frequency regulation and rotational inertia provision in power systems. 52nd IEEE Conference on Decision and Control, IEEE; 2013, pp. 2946-53. doi:10.1109/CDC.2013.6760331

[7] Vandoorn TL, Vasquez JC, De Kooning J, Guerrero JM, Vandevelde L. Microgrids: Hierarchical control and an overview of the control and reserve management strategies. IEEE Industrial Electronics Magazine. 2013;**7**:42-55. doi:10.1109/MIE.2013.2279306

[8] Natarajan V, Weiss G. Synchronverters with better stability due to virtual inductors, virtual capacitors and anti-windup. IEEE Transactions of Industrial Electronics. 2017;**46**:1-1. doi:10.1109/TIE.2017.2674611

[9] Zhong Q-C. Power electronics-enabled autonomous power systems: Architecture and technical routes. IEEE Transactions of Industrial Electronics. 2017;**46**:1-1. doi:10.1109/TIE.2017.2677339

[10] Chen D, Xu Y, Huang AQ. Integration of DC microgrids as virtual synchronous machines into the AC Grid. IEEE Transactions of Industrial Electronics. 2017;**46**:1-1. doi:10.1109/TIE.2017.267462

[11] Wu H, Ruan X, Yang D, Chen X, Zhao W, Lv Z, et al. Small-signal modeling and parameters design for virtual synchronous generators. IEEE Transactions of Industrial Electronics. 2016;**63**:4292-4303. doi:10.1109/TIE.2016.2543181

[12] Suul JA, DArco S, Guidi G. Virtual synchronous machine-based control of a single-phase bi-directional battery charger for providing vehicle-to-grid services. IEEE Transactions of Industrial Application. 2016;**52**:3234-3244. doi:10.1109/TIA.2016.2550588

[13] Zhong Q-C. Virtual synchronous machines: A unified interface for grid integration. IEEE Power Electronics Magazine. 2016;**3**:18-27. doi:10.1109/MPEL.2016.2614906

[14] D'Arco S, Suul JA, Fosso OB. Small-signal modeling and parametric sensitivity of a virtual synchronous machine in islanded operation. International Journal of Electronics Power Energy System. 2015;**72**:3-15. doi:10.1016/j.ijepes.2015.02.005

[15] Zhong QC, Nguyen PL, Ma Z, Sheng W. Self-synchronized synchronverters: Inverters without a dedicated synchronization unit. IEEE Transactions of Power Electronics. 2014;**29**:617-630. doi:10.1109/TPEL.2013.2258684

[16] Unamuno E, Barrena J. Equivalence of primary control strategies for AC and DC microgrids. Energies. 2017;**10**:91. doi:10.3390/en10010091

[17] Liu Y, Hou X, Wang X, Lin C, Guerrero J. A coordinated control for photovoltaic generators and energy storages in low-voltage AC/DC hybrid microgrids under Islanded mode. Energies. 2016;**9**:651. doi:10.3390/en9080651

[18] Gu Y, Li W, He X. Frequency-coordinating virtual impedance for autonomous power management of DC microgrid. IEEE Transactions of Power Electronics. 2015;**30**:2328-2337. doi:10.1109/TPEL.2014.2325856

[19] Loh PC, Li D, Chai YK, Blaabjerg F. Hybrid AC-DC microgrids with energy storages and progressive energy flow tuning. Proceedings of 7th International Power Electronics Motion Control Conference, IEEE; 2012, pp. 120-127. doi:10.1109/IPEMC.2012.6258872

[20] Gu Y, Xiang X, Li W, He X. Mode-adaptive decentralized control for renewable DC microgrid with enhanced reliability and flexibility. IEEE Transactions of Power Electronics. 2014;**29**:5072-5080. doi:10.1109/TPEL.2013.2294204

[21] Peyghami-Akhuleh S, Mokhtari H, Davari P, Loh PC, Blaabjerg F. Smart power management of DC microgrids in future milligrids. 2016 18th European Conference on Power Electronics and Applications. EPE'16 ECCE Eur., IEEE; 2016, pp. 1-10. doi:10.1109/EPE.2016.7695459

[22] Loh PC, Li D, Chai YK, Blaabjerg F. Autonomous operation of hybrid microgrid with ac and dc subgrids. IEEE Transactions of Power Electronics. 2013;**28**:2214-2223. doi:10.1109/TPEL.2012.2214792

[23] Palensky P, Dietrich D. Demand side management: Demand response, intelligent energy systems, and smart loads. IEEE Transactions of Industrial Informatics. 2011;**7**:381-388. doi:10.1109/TII.2011.2158841

[24] Baharizadeh M, Karshenas HR, Guerrero JM. Control strategy of interlinking converters as the key segment of hybrid AC-DC Microgrids. IET Generation, Transmission & Distribution. 2016;**10**:1671-1681. doi:10.1049/iet-gtd.2015.1014